Student Solutions Manual

Differential Equations & Linear Algebra

C. Henry Edwards

David E. Penney

Prentice Hall

Upper Saddle River, NJ 07458

Executive Editor: George Lobell
Supplement Editor: Melanie Van Benthuysen
Assistant Managing Editor: John Matthews
Production Editor: Barbara A. Till
Supplement Cover Manager: Paul Gourhan
Supplement Cover Designer: PM Workshop Inc.
Manufacturing Buyer: Lisa McDowell

© 2001 by Prentice Hall
Upper Saddle River, NJ 07458

All rights reserved. No part of this book may be
reproduced, in any form or by any means,
without permission in writing from the publisher.

Printed in the United States of America

10 9 8 7 6 5 4 3 2 1

ISBN 0-13-091051-1

Prentice-Hall International (UK) Limited, London
Prentice-Hall of Australia Pty. Limited, Sydney
Prentice-Hall Canada, Inc., Toronto
Prentice-Hall Hispanoamericana, S.A., Mexico
Prentice-Hall of India Private Limited, New Delhi
Pearson Education Asia Pte. Ltd., Singapore
Prentice-Hall of Japan, Inc., Tokyo
Editora Prentice-Hall do Brazil, Ltda., Rio de Janeiro

CONTENTS

1 FIRST-ORDER DIFFERENTIAL EQUATIONS
- **1.1** Differential Equations and Mathematical Modeling — 1
- **1.2** Integrals as General and Particular Solutions — 3
- **1.3** Slope Fields and Solution Curves — 6
- **1.4** Separable Equations and Applications — 7
- **1.5** Linear First-Order Equations — 13
- **1.6** Substitution Methods and Exact Equations — 16
- Chapter 1 Review Problems — 21

2 MATHEMATICAL MODELS AND NUMERICAL METHODS
- **2.1** Population Models — 23
- **2.2** Equilibrium Solutions and Stability — 26
- **2.3** Acceleration-Velocity Models — 29
- **2.4** Numerical Approximation: Euler's Method — 32
- **2.5** A Closer Look at the Euler Method — 37
- **2.6** The Runge-Kutta Method — 43

3 LINEAR SYSTEMS AND MATRICES
- **3.1** Introduction to Linear Systems — 49
- **3.2** Matrices and Gaussian Elimination — 51
- **3.3** Reduced Row-Echelon Matrices — 54
- **3.4** Matrix Operations — 59
- **3.5** Inverses of Matrices — 62
- **3.6** Determinants — 67
- **3.7** Linear Equations and Curve Fitting — 72

4 VECTOR SPACES
- **4.1** The Vector Space $\mathbf{R}^3$ — 78
- **4.2** The Vector Space $\mathbf{R}^n$ and Subspaces — 81
- **4.3** Linear Combinations and Independence of Vectors — 85
- **4.4** Bases and Dimension for Vector Spaces — 88
- **4.5** General Vector Spaces — 92

5 LINEAR EQUATIONS OF HIGHER ORDER
- **5.1** Introduction: Second-Order Linear Equations — 96
- **5.2** General Solutions of Linear Equations — 98
- **5.3** Homogeneous Equations with Constant Coefficients — 101
- **5.4** Mechanical Vibrations — 104
- **5.5** Nonhomogeneous Equations and Variation of Parameters — 107
- **5.6** Forced Oscillations and Resonance — 112

6 EIGENVALUES AND EIGENVECTORS
6.1	Introduction to Eigenvalues	118
6.2	Diagonalization of Matrices	125
6.3	Applications Involving Powers of Matrices	132

7 LINEAR SYSTEMS OF DIFFERENTIAL EQUATIONS
7.1	First-Order Systems and Applications	141
7.2	Matrices and Linear Systems	144
7.3	The Eigenvalue Method for Linear Systems	148
7.4	Second-Order Systems and Mechanical Applications	158
7.5	Multiple Eigenvalue Solutions	164
7.6	Numerical Methods for Systems	173

8 MATRIX EXPONENTIAL METHODS
8.1	Matrix Exponentials and Linear Systems	179
8.2	Nonhomogeneous Linear Systems	184
8.3	Spectral Decomposition Methods	189

9 NONLINEAR SYSTEMS AND PHENOMENA
9.1	Stability and the Phase Plane	201
9.2	Linear and Almost Linear Systems	203
9.3	Ecological Applications: Predators and Competitors	207
9.4	Nonlinear Mechanical Systems	210

10 LAPLACE TRANSFORM METHODS
10.1	Laplace Transforms and Inverse Transforms	214
10.2	Transformation of Initial Value Problems	216
10.3	Translation and Partial Fractions	221
10.4	Derivatives, Integrals, and Products of Transforms	225
10.5	Periodic and Piecewise Continuous Forcing Functions	228

11 POWER SERIES METHODS
11.1	Introduction and Review of Power Series	234
11.2	Power Series Solutions	237
11.3	Frobenius Series Solutions	242
11.5	Bessel's Equation	249

APPENDIX A
Existence and Uniqueness of Solutions	254

PREFACE

This is a solutions manual to accompany the textbook **DIFFERENTIAL EQUATIONS AND LINEAR ALGEBRA** by C. Henry Edwards and David E. Penney. We include solutions to most of the odd-numbered problems in the text.

Our goal is to support teaching of the subject of elementary differential equations in every way that we can. We therefore invite comments and suggested improvements for future printings of this manual, as well as advice regarding features that might be added to increase its usefulness in subsequent editions. Additional supplementary material can be found at our textbook Web site listed below.

Henry Edwards & David Penney

hedwards@math.uga.edu
dpenney@math.uga.edu

www.prenhall.com/edwards

CHAPTER 1

FIRST-ORDER DIFFERENTIAL EQUATIONS

SECTION 1.1

DIFFERENTIAL EQUATIONS AND MATHEMATICAL MODELING

The main purpose of Section 1.1 is simply to introduce the basic notation and terminology of differential equations, and to show the student what is meant by a solution of a differential equation. Also, the use of differential equations in the mathematical modeling of real-world phenomena is outlined.

Problems 1-12 are routine verifications by direct substitution of the suggested solutions into the given differential equations. We include here just some typical examples of such verifications.

3. If $y_1 = \cos 2x$ and $y_2 = \sin 2x$, then $y_1' = -2\sin 2x$ and $y_2' = 2\cos 2x$ so

    $$y_1'' = -4\cos 2x = -4y_1 \quad \text{and} \quad y_2'' = -4\sin 2x = -4y_2.$$

    Thus $y_1'' + 4y_1 = 0$ and $y_2'' + 4y_2 = 0$.

5. If $y = e^x - e^{-x}$, then $y' = e^x + e^{-x}$ so $y' - y = (e^x + e^{-x}) - (e^x - e^{-x}) = 2e^{-x}$. Thus $y' = y + 2e^{-x}$.

11. If $y = y_1 = x^{-2}$ then $y' = -2x^{-3}$ and $y'' = 6x^{-4}$, so

    $$x^2 y'' + 5x y' + 4y = x^2(6x^{-4}) + 5x(-2x^{-3}) + 4(x^{-2}) = 0.$$

    If $y = y_2 = x^{-2} \ln x$ then $y' = x^{-3} - 2x^{-3} \ln x$ and $y'' = -5x^{-4} + 6x^{-4} \ln x$, so

    $$x^2 y'' + 5x y' + 4y = x^2(-5x^{-4} + 6x^{-4} \ln x) + 5x(x^{-3} - 2x^{-3} \ln x) + 4(x^{-2} \ln x)$$
    $$= (-5x^{-2} + 5x^{-2}) + (6x^{-2} - 10x^{-2} + 4x^{-2})\ln x = 0.$$

13. Substitution of $y = e^{rx}$ into $3y' = 2y$ gives the equation $3re^{rx} = 2e^{rx}$ that simplifies to $3r = 2$. Thus $r = 2/3$.

15. Substitution of $y = e^{rx}$ into $y'' + y' - 2y = 0$ gives the equation $r^2 e^{rx} + r e^{rx} - 2 e^{rx} = 0$ that simplifies to $r^2 + r - 2 = (r+2)(r-1) = 0$. Thus $r = -2$ or $r = 1$.

The verifications of the suggested solutions in Problems 17-36 are similar to those in Problems 1-12. We illustrate the determination of the value of C only in some typical cases.

17. $C = 2$

19. If $y(x) = Ce^x - 1$ then $y(0) = 5$ gives $C - 1 = 5$, so $C = 6$.

21. $C = 7$

23. If $y(x) = \frac{1}{4}x^5 + Cx^{-2}$ then $y(2) = 1$ gives the equation $\frac{1}{4} \cdot 32 + C \cdot \frac{1}{8} = 1$ with solution $C = -56$.

25. If $y(x) = \tan(x^2 + C)$ then $y(0) = 1$ gives the equation $\tan C = 1$. Hence one value of C is $C = \pi/4$ (as is this value plus any integral multiple of π).

27. $y' = x + y$

29. If $m = y'$ is the slope of the tangent line and m' is the slope of the normal line at (x, y), then the relation $mm' = -1$ yields $m' = 1/y' = (y-1)/(x-0)$. Solution for y' then gives the differential equation $(1-y)y' = x$.

31. The slope of the line through (x, y) and $(-y, x)$ is $y' = (x-y)/(-y-x)$, so the differential equation is $(x+y)y' = y - x$.

In Problems 32-36 we get the desired differential equation when we replace the "time rate of change" of the dependent variable with its derivative, the word "is" with the = sign, the phrase "proportional to" with k, and finally translate the remainder of the given sentence into symbols.

33. $dv/dt = kv^2$

35. $dN/dt = k(P - N)$

37. $y(x) = 1$ or $y(x) = x$

39. $y(x) = x^2$

41. $y(x) = e^x/2$

43. **(a)** $y(10) = 10$ yields $10 = 1/(C-10)$, so $C = 101/10$.

(b) There is no such value of C, but the constant function $y(x) \equiv 0$ satisfies the conditions $y' = y^2$ and $y(0) = 0$.

(c) It is obvious visually that one and only one solution curve passes through each point (a,b) of the xy-plane, so it follows that there exists a unique solution to the initial value problem $y' = y^2$, $y(a) = b$.

SECTION 1.2

INTEGRALS AS GENERAL AND PARTICULAR SOLUTIONS

This section introduces **general solutions** and **particular solutions** in the very simplest situation — a differential equation of the form $y' = f(x)$ — where only direct integration and evaluation of the constant of integration are involved. Students should review carefully the elementary concepts of velocity and acceleration, as well as the fps and mks unit systems.

1. Integration of $y' = 2x+1$ yields $y(x) = \int (2x+1)\,dx = x^2 + x + C$. Then substitution of $x=0$, $y=3$ gives $3 = 0 + 0 + C = C$, so $y(x) = x^2 + x + 3$.

3. Integration of $y' = \sqrt{x}$ yields $y(x) = \int \sqrt{x}\,dx = \frac{2}{3}x^{3/2} + C$. Then substitution of $x=4$, $y=0$ gives $0 = \frac{16}{3} + C$, so $y(x) = \frac{2}{3}(x^{3/2} - 8)$.

5. Integration of $y' = (x+2)^{-1/2}$ yields $y(x) = \int (x+2)^{-1/2}\,dx = 2\sqrt{x+2} + C$. Then substitution of $x=2$, $y=-1$ gives $-1 = 2 \cdot 2 + C$, so $y(x) = 2\sqrt{x+2} - 5$.

7. Integration of $y' = 10/(x^2+1)$ yields $y(x) = \int 10/(x^2+1)\,dx = 10\tan^{-1} x + C$. Then substitution of $x=0$, $y=0$ gives $0 = 10 \cdot 0 + C$, so $y(x) = 10\tan^{-1} x$.

9. Integration of $y' = 1/\sqrt{1-x^2}$ yields $y(x) = \int 1/\sqrt{1-x^2}\,dx = \sin^{-1} x + C$. Then substitution of $x=0$, $y=0$ gives $0 = 0 + C$, so $y(x) = \sin^{-1} x$.

11. If $a(t) = 50$ then $v(t) = \int 50\,dt = 50t + v_0 = 50t + 10$. Hence

$$x(t) = \int (50t+10)\,dt = 25t^2 + 10t + x_0 = 25t^2 + 10t + 10.$$

13. If $a(t) = 3t$ then $v(t) = \int 3t\,dt = \frac{3}{2}t^2 + v_0 = \frac{3}{2}t^2 + 5$. Hence

$$x(t) = \int (\tfrac{3}{2}t^2 + 5)\,dt = \tfrac{1}{2}t^3 + 5t + x_0 = \tfrac{1}{2}t^3 + 5t.$$

15. If $a(t) = 2t+1$ then $v(t) = \int (2t+1)\,dt = t^2 + t + v_0 = t^2 + t - 7$. Hence

$$x(t) = \int (t^2 + t - 7)\,dt = \tfrac{1}{3}t^3 + \tfrac{1}{2}t - 7t + x_0 = \tfrac{1}{3}t^3 + \tfrac{1}{2}t - 7t + 4.$$

17. If $a(t) = (t+1)^{-3}$ then $v(t) = \int (t+1)^{-3}\,dt = -\tfrac{1}{2}(t+1)^{-2} + C = -\tfrac{1}{2}(t+1)^{-2} + \tfrac{1}{2}$ (taking $C = \tfrac{1}{2}$ so that $v(0) = 0$). Hence

$$x(t) = \int \left[-\tfrac{1}{2}(t+1)^{-2} + \tfrac{1}{2} \right] dt = \tfrac{1}{2}(t+1)^{-1} + \tfrac{1}{2}t + C = \tfrac{1}{2}\left[(t+1)^{-1} + t - 1 \right]$$

(taking $C = -\tfrac{1}{2}$ so that $x(0) = 0$).

19. $v = -9.8t + 49$, so the ball reaches its maximum height ($v = 0$) after $t = 5$ seconds. Its maximum height then is $y(5) = -4.9(5)^2 + 49(5) = 122.5$ meters.

21. $a = -10$ m/s^2 and $v_0 = 100$ km/h ≈ 27.78 m/s, so $v = -10t + 27.78$, and hence $x(t) = -5t^2 + 27.78t$. The car stops when $v = 0$, $t \approx 2.78$, and thus the distance traveled before stopping is $x(2.78) \approx 38.59$ meters.

23. $a = -9.8$ m/s2 so $v = -9.8\,t - 10$ and

$$y = -4.9\,t^2 - 10\,t + y_0.$$

The ball hits the ground when $y = 0$ and

$$v = -9.8\,t - 10 = -60,$$

so $t \approx 5.10$ s. Hence

$$y_0 = 4.9(5.10)^2 + 10(5.10) \approx 178.57 \text{ m}.$$

25. Integration of $dv/dt = 0.12\,t^3 + 0.6\,t$, $v(0) = 0$ gives $v(t) = 0.3\,t^2 + 0.04\,t^3$. Hence $v(10) = 70$. Then integration of $dx/dt = 0.3\,t^2 + 0.04\,t^3$, $x(0) = 0$ gives $x(t) = 0.1\,t^3 + 0.04\,t^4$, so $x(10) = 200$. Thus after 10 seconds the car has gone 200 ft and is traveling at 70 ft/sec.

27. If $a = -20$ m/sec^2 and $x_0 = 0$ then the car's velocity and position at time t are given by

$$v = -20t + v_0, \quad x = -10\,t^2 + v_0 t.$$

It stops when $v = 0$ (so $v_0 = 20t$), and hence when

$$x = 75 = -10\,t^2 + (20t)t = 10\,t^2.$$

Thus $t = \sqrt{7.5}$ sec so

$$v_0 = 20\sqrt{7.5} \approx 54.77 \text{ m/sec} \approx 197 \text{ km/hr}.$$

29. If $v_0 = 0$ and $y_0 = 20$ then

$$v = -at \quad \text{and} \quad y = -\tfrac{1}{2}at^2 + 20.$$

Substitution of $t = 2$, $y = 0$ yields $a = 10$ ft/sec^2. If $v_0 = 0$ and $y_0 = 200$ then

$$v = -10t \quad \text{and} \quad y = -5t^2 + 200.$$

Hence $y = 0$ when $t = \sqrt{40} = 2\sqrt{10}$ sec and $v = -20\sqrt{10} \approx -63.25$ ft/sec.

31. If $v_0 = 0$ and $y_0 = h$ then the stone's velocity and height are given by

$$v = -gt, \quad y = -0.5\,gt^2 + h.$$

Hence $y = 0$ when $t = \sqrt{2h/g}$ so

$$v = -g\sqrt{2h/g} = -\sqrt{2gh}.$$

33. We use units of miles and hours. If $x_0 = v_0 = 0$ then the car's velocity and position after t hours are given by

$$v = at, \quad x = \tfrac{1}{2}t^2.$$

Since $v = 60$ when $t = 5/6$, the velocity equation yields $a = 72$ mi/hr^2. Hence the distance traveled by 12:50 pm is

$$x = (0.5)(72)(5/6)^2 = 25 \text{ miles}.$$

35. Integration of $y' = (9/v_s)(1 - 4x^2)$ yields

$$y = (3/v_s)(3x - 4x^3) + C,$$

and the initial condition $y(-1/2) = 0$ gives $C = 3/v_s$. Hence the swimmer's trajectory is

$$y(x) = (3/v_s)(3x - 4x^3 + 1).$$

Substitution of $y(1/2) = 1$ now gives $v_s = 6$ mph.

SECTION 1.3

SLOPE FIELDS AND SOLUTION CURVES

As pointed out in the textbook, the instructor may choose to delay covering Section 1.3 until later in Chapter 1. However, before proceeding to Chapter 2, it is important that students come to grips at some point with the question of the existence of a unique solution of a differential equation — and realize that it makes no sense to look for the solution without knowing in advance that it exists. The instructor may prefer to combine existence and uniqueness by simplifying the statement of the existence-uniqueness theorem as follows:

> Suppose that the function $f(x, y)$ and the partial derivative $\partial f / \partial y$ are both continuous in some neighborhood of the point (a, b). Then the initial value problem
> $$\frac{dy}{dx} = f(x, y), \qquad y(a) = b$$
> has a unique solution in some neighborhood of the point a.

Slope fields and geometrical solution curves are introduced in this section as a concrete aid in visualizing solutions and existence-uniqueness questions. Solution curves corresponding to the slope fields in Problems 1–10 are shown in the answers section of the textbook and will not be duplicated here.

11. Each isocline $x - 1 = C$ is a vertical straight line.

13. Each isocline $y^2 = C \geq 0$, that is, $y = \sqrt{C}$ or $y = -\sqrt{C}$, is a horizontal straight line.

15. Each isocline $y/x = C$, or $y = Cx$, is a straight line through the origin.

17. Each isocline $xy = C$ is a rectangular hyperbola that opens along the line $y = x$ if $C > 0$, along $y = -x$ if $C < 0$.

19. Each isocline $y - x^2 = C$, or $x^2 = y - C$, is a translated parabola that opens along the y–axis.

21. Because both $f(x, y) = 2x^2y^2$ and $\partial f / \partial y = 4x^2y$ are continuous everywhere, the existence-uniqueness theorem of Section 1.3 in the textbook guarantees the existence of a unique solution in some neighborhood of $x = 1$.

23. Both $f(x, y) = y^{1/3}$ and $\partial f / \partial y = (1/3)y^{-2/3}$ are continuous near $(0, 1)$, so the theorem guarantees the existence of a unique solution in some neighborhood of $x = 0$.

25. $f(x, y) = (x - y)^{1/2}$ is not continuous at $(2, 2)$ because it is not even defined if $y > x$. Hence the theorem guarantees neither existence nor uniqueness in any neighborhood of the point $x = 2$.

27. Both $f(x, y) = (x - 1)/y$ and $\partial f / \partial y = -(x - 1)/y^2$ are continuous near $(0, 1)$, so the theorem guarantees both existence and uniqueness of a solution in some neighborhood of $x = 0$.

29. Both $f(x, y) = \ln(1 + y^2)$ and $\partial f / \partial y = 2y/(1 + y^2)$ are continuous near $(0, 0)$, so the theorem guarantees the existence of a unique solution near $x = 0$.

31. If $f(x, y) = -(1 - y^2)^{1/2}$ then $\partial f / \partial y = y(1 - y^2)^{-1/2}$ is not continuous when $y = 1$, so the theorem does not guarantee uniqueness.

35. The isoclines of $y' = y/x$ are the straight lines $y = Cx$ through the origin, and $y' = C$ at points of $y = Cx$, so it appears that these same straight lines are the solution curves of $xy' = y$. Then we observe that there is

(i) a unique one of these lines through any point not on the y-axis;
(ii) no such line through any point on the y-axis other than the origin; and
(iii) infinitely many such lines through the origin.

SECTION 1.4

SEPARABLE EQUATIONS AND APPLICATIONS

Of course it should be emphasized to students that the possibility of separating the variables is the first one you look for. The general concept of natural growth and decay is important for all differential equations students, but the particular applications in this section are optional. Torricelli's law in the form of Equation (24) in the text leads to some nice concrete examples and problems.

1. $\int \dfrac{dy}{y} = -\int 2x\,dx; \quad \ln y = -x^2 + c; \quad y(x) = e^{-x^2 + c} = Ce^{-x^2}$

3. $\int \dfrac{dy}{y} = \int \sin x\,dx; \quad \ln y = -\cos x + c; \quad y(x) = e^{-\cos x + c} = Ce^{-\cos x}$

5. $\int \dfrac{dy}{\sqrt{1 - y^2}} = \int \dfrac{dx}{2\sqrt{x}}; \quad \sin^{-1} y = \sqrt{x} + C; \quad y(x) = \sin\left(\sqrt{x} + C\right)$

7. $\int \dfrac{dy}{y^{1/3}} = \int 4x^{1/3}\,dx;\quad \tfrac{3}{2}y^{2/3} = 3x^{4/3} + \tfrac{3}{2}C;\quad y(x) = \left(2x^{4/3} + C\right)^{3/2}$

9. $\int \dfrac{dy}{y} = \int \dfrac{2\,dx}{1-x^2} = \int\left(\dfrac{1}{1+x} + \dfrac{1}{1-x}\right)dx$ (partial fractions)

$\ln y = \ln(1+x) - \ln(1-x) + \ln C;\quad y(x) = C\dfrac{1+x}{1-x}$

11. $\int \dfrac{dy}{y^3} = \int x\,dx;\quad -\dfrac{1}{2y^2} = \dfrac{x^2}{2} - \dfrac{C}{2};\quad y(x) = \left(C - x^2\right)^{-1/2}$

13. $\int \dfrac{y^3\,dy}{y^4+1} = \int \cos x\,dx;\quad \tfrac{1}{4}\ln\left(y^4+1\right) = \sin x + C$

15. $\int\left(\dfrac{2}{y^2} - \dfrac{1}{y^4}\right)dy = \int\left(\dfrac{1}{x} - \dfrac{1}{x^2}\right)dx;\quad -\dfrac{2}{y} + \dfrac{1}{3y^3} = \ln|x| + \dfrac{1}{x} + C$

17. $y' = 1 + x + y + xy = (1+x)(1+y)$

$\int \dfrac{dy}{1+y} = \int (1+x)\,dx;\quad \ln|1+y| = x + \tfrac{1}{2}x^2 + C$

19. $\int \dfrac{dy}{y} = \int e^x\,dx;\quad \ln y = e^x + \ln C;\quad y(x) = C\exp(e^x)$

$y(0) = 2e$ implies $C = 2$ so $y(x) = 2\exp(e^x)$

21. $\int 2y\,dy = \int \dfrac{x\,dx}{\sqrt{x^2-16}};\quad y^2 = \sqrt{x^2-16} + C$

$y(5) = 2$ implies $C = 1$ so $y^2 = 1 + \sqrt{x^2-16}$

23. $\int \dfrac{dy}{2y-1} = \int dx;\quad \tfrac{1}{2}\ln(2y-1) = x + \tfrac{1}{2}\ln C;\quad 2y - 1 = Ce^{2x}$

$y(1) = 1$ implies $C = e^{-2}$ so $y(x) = \tfrac{1}{2}\left(1 + e^{2x-2}\right)$

25. $\int \dfrac{dy}{y} = \int\left(\dfrac{1}{x} + 2x\right);\quad \ln y = \ln x + x^2 + \ln C;\quad y(x) = Cx\exp(x^2)$

$y(1) = 1$ implies $C = e^{-1}$ so $y(x) = x\exp(x^2 - 1)$

27. $\int e^y \, dy = \int 6e^{2x} \, dx; \quad e^y = 3e^{2x} + C; \quad y(x) = \ln\left(3e^{2x} + C\right)$

$y(0) = 0$ implies $C = -2$ so $y(x) = \ln\left(3e^{2x} - 2\right)$

29. The population growth rate is $k = \ln(30000/25000)/10 \approx 0.01823$, so the population of the city t years after 1960 is given by $P(t) = 25000 e^{0.01823t}$. The expected year 2000 population is then $P(40) = 25000 e^{0.01823 \times 40} \approx 51840$.

31. As in the textbook discussion of radioactive decay, the number of ^{14}C atoms after t years is given by $N(t) = N_0 e^{-0.0001216 t}$. Hence we need only solve the equation $\frac{1}{6} N_0 = N_0 e^{-0.0001216 t}$ for $t = (\ln 6)/0.0001216 \approx 14735$ years to find the age of the skull.

33. The amount in the account after t years is given by $A(t) = 5000 e^{0.08t}$. Hence the amount in the account after 18 years is given by $A(20) = 5000 e^{0.08 \times 20} \approx 21{,}103.48$ dollars.

35. To find the decay rate of this drug in the dog's blood stream, we solve the equation $\frac{1}{2} = e^{-5k}$ (half-life 5 hours) for $k = (\ln 2)/5 \approx 0.13863$. Thus the amount in the dog's bloodstream after t hours is given by $A(t) = A_0 e^{-0.13863 t}$. We therefore solve the equation $A(1) = A_0 e^{-0.13863} = 50 \times 45 = 2250$ for $A_0 \approx 2585$ mg, the amount to anesthetize the dog properly.

37. Taking $t = 0$ when the body was formed and $t = T$ now, the amount $Q(t)$ of ^{238}U in the body at time t (in years) is given by $Q(t) = Q_0 e^{-kt}$, where $k = (\ln 2)/(4.51 \times 10^9)$. The given information tells us that

$$\frac{Q(T)}{Q_0 - Q(T)} = 0.9.$$

After substituting $Q(T) = Q_0 e^{-kT}$, we solve readily for $e^{kT} = 19/9$, so $T = (1/k)\ln(19/9) \approx 4.86 \times 10^9$. Thus the body was formed approximately 4.86 billion years ago.

39. Because $A = 0$ the differential equation reduces to $T' = kT$, so $T(t) = 25 e^{-kt}$. The fact that $T(20) = 15$ yields $k = (1/20)\ln(5/3)$, and finally we solve

$$5 = 25 e^{-kt} \quad \text{for} \quad t = (\ln 5)/k \approx 63 \text{ min}.$$

41. (a) The light intensity at a depth of x meters is given by $I(x) = I_0 e^{-1.4x}$. We solve the equation $I(x) = I_0 e^{-1.4x} = \frac{1}{2} I_0$ for $x = (\ln 2)/1.4 \approx 0.495$ meters.

(b) At depth 10 meters the intensity is $I(10) = I_0 e^{-1.4 \times 10} \approx (8.32 \times 10^{-7}) I_0$.

(c) We solve the equation $I(x) = I_0 e^{-1.4x} = 0.01 I_0$ for $x = (\ln 100)/1.4 \approx 3.29$ meters.

43. (a) $A' = rA + Q$

(b) The solution of the differential equation with $A(0) = 0$ is given by

$$rA + Q = Qe^{rt}.$$

When we substitute $A = 40$ (thousand), $r = 0.11$, and $t = 18$, we find that $Q = 0.70482$, that is, $\$704.82$ per year.

45. The cake's temperature will be $100°$ after 66 min 40 sec; this problem is just like Example 6 in the text.

47. If $N(t)$ denotes the number of people (in thousands) who have heard the rumor after t days, then the initial value problem is

$$N' = k(100 - N), \quad N(0) = 0$$

and we are given that $N(7) = 10$. When we separate variables ($dN/(100-N) = k\, dt$) and integrate, we get $\ln(100-N) = -kt + C$, and the initial condition $N(0) = 0$ gives $C = \ln 100$. Then $100 - N = 100 e^{-kt}$, so $N(t) = 100(1 - e^{-kt})$. We substitute $t = 7$, $N = 10$ and solve for the value $k = \ln(100/90)/7 \approx 0.01505$. Finally, 50 thousand people have heard the rumor after $t = (\ln 2)/k \approx 46.05$ days.

49. With $A = \pi(3)^2$ and $a = \pi(1/12)^2$, and taking $g = 32$ ft/sec^2, Equation (20) reduces to $162\, y' = -\sqrt{y}$. The solution such that $y = 9$ when $t = 0$ is given by $324\sqrt{y} = -t + 972$. Hence $y = 0$ when $t = 972$ sec $= 16$ min 12 sec.

51. The solution of $y' = -k\sqrt{y}$ is given by

$$2\sqrt{y} = -kt + C.$$

The initial condition $y(0) = h$ (the height of the cylinder) yields $C = 2\sqrt{h}$. Then substitution of $t = T$, $y = 0$ gives $k = (2\sqrt{h})/T$. It follows that

$$y = h(1 - t/T)^2.$$

If r denotes the radius of the cylinder, then

$$V(y) = \pi r^2 y = \pi r^2 h(1-t/T)^2 = V_0(1-t/T)^2.$$

53. **(a)** Since $x^2 = by$, the cross-sectional area is $A(y) = \pi x^2 = \pi by$. Hence the equation $A(y)y' = -a\sqrt{2gy}$ reduces to the differential equation

$$y^{1/2}y' = -k = -(a/\pi b)\sqrt{2g}$$

with the general solution

$$(2/3)y^{3/2} = -kt + C.$$

The initial condition $y(0) = 4$ gives $C = 16/3$, and then $y(1) = 1$ yields $k = 14/3$. It follows that the depth at time t is

$$y(t) = (8 - 7t)^{2/3}.$$

(b) The tank is empty after $t = 8/7$ hr, that is, at 1:08:34 p.m.

(c) We see above that $k = (a/\pi b)\sqrt{2g} = 14/3$. Substitution of $a = \pi r^2$, $b = 1$, $g = (32)(3600)^2$ ft/hr² yields $r = (1/60)\sqrt{7/12}$ ft ≈ 0.15 in for the radius of the bottom-hole.

55. $A(y) = \pi(8y - y^2)$ as in Example 7 in the text, but now $a = \pi/144$ in Equation (24), so the initial value problem is

$$18(8y - y^2)y' = -\sqrt{y}, \qquad y(0) = 8.$$

We seek the value of t when $y = 0$. The answer is $t \approx 869$ sec $= 14$ min 29 sec.

57. **(a)** As in Example 8, the initial value problem is

$$\pi(8y - y^2)\frac{dy}{dt} = -\pi k\sqrt{y}, \qquad y(0) = 4$$

where $k = 0.6 r^2 \sqrt{2g} = 4.8 r^2$. Integrating and applying the initial condition just in the Example 8 solution in the text, we find that

$$\frac{16}{3}y^{3/2} - \frac{2}{5}y^{5/2} = -kt + \frac{448}{15}.$$

When we substitute $y = 2$ (ft) and $t = 1800$ (sec, that is, 30 min), we find that $k \approx 0.009469$. Finally, $y = 0$ when

$$t = \frac{448}{15k} \approx 3154 \text{ sec} = 53 \text{ min } 34 \text{ sec}.$$

Thus the tank is empty at 1:53:34 pm.

(b) The radius of the bottom-hole is

$$r = \sqrt{k/4.8} \approx 0.04442 \text{ ft} \approx 0.53 \text{ in, thus about a half inch}.$$

59. Let $t = 0$ at the time of death. Then the solution of the initial value problem

$$T' = k(70 - T), \qquad T(0) = 98.6$$

is

$$T(t) = 70 + 28.6\,e^{-kt}.$$

If $t = a$ at 12 noon, then we know that

$$T(t) = 70 + 28.6\,e^{-ka} = 80,$$

$$T(a+1) = 70 + 28.6\,e^{-k(a+1)} = 75.$$

Hence

$$28.6\,e^{-ka} = 10 \quad \text{and} \quad 28.6\,e^{-ka}e^{-k} = 5.$$

It follows that $e^{-k} = 1/2$, so $k = \ln 2$. Finally the first of the previous two equations yields

$$a = (\ln 2.86)/(\ln 2) \approx 1.516 \text{ hr} \approx 1 \text{ hr } 31 \text{ min},$$

so the death occurred at 10:29 a.m.

61. Let $t = 0$ when it began to snow, and $t = t_0$ at 7:00 a.m. Let x denote distance along the road, with $x = 0$ where the snowplow begins at 7:00 a.m. If $y = ct$ is the snow depth at time t, w is the width of the road, and $v = dx/dt$ is the plow's velocity, then "plowing at a constant rate" means that the product wyv is constant. Hence our differential equation is of the form

$$k\frac{dx}{dt} = \frac{1}{t}.$$

The solution with $x = 0$ when $t = t_0$ is

$$t = t_0 e^{kx}.$$

We are given that $x = 4$ when $t = t_0 + 1$ and $x = 7$ when $t = t_0 + 2$, so it follows that

$$t_0 + 1 = t_0 e^{4k} \quad \text{and} \quad t_0 + 2 = t_0 e^{7k}$$

at 8 a.m. and 9 a.m., respectively. Elimination of t_0 gives the equation

$$2e^{4k} - e^{7k} - 1 = 0,$$

which we solve numerically for $k = 0.08276$. Using this value, we finally solve one of the preceding pair of equations for $t_0 = 2.5483$ hr $\approx$ 2 hr 33 min. Thus it began to snow at 4:27 a.m.

SECTION 1.5

LINEAR FIRST-ORDER EQUATIONS

1. $\rho = \exp\left(\int 1\,dx\right) = e^x;\quad D_x(y \cdot e^x) = 2e^x;\quad y \cdot e^x = 2e^x + C;\quad y(x) = 2 + Ce^{-x}$

   $y(0) = 0$ implies $C = -2$ so $y(x) = 2 - 2e^{-x}$

3. $\rho = \exp\left(\int 3\,dx\right) = e^{3x};\quad D_x(y \cdot e^{3x}) = 2x;\quad y \cdot e^{3x} = x^2 + C;\quad y(x) = (x^2 + C)e^{-3x}$

5. $\rho = \exp\left(\int (2/x)\,dx\right) = e^{2\ln x} = x^2;\quad D_x(y \cdot x^2) = 3x^2;\quad y \cdot x^2 = x^3 + C$

   $y(x) = x + C/x^2;\quad y(1) = 5$ implies $C = 4$ so $y(x) = x + 4/x^2$

7. $\rho = \exp\left(\int (1/2x)\,dx\right) = e^{(\ln x)/2} = \sqrt{x};\quad D_x(y \cdot \sqrt{x}) = 5;\quad y \cdot \sqrt{x} = 5x + C$

   $y(x) = 5\sqrt{x} + C/\sqrt{x}$

9. $\rho = \exp\left(\int (-1/x)\,dx\right) = e^{-\ln x} = 1/x;\quad D_x(y \cdot 1/x) = 1/x;\quad y \cdot 1/x = \ln x + C$

   $y(x) = x\ln x + Cx;\quad y(1) = 7$ implies $C = 7$ so $y(x) = x\ln x + 7x$

11. $\rho = \exp\left(\int (1/x - 3)\,dx\right) = e^{\ln x - 3x} = xe^{-3x}$; $D_x(y \cdot xe^{-3x}) = 0$; $y \cdot xe^{-3x} = C$

$y(x) = Cx^{-1}e^{3x}$; $y(1) = 0$ implies $C = 0$ so $y(x) \equiv 0$ (constant)

13. $\rho = \exp\left(\int 1\,dx\right) = e^x$; $D_x(y \cdot e^x) = e^{2x}$; $y \cdot e^x = \tfrac{1}{2}e^{2x} + C$

$y(x) = \tfrac{1}{2}e^x + Ce^{-x}$; $y(0) = 1$ implies $C = \tfrac{1}{2}$ so $y(x) = \tfrac{1}{2}e^x + \tfrac{1}{2}e^{-x}$

15. $\rho = \exp\left(\int 2x\,dx\right) = e^{x^2}$; $D_x(y \cdot e^{x^2}) = xe^{x^2}$; $y \cdot e^{x^2} = \tfrac{1}{2}e^{x^2} + C$

$y(x) = \tfrac{1}{2} + Ce^{-x^2}$; $y(0) = -2$ implies $C = -\tfrac{5}{2}$ so $y(x) = \tfrac{1}{2} - \tfrac{5}{2}e^{-x^2}$

17. $\rho = \exp\left(\int 1/(1+x)\,dx\right) = e^{\ln(1+x)} = 1+x$; $D_x(y \cdot (1+x)) = \cos x$; $y \cdot (1+x) = \sin x + C$

$y(x) = \dfrac{C + \sin x}{1 + x}$; $y(0) = 1$ implies $C = 1$ so $y(x) = \dfrac{1 + \sin x}{1 + x}$

19. $\rho = \exp\left(\int \cot x\,dx\right) = e^{\ln(\sin x)} = \sin x$; $D_x(y \cdot \sin x) = \sin x \cos x$

$y \cdot \sin x = \tfrac{1}{2}\sin^2 x + C$; $y(x) = \tfrac{1}{2}\sin x + C\csc x$

21. $\rho = \exp\left(\int (-3/x)\,dx\right) = e^{-3\ln x} = x^{-3}$; $D_x(y \cdot x^{-3}) = \cos x$; $y \cdot x^{-3} = \sin x + C$

$y(x) = x^3 \sin x + Cx^3$; $y(2\pi) = 0$ implies $C = 0$ so $y(x) = x^3 \sin x$

23. $\rho = \exp\left(\int (2 - 3/x)\,dx\right) = e^{2x - 3\ln x} = x^{-3}e^{2x}$; $D_x(y \cdot x^{-3}e^{2x}) = 4e^{2x}$

$y \cdot x^{-3}e^{2x} = 2e^{2x} + C$; $y(x) = 2x^3 + Cx^3 e^{-2x}$

25. First we calculate

$$\int \frac{3x^3\,dx}{x^2+1} = \int \left[3x - \frac{3x}{x^2+1}\right]dx = \frac{3}{2}\left[x^2 - \ln(x^2+1)\right].$$

It follows that $\rho = (x^2+1)^{-3/2}\exp(3x^2/2)$ and thence that

$D_x\left(y \cdot (x^2+1)^{-3/2}\exp(3x^2/2)\right) = 6x(x^2+4)^{-5/2}$,

$y \cdot (x^2+1)^{-3/2}\exp(3x^2/2) = -2(x^2+4)^{-3/2} + C$,

$y(x) = -2\exp(3x^2/2) + C(x^2+1)^{3/2}\exp(-3x^2/2)$.

Finally, $y(0) = 1$ implies that $C = 3$ so the desired particular solution is

$$y(x) = -2\exp(3x^2/2) + 3(x^2+1)^{3/2}\exp(-3x^2/2).$$

27. With $x' = dx/dy$, the differential equation is $x' - x = ye^y$. Then with y as the independent variable we calculate

$$\rho(y) = \exp\left(\int(-1)dy\right) = e^{-y}; \quad D_y\left(x \cdot e^{-y}\right) = y$$

$$x \cdot e^{-y} = \tfrac{1}{2}y^2 + C; \quad x(y) = \left(\tfrac{1}{2}y^2 + C\right)e^y$$

29. $\rho = \exp\left(\int(-2x)dx\right) = e^{-x^2}; \quad D_x\left(y \cdot e^{-x^2}\right) = e^{-x^2}; \quad y \cdot e^{-x^2} = C + \int_0^x e^{-t^2}\,dt$

$$y(x) = e^{x^2}\left(C + \tfrac{\sqrt{\pi}}{2}\text{erf}(x)\right)$$

31. (a) $y_c' = Ce^{-\int P\,dx}(-P) = -P\,y_c$, so $y_c' + P\,y_c = 0$.

(b) $y_p' = (-P)e^{-\int P\,dx} \cdot \left[\int\left(Qe^{\int P\,dx}\right)dx\right] + e^{-\int P\,dx} \cdot Qe^{\int P\,dx} = -Py_p + Q$

33. The amount $x(t)$ of salt (in kg) after t seconds satisfies the differential equation $x' = -x/200$, so $x(t) = 100e^{-t/200}$. Hence we need only solve the equation $10 = 100e^{-t/200}$ for $t = 461$ sec $= 7$ min 41 sec (approximately).

35. The only difference from the Example 4 solution in the textbook is that $V = 1640$ km³ and $r = 410$ km³/yr for Lake Ontario, so the time required is

$$t = \frac{V}{r}\ln 4 = 4\ln 4 \approx 5.5452 \text{ years}.$$

37. The volume of brine in the tank after t min is $V(t) = 100 + 2t$ gal, so the initial value problem is

$$\frac{dx}{dt} = 5 - \frac{3x}{100+2t}, \quad x(0) = 50.$$

The integrating factor $\rho(t) = (100+2t)^{3/2}$ leads to the solution

$$x(t) = (100+2t) - \frac{50000}{(100+2t)^{3/2}}.$$

such that $x(0) = 50$. The tank is full after $t = 150$ min, at which time $x(150) = 393.75$ lb.

39. **(a)** The initial value problem

$$\frac{dx}{dt} = -\frac{x}{10}, \qquad x(0) = 100$$

for Tank 1 has solution $x(t) = 100e^{-t/10}$. Then the initial value problem

$$\frac{dy}{dt} = \frac{x}{10} - \frac{y}{10} = 10e^{-t/10} - \frac{y}{10}, \qquad y(0) = 0$$

for Tank 2 has solution $y(t) = 10te^{-t/10}$.

(b) The maximum value of y occurs when

$$y'(t) = 10e^{-t/10} - te^{-t/10} = 0$$

and thus when $t = 10$. We find that $y_{\max} = y(10) = 100e^{-1} \approx 36.79$ gal.

41. **(a)** $A'(t) = 0.06A + 0.12S = 0.06A + 3.6e^{0.05t}$

(b) The solution with $A(0) = 0$ is

$$A(t) = 360(e^{0.06t} - e^{0.05t}),$$

so $A(40) \approx 1308.283$ thousand dollars.

43. The solution of the initial value problem $y' = x - y$, $y(-5) = y_0$ is

$$y(x) = x - 1 + (y_0 + 6)e^{-x-5}.$$

Substituting $x = 5$, we therefore solve the equation $4 + (y_0 + 6)e^{-10} = y_1$ with $y_1 = 3.998, 3.999, 4, 4.001, 4.002$ for the desired initial values $y_0 = -50.0529, -28.0265, -6.0000, 16.0265, 38.0529$, respectively.

SECTION 1.6

SUBSTITUTION METHODS AND EXACT EQUATIONS

It is traditional for every elementary differential equations text to include the particular types of equations that are found in this section. However, no one of them is vitally important solely in its own right. Their real purpose (at this point in the course) is to familiarize students with the technique of transforming a differential equation by substitution. The subsection on airplane flight trajectories (together with Problems 56–59) is optional material and may be omitted if the instructor desires.

The differential equations in Problems 1–15 are homogeneous, so we make the substitutions

$$v = \frac{y}{x}, \quad y = vx, \quad \frac{dy}{dx} = v + x\frac{dv}{dx}.$$

For each problem we give the differential equation in x, $v(x)$, and $v' = dv/dx$ that results, together with the principal steps in its solution.

1. $x(v+1)v' = -(v^2+2v-1);$ $\quad \int \frac{2(v+1)dv}{v^2+2v-1} = -\int 2x\,dx;\quad \ln(v^2+2v-1) = -2\ln x + \ln C$

$x^2(v^2+2v-1) = C;\quad y^2+2xy-x^2 = C$

3. $xv' = 2\sqrt{v};\quad \int \frac{dv}{2\sqrt{v}} = \int \frac{dx}{x};\quad \sqrt{v} = \ln x + C;\quad y = x(\ln x + C)^2$

5. $x(v+1)v' = -2v^2;\quad \int\left(\frac{1}{v}+\frac{1}{v^2}\right)dv = -\int\frac{2\,dx}{x};\quad \ln v - \frac{1}{v} = -2\ln x + C$

$\ln y - \ln x - \frac{x}{y} = -2\ln x + C;\quad \ln(xy) = C + \frac{x}{y}$

7. $xv^2v' = 1;\quad \int 3v^2\,dv = \int \frac{3\,dx}{x};\quad v^3 = 3\ln x + C;\quad y^3 = x^3(3\ln x + C)$

9. $xv' = v^2;\quad -\int \frac{dv}{v^2} = -\int \frac{dx}{x};\quad \frac{1}{v} = -\ln x + C;\quad x = y(C - \ln x)$

11. $x(1-v^2)v' = v + v^3;\quad \int \frac{1-v^2}{v^3+v}dv = \int \frac{dx}{x};\quad \int\left(\frac{1}{v}-\frac{2v}{v^2+1}\right)dv = \int\frac{dx}{x}$

$\ln v - \ln(v^2+1) = \ln x + \ln C;\quad v = Cx(v^2+1);\quad y = C(x^2+y^2)$

13. $xv' = \sqrt{v^2+1};\quad \int \frac{dv}{\sqrt{v^2+1}} = \int \frac{dx}{x};\quad \ln(v+\sqrt{v^2+1}) = \ln x + \ln C$

$v + \sqrt{v^2+1} = Cx;\quad y + \sqrt{x^2+y^2} = Cx^2$

Section 1.6

15. $x(v+1)v' = -2(v^2+2v);$ $\quad \int \frac{2(v+1)\,dv}{v^2+2v} = -\int \frac{4\,dx}{x};$ $\quad \ln(v^2+2v) = -4\ln x + \ln C$

$v^2 + 2v = C/x^4;$ $\quad x^2y^2 + 2x^3y = C$

17. $v = 4x + y;$ $\quad v' = v^2 + 4;$ $\quad x = \int \frac{dv}{v^2+4} = \frac{1}{2}\tan^{-1}\frac{v}{2} + \frac{C}{2}$

$v = 2\tan(2x-C);$ $\quad y = 2\tan(2x-C) - 4x$

Problems 19–25 are Bernoulli equations. For each, we indicate the appropriate substitution as specified in Equation (10) of this section, the resulting linear differential equation in v, its integrating factor ρ, and finally the resulting solution of the original Bernoulli equation.

19. $v = y^{-2};$ $\quad v' - 4v/x = -10/x^2;$ $\quad \rho = 1/x^4;$ $\quad y^2 = x/(Cx^5 + 2)$

21. $v = y^{-2};$ $\quad v' + 2v = -2;$ $\quad \rho = e^{2x};$ $\quad y^2 = 1/(Ce^{-2x} - 1)$

23. $v = y^{-1/3};$ $\quad v' - 2v/x = -1;$ $\quad \rho = x^{-2};$ $\quad y^3 = 1/(x + Cx^2)$

25. $v = y^3;$ $\quad v' + 3v/x = 3/\sqrt{1+x^4};$ $\quad \rho = x^3;$ $\quad y^3 = (C + 3\sqrt{1+x^4})/(2x^3)$

27. The substitution $v = y^3$ yields the linear equation $xv' - v = 3x^4$ with integrating factor $\rho = 1/x$. Solution: $y = (x^4 + Cx)^{1/3}$

29. The substitution $v = \sin y$ yields the homogeneous equation $2xv\,v' = 4x^2 + v^2$. Solution: $\sin^2 y = 4x^2 - Cx$

Each of the differential equations in Problems 31–42 is of the form $M\,dx + N\,dy = 0$, and the exactness condition $\partial M/\partial y = \partial N/\partial x$ is routine to verify. For each problem we give the principal steps in the calculation corresponding to the method of Example 9 in this section.

31. $F = \int (2x+3y)\,dx = x^2 + 3xy + g(y);$ $\quad F_y = 3x + g'(y) = 3x + 2y = N$

$g'(y) = 2y;$ $\quad g(y) = y^2;$ $\quad x^2 + 3xy + y^2 = C$

33. $F = \int (3x^2 + 2y^2)\,dx = x^3 + xy^2 + g(y);$ $\quad F_y = 4xy + g'(y) = 4xy + 6y^2 = N$

$g'(y) = 6y^2;$ $\quad g(y) = 2y^3;$ $\quad x^3 + 2xy^2 + 2y^3 = C$

35. $F = \int (x^3 + y/x)\,dx = \frac{1}{4}x^4 + y\ln x + g(y);$ $\quad F_y = \ln x + g'(y) = y^2 + \ln x = N$

$g'(y) = y^2$; $g(y) = \frac{1}{3}y^3$; $\frac{1}{4}x^4 + \frac{1}{3}y^3 + y\ln x = C$

37. $F = \int(\cos x + \ln y)\,dx = \sin x + x\ln y + g(y)$; $F_y = x/y + g'(y) = x/y + e^y = N$

$g'(y) = e^y$; $g(y) = e^y$; $\sin x + x\ln y + e^y = C$

39. $F = \int(3x^2 y^3 + y^4)\,dx = x^3 y^3 + x y^4 + g(y)$;

$F_y = 3x^3 y^2 + 4xy^3 + g'(y) = 3x^3 y^2 + y^4 + 4xy^3 = N$

$g'(y) = y^4$; $g(y) = \frac{1}{5}y^5$; $x^3 y^3 + xy^4 + \frac{1}{5}y^5 = C$

41. $F = \int\left(\dfrac{2x}{y} - \dfrac{3y^2}{x^4}\right)dx = \dfrac{x^2}{y} + \dfrac{y^2}{x^3} + g(y)$;

$F_y = -\dfrac{x^2}{y^2} + \dfrac{2y}{x^3} + g'(y) = -\dfrac{x^2}{y^2} + \dfrac{2y}{x^3} + \dfrac{1}{\sqrt{y}} = N$

$g'(y) = \dfrac{1}{\sqrt{y}}$; $g(y) = 2\sqrt{y}$; $\dfrac{x^2}{y} + \dfrac{y^2}{x^3} + 2\sqrt{y} = C$

43. The substitution $v = ax + by + c$, $y = (v - ax - c)/b$ in $y' = F(ax + by + c)$ yields the separable differential equation $(dv/dx - a)/b = F(v)$, that is, $dv/dx = a + bF(v)$.

45. If $v = \ln y$ then $y = e^v$ so $y' = e^v v'$. Hence the given equation transforms to $e^v v' + P(x)e^v = Q(x) v e^v$. Cancellation of the factor e^v then yields the linear differential equation $v' - Q(x)v = P(x)$.

47. The substitution $x = u - 1$, $y = v - 2$ yields the homogeneous equation

$$\dfrac{dv}{du} = \dfrac{u - v}{u + v}.$$

The substitution $v = pu$ leads to

$$\ln u = -\int \dfrac{(p+1)\,dp}{(p^2 + 2p - 1)} = -\dfrac{1}{2}\left[\ln\left(p^2 + 2p - 1\right) - \ln C\right].$$

We thus obtain the implicit solution

$$u^2\left(p^2 + 2p - 1\right) = C$$

$$u^2\left(\frac{v^2}{u^2}+2\frac{v}{u}-1\right) = v^2+2uv-u^2 = C$$

$$(y+2)^2+2(x+1)(y+2)-(x+1)^2 = C$$

$$y^2+2xy-x^2+2x+6y = C.$$

49. The substitution $v = x-y$ yields the separable equation $v' = 1-\sin v$. With the aid of the identity

$$\frac{1}{1-\sin v} = \frac{1+\sin v}{\cos^2 v} = \sec^2 v + \sec v \tan v$$

we obtain the solution

$$x = \tan(x-y) + \sec(x-y) + C.$$

51. If we substitute $y = y_1+1/v$, $y' = y_1'-v'/v^2$ (primes denoting differentiation with respect to x) into the Riccati equation $y' = Ay^2+By+C$ and use the fact that $y_1' = Ay_1^2+By_1+C$, then we immediately get the linear differential equation $v'+(B+2Ay_1)v = -A$.

In Problems 52 and 53 we outline the application of the method of Problem 51 to the given Riccati equation.

53. The substitution $y = x+1/v$ yields the trivial linear equation $v' = -1$ with immediate solution $v(x) = C-x$. Hence the general solution of our Riccati equation is given by $y(x) = x+1/(C-x)$.

55. Clearly the line $y = Cx - C^2/4$ and the tangent line at $(C/2, C^2/4)$ to the parabola $y = x^2$ both have slope C.

57. With $a = 100$ and $k = 1/10$, Equation (19) in the text is

$$y = 50[(x/100)^{9/10} - (x/100)^{11/10}].$$

The equation $y'(x) = 0$ then yields

$$(x/100)^{1/10} = (9/11)^{1/2},$$

so it follows that

$$y_{\max} = 50[(9/11)^{9/2} - (9/11)^{11/2}] \approx 3.68 \text{ mi.}$$

59. **(a)** With $a = 100$ and $k = w/v_0 = 2/4 = 1/2$, the solution given by equation (19) in the textbook is $y(x) = 50[(x/100)^{1/2} - (x/100)^{3/2}]$. The fact that $y(0) = 0$ means that this trajectory goes through the origin where the tree is located.

(b) With $k = 4/4 = 1$ the solution is $y(x) = 50[1 - (x/100)^2]$ and we see that the swimmer hits the bank at a distance $y(0) = 50$ north of the tree.

(c) With $k = 6/4 = 1$ the solution is $y(x) = 50[(x/100)^{-1/2} - (x/100)^{5/2}]$. This trajectory is asymptotic to the positive x-axis, so we see that the swimmer never reaches the west bank of the river.

CHAPTER 1 Review Problems

The main objective of this set of review problems is practice in the identification of the different types of first-order differential equations discussed in this chapter. In each of Problems 1-36 we identify the type of the given equation and indicate an appropriate method of solution.

1. If we write the equation in the form $y' - (3/x)y = x^2$ we see that it is *linear* with integrating factor $\rho = x^{-3}$. The method of Section 1.5 then yields the general solution $y = x^3(C + \ln x)$.

3. This equation is *homogeneous*. The substitution $y = vx$ of Equation (8) in Section 1.6 leads to the general solution $y = x/(C - \ln x)$.

5. We write this equation in the *separable* form $y'/y^2 = (2x-3)/x^4$. Then separation of variables and integration as in Section 1.4 yields the general solution $y = C \exp[(1-x)/x^3]$.

7. If we write the equation in the form $y' + (2/x)y = 1/x^3$ we see that it is *linear* with integrating factor $\rho = x^2$. The method of Section 1.5 then yields the general solution $y = x^{-2}(C + \ln x)$.

9. If we write the equation in the form $y' + (2/x)y = 6x\sqrt{y}$ we see that it is a *Bernoulli equation* with $n = 1/2$. The substitution $v = y^{-1/2}$ of Eq. (10) in Section 1.6 then yields the general solution $y = (x^2 + C/x)^2$.

11. This equation is *homogeneous*. The substitution $y = vx$ of Equation (8) in Section 1.6 leads to the general solution $y = x/(C - 3 \ln x)$.

13. We write this equation in the *separable* form $y'/y^2 = 5x^4 - 4x$. Then separation of variables and integration as in Section 1.4 yields the general solution $y = 1/(C + 2x^2 - x^5)$.

15. This is a *linear* differential equation with integrating factor $\rho = e^{3x}$. The method of Section 1.5 yields the general solution $y = (x^3 + C)e^{-3x}$.

17. We note that $D_y\left(e^x + ye^{xy}\right) = D_x\left(e^y + xe^{xy}\right) = e^{xy} + xye^{xy}$, so the given equation is *exact*. The method of Example 9 in Section 1.6 yields the implicit general solution $e^x + e^y + e^{xy} = C.$

19. We write this equation in the *separable* form $y'/y^2 = (2 - 3x^5)/x^3$. Then separation of variables and integration as in Section 1.4 yields the general solution $y = x^2/(x^5 + Cx^2 + 1)$.

21. If we write the equation in the form $y' + (1/(x+1))y = 1/(x^2 - 1)$ we see that it is *linear* with integrating factor $\rho = x + 1$. The method of Section then 1.5 yields the general solution $y = [C + \ln(x - 1)]/(x + 1)$.

23. We note that $D_y\left(e^y + y\cos x\right) = D_x\left(xe^y + \sin x\right) = e^y + \cos x$, so the given equation is *exact*. The method of Example 9 in Section 1.6 yields the implicit general solution $xe^y + y\sin x = C$

25. If we write the equation in the form $y' + (2/(x+1))y = 3$ we see that it is *linear* with integrating factor $\rho = (x+1)^2$. The method of Section 1.5 then yields the general solution $y = x + 1 + C(x + 1)^{-2}$.

27. If we write the equation in the form $y' + (1/x)y = -x^2y^4/3$ we see that it is a *Bernoulli* equation with $n = 4$. The substitution $v = y^{-3}$ of Eq. (10) in Section 1.6 then yields the general solution $y = x^{-1}(C + \ln x)^{-1/3}$.

29. If we write the equation in the form $y' + (1/(2x+1))y = (2x+1)^{1/2}$ we see that it is *linear* with integrating factor $\rho = (2x+1)^{1/2}$. The method of Section 1.5 then yields the general solution $y = (x^2 + x + C)(2x + 1)^{-1/2}$.

31. $dy/(y+7) = 3x^2 dx$ is separable; $y' + 3x^2 y = 21x^2$ is linear.

33. $(3x^2 + 2y^2)dx + 4xy\, dy = 0$ is exact; $y' = -\frac{1}{4}(3x/y + 2y/x)$ is homogeneous.

35. $dy/(y+1) = 2x\,dx/(x^2+1)$ is separable; $y' - (2x/(x^2+1))y = 2x/(x^2+1)$ is linear.

CHAPTER 2

MATHEMATICAL MODELS AND NUMERICAL METHODS

SECTION 2.1

POPULATION MODELS

Section 2.1 introduces the first of the two major classes of mathematical models studied in the textbook, and is a prerequisite to the discussion of equilibrium solutions and stability in Section 2.2.

In Problems 1-4 we outline the derivation of the desired particular solution.

1. Noting that $x > 5$ because $x(0) = 8$, we write

$$\int \frac{dx}{x(x-5)} = \int (-3)\, dt; \qquad \int \left(\frac{1}{x} - \frac{1}{x-5}\right) dx = \int 15\, dt$$

$$\ln x - \ln(x-5) = 15t + \ln C; \qquad \frac{x}{x-5} = C e^{15t}$$

$$x(0) = 8 \text{ implies } C = 8/3; \qquad 3x = 8(x-5)e^{15t}$$

$$x(t) = \frac{-40 e^{15t}}{3 - 8 e^{15t}} = \frac{40}{8 - 3 e^{-15t}}$$

3. Noting that $x > 7$ because $x(0) = 11$, we write

$$\int \frac{dx}{x(x-7)} = \int (-4)\, dt; \qquad \int \left(\frac{1}{x} - \frac{1}{x-7}\right) dx = \int 28\, dt$$

$$\ln x - \ln(x-7) = 28t + \ln C; \qquad \frac{x}{x-7} = C e^{28t}$$

$$x(0) = 11 \text{ implies } C = 11/4; \qquad 4x = 11(x-7)e^{28t}$$

$$x(t) = \frac{-77 e^{28t}}{4 - 11 e^{28t}} = \frac{77}{11 - 4 e^{-28t}}$$

5. Substitution of $P(0)=100$ and $P'(0)=20$ into $P'=k\sqrt{P}$ yields $k=2$, so the differential equation is $P'=2\sqrt{P}$. Separation of variables and integration, $\int dP/2\sqrt{P} = \int dt$, gives $\sqrt{P} = t+C$. Then $P(0)=100$ implies $C=10$, so $P(t) = (t+10)^2$. Hence the number of rabbits after one year is $P(12) = 484$.

7. (a) Starting with $dP/dt = k\sqrt{P}$, $dP/dt = k\sqrt{P}$, we separate the variables and integrate to get $P(t) = (kt/2 + C)^2$. Clearly $P(0) = P_0$ implies $C = \sqrt{P_0}$.

(b) If $P(t) = (kt/2 + 10)^2$, then $P(6) = 169$ implies that $k = 1$. Hence $P(t) = (t/2 + 10)^2$, so there are 256 fish after 12 months.

9. (a) If the birth and death rates both are proportional to P^2 and $\beta > \delta$, then Eq. (1) in this section gives $P' = kP^2$ with k positive. Separating variables and integrating as in Problem 8, we find that $P(t) = 1/(C-kt)$. The initial condition $P(0) = P_0$ then gives $C = 1/P_0$, so $P(t) = 1/(1/P_0 - kt) = P_0/(1-kP_0 t)$.

(b) If $P_0 = 6$ then $P(t) = 6/(1-6kt)$. Now the fact that $P(10) = 9$ implies that $k = 180$, so $P(t) = 6/(1-t/30) = 180/(30-t)$. Hence it is clear that $P \to \infty$ as $t \to 30$ (doomsday).

11. If we write $P' = bP(a/b - P)$ we see that $M = a/b$. Hence

$$\frac{B_0 P_0}{D_0} = \frac{(aP_0)P_0}{bP_0^2} = \frac{a}{b} = M.$$

Note also (for Problems 12 and 13) that $a = B_0/P_0$ and $b = D_0/P_0^2 = k$.

13. The relations in Problem 11 give $k = 1/2400$ and $M = 180$. The solution is $P(t) = 43200/(240 - 60e^{-3t/80})$. We find that $P = 1.05M$ after about 44.22 months.

15. The relations in Problem 14 give $k = 1/1000$ and $M = 90$. The solution is $P(t) = 9000/(100 - 10e^{9t/100})$. We find that $P = 10M$ after about 24.41 months.

17. The only difference is that, if $P > M$, then

$$\int \frac{dP}{M-P} = -\ln|M-P| = -\ln(P-M)$$

so integration of the separated equation yields

$$\frac{P}{P-M} = Ae^{kMt}.$$

This gives $A = P_0/(P_0 - M)$, so each side of the equation preceding Equation (4) in this section is simply multiplied by -1, and the result in Equation (4) is unchanged.

19. **(a)** $x' = 0.8x - 0.004x^2 = 0.004x(200 - x)$, so the maximum amount that will dissolve is $M = 200$ g.

    (b) With $M = 200$, $P_0 = 50$, and $k = 0.004$, Equation (4) in the text yields the solution

    $$x(t) = \frac{10000}{50 + 150 e^{-0.08t}}.$$

    Substituting $x = 100$ on the left, we solve for $t = 1.25 \ln 3 \approx 1.37$ sec.

21. Proceeding as in Example 4 in the text, we solve the equations

    $$25.00k(M - 25.00) = 3/8, \qquad 47.54k(M - 47.54) = 1/2$$

    for $M = 100$ and $k = 0.0002$. Then Equation (4) gives the population function

    $$P(t) = \frac{2500}{25 + 75e^{-0.02t}}.$$

    We find that $P = 75$ when $t = 50 \ln 9 \approx 110$, that is, in 2035 A. D.

23. We are given that
    $$P' = kP^2 - 0.01P,$$

    and the fact that $P = 200$ and $P' = 2$ when $t = 0$ implies that $k = 0.0001$, so

    $$P' = 10^{-4}(P^2 - P).$$

    The solution with $P(0) = 200$ is $P(t) = 100/\left(1 - 0.5 e^{0.01t}\right)$.

    (a) $P = 1000$ when $t = 100 \ln(9/5) \approx 58.78$.

    (b) $P \to \infty$ as $t \to 100 \ln 2 \approx 69.31$.

25. The equation is separable, so we have

    $$\int \frac{dP}{P} = \int \beta_0 e^{-\alpha t} dt, \quad \text{so} \quad \ln P = -\frac{\beta_0}{\alpha} e^{-\alpha t} + C.$$

The initial condition $P(0) = P_0$ gives $C = \ln P_0 + \beta_0/\alpha$, so

$$P(t) = P_0 \exp\left[\frac{\beta_0}{\alpha}\left(1 - e^{-\alpha t}\right)\right].$$

27. Any way you look at it, you should see that, the larger the parameter $k > 0$ is, the faster the logistic population $P(t)$ approaches its limiting population M.

In Problems 29 and 30 we give just the values of k and M calculated using Eqs. (ii) and (iii) in Problem 28 above, the resulting logistic solution, and the predicted year 2000 population.

29. $k = 0.0000668717$ and $M = 338.027$, so $P(t) = \dfrac{25761.7}{76.212 + 261.815\,e^{-0.0226045\,t}}$,

predicting $P = 192.525$ in the year 2000.

SECTION 2.2

EQUILIBRIUM SOLUTIONS AND STABILITY

In Problems 1-12 we identify the stable and unstable critical points as well as the funnels and spouts along the equilibrium solutions. In each problem the indicated solution satisfying $x(0) = x_0$ is derived fairly routinely by separation of variables. In some cases, various signs in the solution depend on the initial value, and we give a typical solution.

1. Unstable critical point: $x = 4$
Spout: Along the equilibrium solution $x(t) = 4$

Solution: If $x_0 > 4$ then

$$\int \frac{dx}{x-4} = \int dt; \quad \ln(x-4) = t + C; \quad C = \ln(x_0 - 4)$$

$$x - 4 = (x_0 - 4)e^t; \quad x(t) = 4 + (x_0 - 4)e^t$$

3. Stable critical point: $x = 0$
Unstable critical point: $x = 4$
Funnel: Along the equilibrium solution $x(t) = 0$
Spout: Along the equilibrium solution $x(t) = 4$

Solution: If $x_0 > 4$ then

$$\int 4\,dt = \int \frac{4\,dx}{x(x-4)} = \int \left(\frac{1}{x-4} - \frac{1}{x}\right)dx$$

$$4t + C = \ln\frac{x-4}{x}; \quad C = \ln\frac{x_0 - 4}{x_0}$$

$$4t = \ln\frac{x_0(x-4)}{x(x_0 - 4)}; \quad e^{4t} = \frac{x_0(x-4)}{x(x_0 - 4)}$$

$$x(t) = \frac{4x_0}{x_0 + (4 - x_0)e^{4t}}$$

5. Stable critical point: $x = -2$
   Unstable critical point: $x = 2$
   Funnel: Along the equilibrium solution $x(t) = -2$
   Spout: Along the equilibrium solution $x(t) = 2$

   Solution: If $x_0 > 2$ then

   $$\int 4\,dt = \int \frac{4\,dx}{x^2 - 4} = \int \left(\frac{1}{x-2} - \frac{1}{x+2}\right) dx$$

   $$4t + C = \ln\frac{x-2}{x+2}; \quad C = \ln\frac{x_0 - 2}{x_0 + 2}$$

   $$4t = \ln\frac{(x-2)(x_0 + 2)}{(x+2)(x_0 - 2)}; \quad e^{4t} = \frac{(x-2)(x_0 + 2)}{(x+2)(x_0 - 2)}$$

   $$x(t) = \frac{2\left[(x_0 + 2) + (x_0 - 2)e^{4t}\right]}{(x_0 + 2) - (x_0 - 2)e^{4t}}$$

7. Critical point: $x = 2$

   This single critical point is *semi-stable*, meaning that solutions with $x_0 > 2$ go to infinity as t increases, while solutions with $x_0 < 2$ approach 2.

   Solution: If $x_0 > 2$ then

   $$\int \frac{-dx}{(x-2)^2} = \int (-1)\,dt; \quad \frac{1}{x-2} = -t + C; \quad C = \frac{1}{x_0 - 2}$$

   $$\frac{1}{x-2} = -t + \frac{1}{x_0 - 2} = \frac{1 - t(x_0 - 2)}{x_0 - 2}$$

   $$x(t) = 2 + \frac{x_0 - 2}{1 - t(x_0 - 2)} = \frac{x_0(2t - 1) - 4t}{t x_0 - 2t - 1}$$

9. Stable critical point: $x = 1$
   Unstable critical point: $x = 4$
   Funnel: Along the equilibrium solution $x(t) = 1$

Spout: Along the equilibrium solution $x(t) = 4$

Solution: If $x_0 > 4$ then

$$\int 3 dt = \int \frac{3 dx}{(x-4)(x-1)} = \int \left(\frac{1}{x-4} - \frac{1}{x-1} \right) dx$$

$$3t + C = \ln \frac{x-4}{x-1}; \quad C = \ln \frac{x_0 - 4}{x_0 - 1}$$

$$3t = \ln \frac{(x-4)(x_0 - 1)}{(x-1)(x_0 - 4)}; \quad e^{3t} = \frac{(x-4)(x_0 - 1)}{(x-1)(x_0 - 4)}$$

$$x(t) = \frac{4(1-x_0) + (x_0 - 4)e^{3t}}{(1-x_0) + (x_0 - 4)e^{3t}}$$

11. Unstable critical point: $x = 1$

Spout: Along the equilibrium solution $x(t) = 1$

Solution: $\int \frac{-2 dx}{(x-1)^3} = \int (-2) dt; \quad \frac{1}{(x-1)^2} = -2t + \frac{1}{(x_0 - 1)^2}$

13. **(a)** If $k = -a^2$ where $h > 0$ then $kx - x^3 = -a^2 x - x^3 = -x(a^2 + x^2)$ is positive if $x < 0$, negative if $x > 0$, and is 0 only if $x = 0$.

(b) If $k = +a^2$ where $h > 0$ then $kx - x^3 = +a^2 x - x^3 = -x(x+a)(x-a)$ is positive if $x < -a$, negative if $-a < x < 0$, positive if $0 < x < a$, and negative if $x > a$.

15. If $x_0 > M$ then

$$\int kM \, dt = \int \frac{M \, dx}{x(x-M)} = \int \left(\frac{1}{x-M} - \frac{1}{x} \right) dx$$

$$kMt + C = \ln \frac{x-M}{x}; \quad C = \ln \frac{x_0 - M}{x_0}$$

$$kMt = \ln \frac{x_0(x-M)}{x(x_0 - M)}; \quad e^{kMt} = \frac{x_0(x-M)}{x(x_0 - M)}$$

$$x(t) = \frac{Mx_0}{x_0 + (M - x_0)e^{kMt}}$$

17. **(i)** In the first alternative form that is given, all of the coefficients within parentheses are positive if $H < x_0 < N$. Hence it is obvious that $x(t) \to N$ as $t \to \infty$.

(ii) In the second alternative form that is given, all of the coefficients within parentheses are positive if $x_0 < H$. Hence the denominator is initially equal to $N - H > 0$, but decreases as t increases, and reaches the value 0 when

$$t = \frac{1}{N-H} \ln \frac{N-x_0}{H-x_0} > 0.$$

19. Separation of variables in the differential equation $x' = -k\left((x-a)^2 + b^2\right)$ yields

$$x(t) = a - b \tan\left(bkt + \tan^{-1}\frac{a-x_0}{b}\right).$$

It therefore follows that $x(t)$ goes to minus infinity in a finite period of time.

21. This is simply a matter of analyzing the signs of x' in the cases $x < a$, $a < x < b$, $b < x < c$, and $c > x$.

SECTION 2.3

ACCELERATION-VELOCITY MODELS

This section consists of three essentially independent subsections that can be studied separately: resistance proportional to velocity, resistance proportional to velocity-squared, and inverse-square gravitational acceleration.

1. Equation: $v' = k(250 - v)$, $v(0) = 0$, $v(10) = 100$

Solution: $\int \frac{(-1)\,dv}{250-v} = -\int k\,dt;$ $\ln(250-v) = -kt + \ln C,$

$v(0) = 0$ implies $C = 250$; $v(t) = 250(1 - e^{-kt})$

$v(10) = 100$ implies $k = \frac{1}{10}\ln(250/150) \approx 0.0511;$

Answer: $v = 200$ when $t = -(\ln 50/250)/k \approx 31.5$ sec

3. Equation: $v' = -kv$, $v(0) = 40;$ $v(10) = 20$ $x' = v$, $x(0) = 0$

Solution: $v(t) = 40\,e^{-kt}$ with $k = (1/10)\ln 2$

$x(t) = (40/k)(1 - e^{-kt})$

Answer: $x(\infty) = \lim_{t \to \infty}(40/k)(1 - e^{-kt}) = 40/k = 400/\ln 2 \approx 577$ ft

5. Equation: $v' = -kv$, $v(0) = 40$; $v(10) = 20$ $x' = v$, $x(0) = 0$

Solution: $v = \dfrac{40}{1+40kt}$ (as in Problem 3)

$v(10) = 20$ implies $40k = 1/10$, so $v(t) = \dfrac{400}{10+t}$

$x(t) = 400 \ln[(10+t)/10]$

Answer: $x(60) = 400 \ln 7 \approx 778$ ft

7. Equation: $v' = 10 - 0.1v$, $x(0) = v(0) = 0$

(a) $\displaystyle\int \dfrac{-0.1\,dv}{10-0.1v} = \int (-0.1)\,dt$; $\ln(10-0.1v) = -t/10 + \ln C$

$v(0) = 0$ implies $C = 10$; $\ln\bigl[(10-0.1v)/10\bigr] = -t/10$

$v(t) = 100(1 - e^{-t/10})$; $v(\infty) = 100$ ft/sec (limiting velocity)

(b) $x(t) = 100t - 1000(1 - e^{-t/10})$

$v = 90$ ft/sec when $t = 23.0259$ sec and $x = 1402.59$ ft

9. The solution of the initial value problem

$$1000 v' = 5000 - 100 v, \qquad v(0) = 0$$

is

$$v(t) = 50(1 - e^{-t/10}).$$

Hence, as $t \to \infty$, we see that $v(t)$ approaches $v_{\max} = 50$ ft/sec ≈ 34 mph.

11. If the paratrooper's terminal velocity was 100 mph = 440/3 ft/sec, then Equation (7) in the text yields $\rho = 12/55$. Then we find by solving Equation (9) numerically with $y_0 = 1200$ and $v_0 = 0$ that $y = 0$ when $t \approx 12.5$ sec. Thus the newspaper account is inaccurate.

Given the hints and integrals provided in the text, Problems 13-16 are fairly straightforward (and fairly tedious) integration problems.

17. To solve the initial value problem $v' = -9.8 - 0.0011v^2$, $v(0) = 49$ we write

$$\int \dfrac{dv}{9.8+0.0011v^2} = -\int dt; \qquad \int \dfrac{0.010595\,dv}{1+(0.010595v)^2} = -\int 0.103827\,dt$$

$$\tan^{-1}(0.010595\,v) = -0.103827\,t + C; \quad v(0) = 49 \text{ implies } C = 0.478854$$

$$v(t) = 94.3841 \tan(0.478854 - 0.103827\,t)$$

Integration with $y(0) = 0$ gives

$$y(t) = 108.468 + 909.052 \ln(\cos(0.478854 - 0.103827\,t)).$$

We solve $v(0) = 0$ for $t = 4.612$, and then calculate $y(4.612) = 108.468$.

19. Equation: $v' = 4 - (1/400)v^2, \quad v(0) = 0$

Solution: $\displaystyle\int \frac{dv}{4 - (1/400)v^2} = \int dt; \quad \int \frac{(1/40)\,dv}{1 - (v/40)^2} = \int \frac{1}{10}\,dt$

$\tanh^{-1}(v/40) = t/10 + C; \quad C = 0; \quad v(t) = 40 \tanh(t/10)$

Answer: $v(10) \approx 30.46$ ft/sec, $\quad v(\infty) = 40$ ft/sec

21. Equation: $v' = -g - \rho v^2, \quad v(0) = v_0, \quad y(0) = 0$

Solution: $\displaystyle\int \frac{dv}{g + \rho v^2} = -\int dt; \quad \int \frac{\sqrt{\rho/g}\,dv}{1 + \left(\sqrt{\rho/g}\,v\right)^2} = -\int \sqrt{g\rho}\,dt;$

$\tan^{-1}\!\left(\sqrt{\rho/g}\,v\right) = -\sqrt{g\rho}\,t + C; \quad v(0) = v_0 \text{ implies } C = \tan^{-1}\!\left(\sqrt{\rho/g}\,v_0\right)$

$$v(t) = -\sqrt{\frac{g}{\rho}} \tan\!\left(t\sqrt{g\rho} - \tan^{-1}\!\left(v_0\sqrt{\frac{\rho}{g}}\right)\right)$$

We solve $v(t) = 0$ for $t = \dfrac{1}{\sqrt{g\rho}} \tan^{-1}\!\left(v_0\sqrt{\dfrac{\rho}{g}}\right)$ and substitute in Eq. (17) for $y(t)$:

$$y_{max} = \frac{1}{\rho} \ln\left|\frac{\cos\!\left(\tan^{-1} v_0\sqrt{\rho/g} - \tan^{-1} v_0\sqrt{\rho/g}\right)}{\cos\!\left(\tan^{-1} v_0\sqrt{\rho/g}\right)}\right|$$

$$= \frac{1}{\rho} \ln\!\left(\sec\!\left(\tan^{-1} v_0\sqrt{\rho/g}\right)\right) = \frac{1}{\rho} \ln\sqrt{1 + \frac{\rho v_0^2}{g}}$$

$$y_{max} = \frac{1}{2\rho} \ln\!\left(1 + \frac{\rho v_0^2}{g}\right)$$

23. Before opening parachute:

$$v' = -32 + 0.00075\,v^2, \quad v(0) = 0, \quad y(0) = 10000$$
$$v(t) = -206.559\tanh(0.154919\,t) \quad v(30) = -206.521 \text{ ft/sec}$$
$$y(t) = 10000 - 1333.33\ln(\cosh(0.154919\,t)), \quad y(30) = 4727.30 \text{ ft}$$

After opening parachute:

$$v' = -32 + 0.075\,v^2, \quad v(0) = -206.521, \quad y(0) = 4727.30$$
$$v(t) = -20.6559\tanh(1.54919\,t + 0.00519595)$$
$$y(t) = 4727.30 - 13.3333\ln(\cosh(1.54919\,t + 0.00519595))$$
$$y = 0 \text{ when } t = 229.304$$

Thus she opens her parachute after 30 sec at a height of 4727 feet, and the total time of descent is $30 + 229.304 = 259.304$ sec, about 4 minutes and 19.3 seconds.

25. We get the desired formula when we set $v = 0$ in Eq. (23) and solve for r.

27. Integration of $v\dfrac{dv}{dy} = -\dfrac{GM}{(y+R)^2}$, $y(0)=0$, $v(0)=v_0$ gives

$$\frac{1}{2}v^2 = \frac{GM}{y+R} - \frac{GM}{R} + \frac{1}{2}v_0^2$$

which simplifies to the desired formula for v^2. Then substitution of $G = 6.6726\times10^{-11}$, $M = 5.975\times10^{24}$ kg, $R = 6.378\times10^6$ m, $v = 0$, and $v_0 = 1$ yields an equation that we easily solve for $y = 51427.3$, that is, about 51.427 km.

SECTION 2.4

NUMERICAL APPROXIMATION: EULER'S METHOD

In each of Problems 1-10 we also give first the iterative formula of Euler's method. These iterations are readily implemented, either manually or with a computer system or graphing calculator (as we illustrate in Problem 1). We give in each problem a table showing the approximate values obtained, as well as the corresponding values of the exact solution.

1. For the differential equation $y' = f(x, y)$ with $f(x, y) = -y$, the iterative formula of Euler's method is $y_{n+1} = y_n + h(-y_n)$, and the exact solution is $y(x) = 2\,e^{-x}$. The TI-83 screen on the left shows a graphing calculator implementation of this iterative formula.

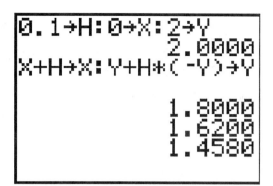
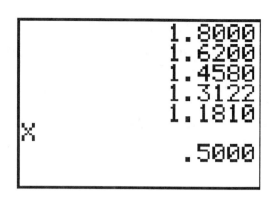

After the variables are initialized (in the first line), and the formula is entered, each press of the enter key carries out an additional step. The screen on the right above shows the results of 5 steps from $x = 0$ to $x = 0.5$ with step size $h = 0.1$ — winding up with $y(0.5) \approx 1.1810$ — and we see the approximate values shown in the second row of the table below. We get the values shown in the next row if we start afresh with $h = 0.05$ and record every other approximation that is obtained.

x	0.0	0.1	0.2	0.3	0.4	0.5
y with $h=0.1$	2.0000	1.8000	1.6200	1.4580	1.3122	1.1810
y with $h=0.05$	2.0000	1.8050	1.6290	1.4702	1.3258	1.1975
y actual	2.0000	1.8097	1.6375	1.4816	1.3406	1.2131

3. Iterative formula: $\quad y_{n+1} = y_n + h(y_n + 1)$

Exact solution: $\quad y(x) = 2e^x - 1$

x	0.0	0.1	0.2	0.3	0.4	0.5
y with $h=0.1$	1.0000	1.2000	1.4200	1.6620	1.9282	2.2210
y with $h=0.05$	1.0000	1.2050	1.4310	1.6802	1.9549	2.2578
y actual	1.0000	1.2103	1.4428	1.6997	1.9837	2.2974

5. Iterative formula: $\quad y_{n+1} = y_n + h(y_n - x_n - 1)$

Exact solution: $\quad y(x) = 2 + x - e^x$

x	0.0	0.1	0.2	0.3	0.4	0.5
y with $h=0.1$	1.0000	1.0000	0.9900	0.9690	0.9359	0.8895
y with $h=0.05$	1.0000	0.9975	0.9845	0.9599	0.9225	0.8711
y actual	1.0000	0.9948	0.9786	0.9501	0.9082	0.8513

7. Iterative formula: $\quad y_{n+1} = y_n + h(-3x_n^2 y_n)$

Exact solution: $\quad y(x) = 3 \exp(-x^3)$

Section 2.4

x	0.0	0.1	0.2	0.3	0.4	0.5
y with h=0.1	3.0000	3.0000	2.9910	2.9551	2.8753	2.7373
y with h=0.05	3.0000	2.9989	2.9843	2.9386	2.8456	2.6930
y actual	3.0000	2.9970	2.9761	2.9201	2.8140	2.6475

9. Iterative formula: $\quad y_{n+1} = y_n + h(1 + y_n^2)/4$

Exact solution: $\quad y(x) = \tan[(x + \pi)/4]$

x	0.0	0.1	0.2	0.3	0.4	0.5
y with h=0.1	1.0000	1.0500	1.1026	1.1580	1.2165	1.2785
y with h=0.05	1.0000	1.0506	1.1039	1.1602	1.2197	1.2828
y actual	1.0000	1.0513	1.1054	1.1625	1.2231	1.2874

The tables of approximate and actual values called for in Problems 11–16 were produced using the following MATLAB script (appropriately altered for each problem).

```
% Section 2.4, Problems 11-16
x0 = 0;    y0 = 1;
% first run:
h = 0.01;
x = x0;   y = y0;   y1 = y0;
for   n = 1:100
    y = y + h*(y-2);
    y1 = [y1,y];
    x = x + h;
    end
% second run:
h = 0.005;
x = x0;   y = y0;   y2 = y0;
for   n = 1:200
    y = y + h*(y-2);
    y2 = [y2,y];
    x = x + h;
    end
% exact values
x = x0 : 0.2 : x0+1;
ye = 2 - exp(x);
% display table
ya = y2(1:40:201);
err = 100*(ye-ya)./ye;
[x; y1(1:20:101); ya; ye; err]
```

11. The iterative formula of Euler's method is $\quad y_{n+1} = y_n + h(y_n - 2)$, and the exact solution is $y(x) = 2 - e^x$. The resulting table of approximate and actual values is

x	0.0	0.2	0.4	0.6	0.8	1.0
y ($h=0.01$)	1.0000	0.7798	0.5111	0.1833	−0.2167	−0.7048
y ($h=0.005$)	1.0000	0.7792	0.5097	0.1806	−0.2211	−0.7115
y actual	1.0000	0.7786	0.5082	0.1779	−0.2255	−0.7183
error	0%	−0.08%	−0.29%	−1.53%	1.97%	0.94%

13. Iterative formula: $y_{n+1} = y_n + 2hx_n^3/y_n$

Exact solution: $y(x) = (8 + x^4)^{1/2}$

x	1.0	1.2	1.4	1.6	1.8	2.0
y ($h=0.01$)	3.0000	3.1718	3.4368	3.8084	4.2924	4.8890
y ($h=0.005$)	3.0000	3.1729	3.4390	3.8117	4.2967	4.8940
y actual	3.0000	3.1739	3.4412	3.8149	4.3009	4.8990
error	0%	0.03%	0.06%	0.09%	0.10%	0.10%

15. Iterative formula: $y_{n+1} = y_n + h(3 - 2y_n/x_n)$

Exact solution: $y(x) = x + 4/x^2$

x	2.0	2.2	2.4	2.6	2.8	3.0
y ($h=0.01$)	3.0000	3.0253	3.0927	3.1897	3.3080	3.4422
y ($h=0.005$)	3.0000	3.0259	3.0936	3.1907	3.3091	3.4433
y actual	3.0000	3.0264	3.0944	3.1917	3.3102	3.4444
error	0%	0.019%	0.028%	0.032%	0.033%	0.032%

The tables of approximate values called for in Problems 17-24 were produced using a MATLAB script similar to the one listed preceding the Problem 11 solution above.

17.

x	0.0	0.2	0.4	0.6	0.8	1.0
y ($h=0.1$)	0.0000	0.0010	0.0140	0.0551	0.1413	0.2925
y ($h=0.02$)	0.0000	0.0023	0.0198	0.0688	0.1672	0.3379
y ($h=0.004$)	0.0000	0.0026	0.0210	0.0717	0.1727	0.3477
y ($h=0.0008$)	0.0000	0.0027	0.0213	0.0723	0.1738	0.3497

These data that $y(1) \approx 0.35$, in contrast with Example 4 in the text, where the initial condition is $y(0) = 1$.

In Problems 18-24 we give only the final approximate values of y obtained using Euler's method with step sizes $h = 0.1$, $h = 0.02$, $h = 0.004$, and $h = 0.0008$.

19. With $x_0 = 0$ and $y_0 = 1$, the approximate values of $y(2)$ obtained are:

h	0.1	0.02	0.004	0.0008
y	6.1831	6.3653	6.4022	6.4096

Section 2.4

21. With $x_0 = 1$ and $y_0 = 2$, the approximate values of $y(2)$ obtained are:

h	0.1	0.02	0.004	0.0008
y	2.8508	2.8681	2.8716	2.8723

23. With $x_0 = 0$ and $y_0 = 0$, the approximate values of $y(1)$ obtained are:

h	0.1	0.02	0.004	0.0008
y	1.2262	1.2300	1.2306	1.2307

25. With step sizes $h = 0.15$, $h = 0.03$, and $h = 0.006$ we get the following results:

x	y with h=0.15	y with h=0.03	y with h=0.006
−1.0	1.0000	1.0000	1.0000
−0.7	1.0472	1.0512	1.0521
−0.4	1.1213	1.1358	1.1390
−0.1	1.2826	1.3612	1.3835
+0.2	0.8900	1.4711	0.8210
+0.5	0.7460	1.2808	0.7192

While the values for $h = 0.15$ alone are not conclusive, a comparison of the values of y for all three step sizes with $x > 0$ suggests some anomaly in the transition from negative to positive values of x.

27. With step sizes $h = 0.1$ and $h = 0.01$ we get the following results:

x	y with h = 0.1	y with h = 0.01
0.0	1.0000	1.0000
0.1	1.2000	1.2200
0.2	1.4428	1.4967
.	.	.
.	.	.
.	.	.
0.7	4.3460	6.4643
0.8	5.8670	11.8425
0.9	8.3349	39.5010

Clearly there is some difficulty near $x = 0.9$.

SECTION 2.5

A CLOSER LOOK AT THE EULER METHOD

In each of Problems 1-10 we give first the predictor formula for u_{n+1}, next the improved Euler corrector for y_{n+1} and the exact solution. These predictor-corrector iterations are readily implemented, either manually or with a computer system or graphing calculator (as we illustrate in Problem 1). We give in each problem a table showing the approximate values obtained, as well as the corresponding values of the exact solution.

1. $u_{n+1} = y_n + h(-y_n)$

    $y_{n+1} = y_n + (h/2)[-y_n - u_{n+1}]$

    $y(x) = 2e^{-x}$

    The TI-83 screen on the left above shows a graphing calculator implementation of this iteration. After the variables are initialized (in the first line), and the formulas are entered, each press of the enter key carries out an additional step. The screen on the right shows the results of 5 steps from $x = 0$ to $x = 0.5$ with step size $h = 0.1$ — winding up with $y(0.5) \approx 1.2142$ — and we see the approximate values shown in the second row of the table below.

x	0.0	0.1	0.2	0.3	0.4	0.5
y with $h=0.1$	2.0000	1.8100	1.6381	1.4824	1.3416	1.2142
y actual	2.0000	1.8097	1.6375	1.4816	1.3406	1.2131

3. $u_{n+1} = y_n + h(y_n + 1)$

    $y_{n+1} = y_n + (h/2)[(y_n + 1) + (u_{n+1} + 1)]$

    $y(x) = 2e^x - 1$

x	0.0	0.1	0.2	0.3	0.4	0.5
y with $h=0.1$	1.0000	1.2100	1.4421	1.6985	1.9818	2.2949
y actual	1.0000	1.2103	1.4428	1.6997	1.9837	2.2974

5. $u_{n+1} = y_n + h(y_n - x_n - 1)$

   $y_{n+1} = y_n + (h/2)[(y_n - x_n - 1) + (u_{n+1} - x_n - h - 1)]$

   $y(x) = 2 + x - e^x$

x	0.0	0.1	0.2	0.3	0.4	0.5
y with $h=0.1$	1.0000	0.9950	0.9790	0.9508	0.9091	0.8526
y actual	1.0000	0.9948	0.9786	0.9501	0.9082	0.8513

7. $u_{n+1} = y_n - 3x_n^2 y_n h$

   $y_{n+1} = y_n - (h/2)[3x_n^2 y_n + 3(x_n + h)^2 u_{n+1}]$

   $y(x) = 3\exp(-x^3)$

x	0.0	0.1	0.2	0.3	0.4	0.5
y with $h=0.1$	3.0000	2.9955	2.9731	2.9156	2.8082	2.6405
y actual	3.0000	2.9970	2.9761	2.9201	2.8140	2.6475

9. $u_{n+1} = y_n + h(1 + y_n^2)/4$

   $y_{n+1} = y_n + h[1 + y_n^2 + 1 + (u_{n+1})^2]/8$

   $y(x) = \tan[(x + \pi)/4]$

x	0.0	0.1	0.2	0.3	0.4	0.5
y with $h=0.1$	1.0000	1.0513	1.1053	1.1625	1.2230	1.2873
y actual	1.0000	1.0513	1.1054	1.1625	1.2231	1.2874

The results given below for Problems 11-16 were computed using the following MATLAB script.

```
% Section 2.5, Problems 11-16
x0 = 0;   y0 = 1;
% first run:
h = 0.01;
x = x0;   y = y0;   y1 = y0;
for  n = 1:100
   u = y + h*f(x,y);                    %predictor
   y = y + (h/2)*(f(x,y)+f(x+h,u));     %corrector
   y1 = [y1,y];
   x = x + h;
   end
% second run:
h = 0.005;
```

```
          x = x0;   y = y0;   y2 = y0;
          for n = 1:200
              u = y + h*f(x,y);                    %predictor
              y = y + (h/2)*(f(x,y)+f(x+h,u));     %corrector
              y2 = [y2,y];
              x = x + h;
          end
          % exact values
          x = x0 : 0.2 : x0+1;

          ye = g(x);
          % display table
          ya = y2(1:40:201);
          err = 100*(ye-ya)./ye;
          x = sprintf('%10.5f',x), sprintf('\n');
          y1 = sprintf('%10.5f',y1(1:20:101)), sprintf('\n');
          ya = sprintf('%10.5f',ya), sprintf('\n');
          ye = sprintf('%10.5f',ye), sprintf('\n');
          err = sprintf('%10.5f',err), sprintf('\n');
          table = [x; y1; ya; ye; err]
```

For each problem the differential equation $y' = f(x,y)$ and the known exact solution $y = g(x)$ are stored in the files **f.m** and **g.m** — for instance, the files

```
          function yp = f(x,y)
          yp = y-2;

          function ye = g(x,y)
          ye = 2-exp(x);
```

for Problem 11. (The exact solutions for Problems 11-16 here are given in the solutions for Problems 11-16 in Section 2.4.)

11.

x	0.0	0.2	0.4	0.6	0.8	1.0
y (h=0.01)	1.00000	0.77860	0.50819	0.17790	−0.22551	−0.71824
y (h=0.005)	1.00000	0.77860	0.50818	0.17789	−0.22553	−0.71827
y actual	1.00000	0.77860	0.50818	0.17788	−0.22554	−0.71828
error	0.000%	−0.000%	−0.001%	−0.003%	0.003%	0.002%

13.

x	1.0	1.2	1.4	1.6	1.8	2.0
y (h=0.01)	3.00000	3.17390	3.44118	3.81494	4.30091	4.89901
y (h=0.005)	3.00000	3.17390	3.44117	3.81492	4.30089	4.89899
y actual	3.00000	3.17389	3.44116	3.81492	4.30088	4.89898
error	0.0000%	−0.0001%	−0.0001%	0.0001%	−0.0002%	−0.0002%

15.

x	2.0	2.2	2.4	2.6	2.8	3.0
y (h=0.01)	3.000000	3.026448	3.094447	3.191719	3.310207	3.444448
y (h=0.005)	3.000000	3.026447	3.094445	3.191717	3.310205	3.444445
y actual	3.000000	3.026446	3.094444	3.191716	3.310204	3.444444
error	0.00000%	–0.00002%	–0.00002%	–0.00002%	–0.00002%	–0.00002%

17. With $h = 0.1$: $y(1) \approx 0.35183$
      With $h = 0.02$: $y(1) \approx 0.35030$
      With $h = 0.004$: $y(1) \approx 0.35023$
      With $h = 0.0008$: $y(1) \approx 0.35023$

The table of numerical results is

x	y with h = 0.1	y with h = 0.02	y with h = 0.004	y with h = 0.0008
0.0	0.00000	0.00000	0.00000	0.00000
0.2	0.00300	0.00268	0.00267	0.00267
0.4	0.02202	0.02139	0.02136	0.02136
0.6	0.07344	0.07249	0.07245	0.07245
0.8	0.17540	0.17413	0.17408	0.17408
1.0	0.35183	0.35030	0.35023	0.35023

In Problems 18-24 we give only the final approximate values of y obtained using the improved Euler method with step sizes $h = 0.1$, $h = 0.02$, $h = 0.004$, and $h = 0.0008$.

19. With $h = 0.1$: $y(2) \approx 6.40834$
      With $h = 0.02$: $y(2) \approx 6.41134$
      With $h = 0.004$: $y(2) \approx 6.41147$
      With $h = 0.0008$: $y(2) \approx 6.41147$

21. With $h = 0.1$: $y(2) \approx 2.87204$
      With $h = 0.02$: $y(2) \approx 2.87245$
      With $h = 0.004$: $y(2) \approx 2.87247$
      With $h = 0.0008$: $y(2) \approx 2.87247$

23. With $h = 0.1$: $y(1) \approx 1.22967$
      With $h = 0.02$: $y(1) \approx 1.23069$
      With $h = 0.004$: $y(1) \approx 1.23073$
      With $h = 0.0008$: $y(1) \approx 1.23073$

In the solutions for Problems 25 and 26 we illustrate the following general MATLAB ode solver.

```
function [t,y] = ode(method, yp, t0,b, y0, n)
%   [t,y] = ode(method, yp, t0,b, y0, n)
%   calls the method described by 'method' for the
%   ODE 'yp' with function header
%
%               y' = yp(t,y)
%
%   on the interval [t0,b] with initial (column)
%   vector y0.  Choices for method are 'euler',
%   'impeuler', 'rk' (Runge-Kutta), 'ode23', 'ode45'.
%   Results are saved at the endPoints of n subintervals,
%   that is, in steps of length h = (b - t0)/n.  The
%   result t is an (n+1)-column vector from b to t1,
%   while y is a matrix with  n+1  rows (one for each
%   t-value) and one column for each dependent variable.

h = (b - t0)/n;             % step size
t = t0 : h : b;
t = t';                     % col. vector of t-values
y = y0';                    % 1st row of result matrix
for i = 2 : n+1             % for i=2 to i=n+1
    t0 = t(i-1);            % old t
    t1 = t(i);              % new t
    y0 = y(i-1,:)';         % old y-row-vector
    [T,Y] = feval(method, yp, t0,t1, y0);
    y = [y;Y'];             % adjoin new y-row-vector
end
```

To use the improved Euler method, we call as **'method'** the following function.

```
function [t,y] = impeuler(yp, t0,t1, y0)
%
%   [t,y] = impeuler(yp, t0,t1, y0)
%   Takes one improved Euler step for
%
%         y' = yprime( t,y ),
%
%   from t0 to t1 with initial value the
%   column vector  y0.

h = t1 - t0;
k1 = feval( yp, t0, y0        );
k2 = feval( yp, t1, y0 + h*k1 );
k  = (k1 + k2)/2;
t = t1;
y = y0 + h*k;
```

25. Here our differential equation is described by the MATLAB function

```
function vp = vpbolt1(t,v)
vp = -0.04*v - 9.8;
```

Section 2.5

Then the commands

```
n = 50;
[t1,v1] = ode('impeuler','vpbolt1',0,10,49,n);
n = 100;
[t2,v2] = ode('impeuler','vpbolt1',0,10,49,n);
t = (0:10)';
ve = 294*exp(-t/25)-245;
[t, v1(1:5:51), v2(1:10:101), ve]
```

generate the table

t	with $n = 50$	with $n = 100$	actual v
0	49.0000	49.0000	49.0000
1	37.4722	37.4721	37.4721
2	26.3964	26.3963	26.3962
3	15.7549	15.7547	15.7546
4	5.5307	5.5304	5.5303
5	-4.2926	-4.2930	-4.2932
6	-13.7308	-13.7313	-13.7314
7	-22.7989	-22.7994	-22.7996
8	-31.5115	-31.5120	-31.5122
9	-39.8824	-39.8830	-39.8832
10	-47.9251	-47.9257	-47.9259

We notice first that the final two columns agree to 3 decimal places (each difference being than 0.0005). Scanning the $n = 100$ column for sign changes, we suspect that $v = 0$ (at the bolt's apex) occurs just after $t = 4.5$ sec. Then interpolation between $t = 4.5$ and $t = 4.6$ in the table

[t2(40:51),v2(40:51)]

3.9000	6.5345
4.0000	5.5304
4.1000	4.5303
4.2000	3.5341
4.3000	2.5420
4.4000	1.5538
4.5000	0.5696
4.6000	-0.4108
4.7000	-1.3872
4.8000	-2.3597
4.9000	-3.3283
5.0000	-4.2930

indicates that $t = 4.56$ at the bolt's apex. Finally, interpolation in

[t2(95:96),v2(95:96)]

```
              9.4000    -43.1387
              9.5000    -43.9445
```

gives the impact velocity $v(9.41) \approx -43.22$ m/s.

SECTION 2.6

THE RUNGE-KUTTA METHOD

The actual solutions in Problems 1-10 are the same as those given in the solutions for Problems 1-10 (respectively) in Section 2.4. Each problem can be solved with a "template" of computations like those listed in Problem 1. We include a table showing the slope values k_1, k_2, k_3, k_4 and the xy-values at the ends of two successive steps of size $h = 0.25$.

1. To make the first step of size $h = 0.25$ we start with the function defined by

```
    f[x_, y_] := -y
```

and the initial values

```
    x = 0;      y = 2;     h = 0.25;
```

and then perform the calculations

```
    k1 = f[x, y]
    k2 = f[x + h/2, y + h*k1/2]
    k3 = f[x + h/2, y + h*k2/2]
    k4 = f[x + h, y + h*k3]
    y  = y + h/6*(k1 + 2*k2 + 2*k3 + k4)
    x  = x + h
```

in turn. Here we are using Mathematica notation that translates transparently to standard mathematical notation describing the corresponding manual computations. A repetition of this same block of calculations carries out a second step of size $h = 0.25$. The following table lists the intermediate and final results obtained in these two steps.

k_1	k_2	k_3	k_4	x	Approx. y	Actual y
−2	−1/75	−1.78125	−1.55469	0.25	1.55762	1.55760
−1.55762	−1.36292	−1.38725	−1.2108	0.5	1.21309	1.21306

3.

k_1	k_2	k_3	k_4	x	Approx. y	Actual y
2	2.25	2.28125	2.57031	0.25	1.56803	1.56805
2.56803	2.88904	2.92916	3.30032	0.5	2.29740	2.29744

5.

k_1	k_2	k_3	k_4	x	Approx. y	Actual y
0	−0.125	−0.14063	−0.28516	0.25	0.96598	0.96597
−28402	−0.44452	−0.46458	−0.65016	0.5	0.85130	0.85128

7.

k_1	k_2	k_3	k_4	x	Approx. y	Actual y
0	−0.14063	−0.13980	−0.55595	0.25	2.95347	2.95349
−0.55378	−1.21679	−1.18183	−1.99351	0.5	2.6475	2.64749

9.

k_1	k_2	k_3	k_4	x	Approx. y	Actual y
0.5	0.53223	0.53437	0.57126	0.25	1.13352	1.13352
0.57122	0.61296	0.61611	0.66444	0.5	1.28743	1.28743

The results given below for Problems 11–16 were computed using the following MATLAB script.

```
% Section 2.6, Problems 11-16
x0 = 0;   y0 = 1;
% first run:
h = 0.2;
x = x0;   y = y0;   y1 = y0;
for   n = 1:5
   k1 = f(x,y);
   k2 = f(x+h/2,y+h*k1/2);
   k3 = f(x+h/2,y+h*k2/2);
   k4 = f(x+h,y+h*k3);
   y = y +(h/6)*(k1+2*k2+2*k3+k4);
   y1 = [y1,y];
   x = x + h;
   end
% second run:
h = 0.1;
x = x0;   y = y0;   y2 = y0;
for   n = 1:10
   k1 = f(x,y);
   k2 = f(x+h/2,y+h*k1/2);
   k3 = f(x+h/2,y+h*k2/2);
   k4 = f(x+h,y+h*k3);
   y = y +(h/6)*(k1+2*k2+2*k3+k4);
   y2 = [y2,y];
   x = x + h;
end
% exact values
x = x0 : 0.2 : x0+1;
ye = g(x);
% display table
y2 = y2(1:2:11);
```

```
err = 100*(ye-y2)./ye;
x  = sprintf('%10.6f',x), sprintf('\n');
y1 = sprintf('%10.6f',y1), sprintf('\n');
y2 = sprintf('%10.6f',y2), sprintf('\n');
ye = sprintf('%10.6f',ye), sprintf('\n');
err = sprintf('%10.6f',err), sprintf('\n');
table = [x;y1;y2;ye;err]
```

For each problem the differential equation $y' = f(x, y)$ and the known exact solution $y = g(x)$ are stored in the files **f.m** and **g.m** — for instance, the files

```
function  yp = f(x,y)
yp = y-2;

function ye = g(x,y)
ye = 2-exp(x);
```

for Problem 11. (The exact solutions for Problems 11-16 here are given in the solutions for Problems 11-16 in Section 2.4.)

11.

x	0.0	0.2	0.4	0.6	0.8	1.0
y (h=0.2)	1.000000	0.778600	0.508182	0.177894	−0.225521	−0.718251
y (h=0.1)	1.000000	0.778597	0.508176	0.177882	−0.225540	−0.718280
y actual	1.000000	0.778597	0.508175	0.177881	−0.225541	−0.718282
error	0.00000%	−0.00002%	−0.00009%	−0.00047%	−0.00061%	−0.00029%

13.

x	1.0	1.2	1.4	1.6	1.8	2.0
y (h=0.2)	3.000000	3.173896	3.441170	3.814932	4.300904	4.899004
y (h=0.1)	3.000000	3.173894	3.441163	3.814919	4.300885	4.898981
y actual	3.000000	3.173894	3.441163	3.814918	4.300884	4.898979
error	0.00000%	−0.00001%	−0.00001%	−0.00002%	−0.00003%	−0.00003%

15.

x	2.0	2.2	2.4	2.6	2.9	3.0
y (h=0.2)	3.000000	3.026448	3.094447	3.191719	3.310207	3.444447
y (h=0.1)	3.000000	3.026446	3.094445	3.191716	3.310204	3.444445
y actual	3.000000	3.026446	3.094444	3.191716	3.310204	3.444444
error	0.000000%	−0.000004%	−0.000005%	−0.000005%	−0.000005%	−0.000004%

17. With h = 0.2: $y(1) \approx 0.350258$
With h = 0.1: $y(1) \approx 0.350234$
With h = 0.05: $y(1) \approx 0.350232$
With h = 0.025: $y(1) \approx 0.350232$

The table of numerical results is

x	y with h = 0.2	y with h = 0.1	y with h = 0.05	y with h = 0.025
0.0	0.000000	0.000000	0.000000	0.000000
0.2	0.002667	0.002667	0.002667	0.002667
0.4	0.021360	0.021359	0.021359	0.021359
0.6	0.072451	0.072448	0.072448	0.072448
0.8	0.174090	0.174081	0.174080	0.174080
1.0	0.350258	0.350234	0.350232	0.350232

In Problems 18-24 we give only the final approximate values of y obtained using the Runge-Kutta method with step sizes $h = 0.2$, $h = 0.1$, $h = 0.05$, and $h = 0.025$.

19. With $h = 0.2$: $y(2) \approx 6.411464$
With $h = 0.1$: $y(2) \approx 6.411474$
With $h = 0.05$: $y(2) \approx 6.411474$
With $h = 0.025$: $y(2) \approx 6.411474$

21. With $h = 0.2$: $y(2) \approx 2.872467$
With $h = 0.1$: $y(2) \approx 2.872468$
With $h = 0.05$: $y(2) \approx 2.872468$
With $h = 0.025$: $y(2) \approx 2.872468$

23. With $h = 0.2$: $y(1) \approx 1.230735$
With $h = 0.1$: $y(1) \approx 1.230731$
With $h = 0.05$: $y(1) \approx 1.230731$
With $h = 0.025$: $y(1) \approx 1.230731$

In the solutions for Problems 25 and 26 we use the general MATLAB solver **ode** that was listed prior to the Problem 25 solution in Section 2.5. To use the Runge-Kutta method, we call as **'method'** the following function.

```
function [t,y] = rk(yp, t0,t1, y0)

%   [t, y] = rk(yp, t0, t1, y0)
%   Takes one Runge-Kutta step for
%
%        y' = yp( t,y ),
%
%   from t0  to  t1  with initial value  the
%   column vector  y0.
h = t1 - t0;
k1 = feval(yp, t0       , y0            );
k2 = feval(yp, t0 + h/2, y0 + (h/2)*k1 );
k3 = feval(yp, t0 + h/2, y0 + (h/2)*k2 );
```

```
k4 = feval(yp, t0 + h    ,y0 +     h *k3 );
k  = (1/6)*(k1 + 2*k2 + 2*k3 + k4);
t = t1;
y = y0 + h*k;
```

25.   Here our differential equation is described by the MATLAB function

```
function  vp = vpbolt1(t,v)
vp = -0.04*v - 9.8;
```

Then the commands

```
n = 100;
[t1,v1] = ode('rk','vpbolt1',0,10,49,n);
n = 200;
[t2,v] = ode('rk','vpbolt1',0,10,49,n);
t = (0:10)';
ve = 294*exp(-t/25)-245;
[t, v1(1:n/20:1+n/2), v(1:n/10:n+1), ve]
```

generate the table

t	with $n = 100$	with $n = 200$	actual v
0	49.0000	49.0000	49.0000
1	37.4721	37.4721	37.4721
2	26.3962	26.3962	26.3962
3	15.7546	15.7546	15.7546
4	5.5303	5.5303	5.5303
5	-4.2932	-4.2932	-4.2932
6	-13.7314	-13.7314	-13.7314
7	-22.7996	-22.7996	-22.7996
8	-31.5122	-31.5122	-31.5122
9	-39.8832	-39.8832	-39.8832
10	-47.9259	-47.9259	-47.9259

We notice first that the final three columns agree to the 4 displayed decimal places. Scanning the last column for sign changes in v, we suspect that $v = 0$ (at the bolt's apex) occurs just after $t = 4.5$ sec. Then interpolation between $t = 4.55$ and $t = 4.60$ in the table

```
[t2(91:95),v(91:95)]
```

4.5000	0.5694
4.5500	0.0788
4.6000	-0.4109
4.6500	-0.8996
4.7000	-1.3873

indicates that $t = 4.56$ at the bolt's apex. Now the commands

```
y = zeros(n+1,1);
h = 10/n;
for j = 2:n+1
   y(j) = y(j-1) + v(j-1)*h +
                   0.5*(-.04*v(j-1) - 9.8)*h^2;
end
ye = 7350*(1 - exp(-t/25)) - 245*t;
[t, y(1:n/10:n+1), ye]
```

generate the table

t	Approx y	Actual y
0	0	0
1	43.1974	43.1976
2	75.0945	75.0949
3	96.1342	96.1348
4	106.7424	106.7432
5	107.3281	107.3290
6	98.2842	98.2852
7	79.9883	79.9895
8	52.8032	52.8046
9	17.0775	17.0790
10	-26.8540	-26.8523

We see at least 2-decimal place agreement between approximate and actual values of y. Finally, interpolation between $t = 9$ and $t = 10$ here suggests that $y = 0$ just after $t = 9.4$. Then interpolation between $t = 9.40$ and $t = 9.45$ in the table

```
[t2(187:191),y(187:191)]
```

```
         9.3000     4.7448
         9.3500     2.6182
         9.4000     0.4713
         9.4500    -1.6957
         9.5000    -3.8829
```

indicates that the bolt is aloft for about 9.41 seconds.

CHAPTER 3

LINEAR SYSTEMS AND MATRICES

SECTION 3.1

INTRODUCTION TO LINEAR SYSTEMS

This initial section takes account of the fact that some students remember only hazily the method of elimination for 2×2 and 3×3 systems. Moreover, high school algebra courses generally emphasize only the case in which a unique solution exists. Here we treat on an equal footing the other two cases — in which either no solution exists or infinitely many solutions exist.

1. Subtraction of twice the first equation from the second equation gives $-5y = -10$, so $y = 2$, and it follows that $x = 3$.

3. Subtraction of 3/2 times the first equation from the second equation gives $\frac{1}{2}y = \frac{3}{2}$, so $y = 3$, and it follows that $x = -4$.

5. Subtraction of twice the first equation from the second equation gives $0 = 1$, so no solution exists.

7. The second equation is -2 times the first equation, so we can choose $y = t$ arbitrarily. The first equation then gives $x = -10 + 4t$.

9. Subtraction of twice the first equation from the second equation gives $-9y - 4z = -3$. Subtraction of the first equation from the third equation gives $2y + z = 1$. Solution of these latter two equations gives $y = -1$, $z = 3$. Finally substitution in the first equation gives $x = 4$.

11. First we interchange the first and second equations. Then subtraction of twice the new first equation from the new second equation gives $y - z = 7$, and subtraction of three times the new first equation from the third equation gives $-2y + 3z = -18$. Solution of these latter two equations gives $y = 3$, $z = -4$. Finally substitution in the (new) first equation gives $x = 1$.

13. First we subtract the second equation from the first equation to get the new first equation $x + 2y + 3z = 0$. Then subtraction of twice the new first equation from the second equation gives $3y - 2z = 0$, and subtraction of twice the new first equation from the

third equation gives $2y - z = 0$. Solution of these latter two equations gives $y = 0$, $z = 0$. Finally substitution in the (new) first equation gives $x = 0$ also.

15. Subtraction of the first equation from the second equation gives $-4y + z = -2$. Subtraction of three times the first equation from the third equation gives (after division by 2) $-4y + z = -5/2$. These latter two equations obviously are inconsistent, so the original system has no solution.

17. First we subtract the first equation from the second equation to get the new first equation $x + 3y - 6z = -4$. Then subtraction of three times the new first equation from the second equation gives $-7y + 16z = 15$, and subtraction of five times the new first equation from the third equation gives (after division by 2) $-7y + 16z = 35/2$. These latter two equations obviously are inconsistent, so the original system has no solution.

19. Subtraction of twice the first equation from the second equation gives $3y - 6z = 9$. Subtraction of the first equation from the third equation gives $y - 2z = 3$. Obviously these latter two equations are scalar multiples of each other, so we can choose $z = t$ arbitrarily. It follows first that $y = 3 + 2t$ and then that $x = 8 + 3t$.

21. Subtraction of three times the first equation from the second equation gives $3y - 6z = 9$. Subtraction of four times the first equation from the third equation gives $-3y + 9z = -6$. Obviously these latter two equations are both scalar multiples of the equation $y - 3z = 2$, so we can choose $z = t$ arbitrarily. It follows first that $y = 2 + 3t$ and then that $x = 3 - 2t$.

23. The initial conditions $y(0) = 3$ and $y'(0) = 8$ yield the equations $A = 3$ and $2B = 8$, so $A = 3$ and $B = 4$. It follows that $y(x) = 3\cos 2x + 4\sin 2x$.

25. The initial conditions $y(0) = 10$ and $y'(0) = 20$ yield the equations $A + B = 10$ and $5A - 5B = 20$ with solution $A = 7$, $B = 3$. Thus $y(x) = 7e^{5x} + 3e^{-5x}$.

27. The initial conditions $y(0) = 40$ and $y'(0) = -16$ yield the equations $A + B = 40$ and $3A - 5B = -16$ with solution $A = 23$, $B = 17$. Thus $y(x) = 23e^{3x} + 17e^{-5x}$.

29. The initial conditions $y(0) = 7$ and $y'(0) = 11$ yield the equations $A + B = 7$ and $\frac{1}{2}A + \frac{1}{3}B = 11$ with solution $A = 52$, $B = -45$. Thus $y(x) = 52e^{x/2} - 45e^{x/3}$.

31. The graph of each of these linear equations in x and y is a straight line through the origin $(0, 0)$ in the xy-plane. If these two lines are distinct then they intersect only at the origin, so the two equations have the unique solution $x = y = 0$. If the two lines coincide, then each of the infinitely many different points (x, y) on this common line provides a solution of the system.

33. **(a)** The three lines have no common point of intersection, so the system has no solution.

   (b) The three lines have a single point of intersection, so the system has a unique solution.

   (c) The three lines — two of them parallel — have no common point of intersection, so the system has no solution.

   (d) The three distinct parallel lines have no common point of intersection, so the system has no solution.

   (e) Two of the lines coincide and intersect the third line in a single point, so the system has a unique solution.

   (f) The three lines coincide, and each of the infinitely many different points (x, y, z) on this common line provides a solution of the system.

SECTION 3.2

MATRICES AND GAUSSIAN ELIMINATION

Because the linear systems in Problems 1-10 are already in echelon form, we need only start at the end of the list of unknowns and work backwards.

1. Starting with $x_3 = 2$ from the third equation, the second equation gives $x_2 = 0$, and then the first equation gives $x_1 = 1$.

3. If we set $x_3 = t$ then the second equation gives $x_2 = 2 + 5t$, and next the first equation gives $x_1 = 13 + 11t$.

5. If we set $x_4 = t$ then the third equation gives $x_3 = 5 + 3t$, next the second equation gives $x_2 = 6 + t$, and finally the first equation gives $x_1 = 13 + 4t$.

7. If we set $x_3 = s$ and $x_4 = t$, then the second equation gives $x_2 = 7 + 2s - 7t$, and next the first equation gives $x_1 = 3 - 8s + 19t$.

9. Starting with $x_4 = 6$ from the fourth equation, the third equation gives $x_3 = -5$, next the second equation gives $x_2 = 3$, and finally the first equation gives $x_1 = 1$.

In each of Problems 11-22, we give just the first two or three steps in the reduction. Then we display a resulting echelon form **E** of the augmented coefficient matrix **A** of the given linear system, and finally list the resulting solution (if any). The student should understand that the echelon matrix **E** is not unique, so a different sequence of elementary row operations may produce a different echelon matrix.

11. Begin by interchanging rows 1 and 2 of **A**. Then subtract twice row 1 both from row 2 and from row 3.

$$\mathbf{E} = \begin{bmatrix} 1 & 3 & 2 & 5 \\ 0 & 1 & 0 & -2 \\ 0 & 0 & 1 & 4 \end{bmatrix}; \quad x_1 = 3, \ x_2 = -2, \ x_3 = 4$$

13. Begin by subtracting twice row 1 of **A** both from row 2 and from row 3. Then add row 2 to row 3.

$$\mathbf{E} = \begin{bmatrix} 1 & 3 & 3 & 13 \\ 0 & 1 & 2 & 3 \\ 0 & 0 & 0 & 0 \end{bmatrix}; \quad x_1 = 4 + 3t, \ x_2 = 3 - 2t, \ x_3 = t$$

15. Begin by interchanging rows 1 and 2 of **A**. Then subtract three times row 1 from row 2, and five times row 1 from row 3.

$$\mathbf{E} = \begin{bmatrix} 1 & 1 & 1 & 1 \\ 0 & 1 & 3 & 3 \\ 0 & 0 & 0 & 1 \end{bmatrix}. \qquad \text{The system has no solution.}$$

17.
$$\begin{aligned} x_1 - 4x_2 - 3x_3 - 3x_4 &= 4 \\ 2x_1 - 6x_2 - 5x_3 - 5x_4 &= 5 \\ 3x_1 - x_2 - 4x_3 - 5x_4 &= -7 \end{aligned}$$

[The first printing of the textbook had a misprinted 2 (instead of 4) as the right-hand side constant in the first equation.] Begin by subtracting twice row 1 from row 2 of **A**, and three times row 1 from row 3.

$$\mathbf{E} = \begin{bmatrix} 1 & -4 & -3 & -3 & 4 \\ 0 & 1 & 0 & -1 & -4 \\ 0 & 0 & 1 & 3 & 5 \end{bmatrix}; \quad x_1 = 3 - 2t, \ x_2 = -4 + t, \ x_3 = 5 - 3t, \ x_4 = t$$

19. Begin by interchanging rows 1 and 2 of **A**. Then subtract three times row 1 from row 2, and four times row 1 from row 3.

$$\mathbf{E} = \begin{bmatrix} 1 & -2 & 5 & -5 & -7 \\ 0 & 1 & -2 & 3 & 5 \\ 0 & 0 & 0 & 0 & 0 \end{bmatrix}; \quad x_1 = 3-s-t, \ x_2 = 5+2s-3t, \ x_3 = s, \ x_4 = t$$

21. Begin by subtracting twice row 1 from row 2, three times row 1 from row 3, and four times row 1 from row 4.

$$\mathbf{E} = \begin{bmatrix} 1 & 1 & 1 & 0 & 6 \\ 0 & 1 & 5 & 1 & 20 \\ 0 & 0 & 1 & 0 & 3 \\ 0 & 0 & 0 & 1 & 4 \end{bmatrix}; \quad x_1 = 2, \ x_2 = 1, \ x_3 = 3, \ x_4 = 4$$

23. If we subtract twice the first row from the second row, we obtain the echelon form

$$\mathbf{E} = \begin{bmatrix} 3 & 2 & 1 \\ 0 & 0 & k-2 \end{bmatrix}$$

of the augmented coefficient matrix. It follows that the given system has no solutions unless $k = 2$, in which case it has infinitely many solutions given by $x_1 = \frac{1}{3}(1-2t), \ x_2 = t$.

25. If we subtract twice the first row from the second row, we obtain the echelon form

$$\mathbf{E} = \begin{bmatrix} 3 & 2 & 11 \\ 0 & k-4 & -1 \end{bmatrix}$$

of the augmented coefficient matrix. It follows that the given system has a unique solution if $k \neq 4$, but no solution if $k = 4$.

27. If we first subtract twice the first row from the second row, then subtract 4 times the first row from the third row, and finally subtract the second row from the third row, we obtain the echelon form

$$\mathbf{E} = \begin{bmatrix} 1 & 2 & 1 & 3 \\ 0 & 5 & 5 & 1 \\ 0 & 0 & 0 & k-11 \end{bmatrix}$$

of the augmented coefficient matrix. It follows that the given system has no solution unless $k = 11$, in which case it has infinitely many solutions with x_3 arbitrary.

29. In each of parts (a)-(c), we start with a typical 2×2 matrix **A** and carry out two row successive operations as indicated, observing that we wind up with the original matrix **A**.

(a) $\quad \mathbf{A} = \begin{bmatrix} s & t \\ u & v \end{bmatrix} \xrightarrow{cR2} \begin{bmatrix} s & t \\ cu & cv \end{bmatrix} \xrightarrow{(1/c)R2} \begin{bmatrix} s & t \\ u & v \end{bmatrix} = \mathbf{A}$

(b) $\quad \mathbf{A} = \begin{bmatrix} s & t \\ u & v \end{bmatrix} \xrightarrow{SWAP(R1,R2)} \begin{bmatrix} u & v \\ s & t \end{bmatrix} \xrightarrow{SWAP(R1,R2)} \begin{bmatrix} s & t \\ u & v \end{bmatrix} = \mathbf{A}$

(c) $\quad \mathbf{A} = \begin{bmatrix} s & t \\ u & v \end{bmatrix} \xrightarrow{cR1+R2} \begin{bmatrix} u & v \\ cu+s & cv+t \end{bmatrix} \xrightarrow{(-c)R1+R2} \begin{bmatrix} s & t \\ u & v \end{bmatrix} = \mathbf{A}$

Since we therefore can "reverse" any single elementary row operation, it follows that we can reverse any finite sequence of such operations — on at a time — so part (d) follows.

SECTION 3.3

REDUCED ROW-ECHELON MATRICES

Each of the matrices in Problems 1-20 can be transformed to reduced echelon form without the appearance of any fractions. The main thing is to get started right. Generally our first goal is to get a 1 in the upper left corner of **A**, then clear out the rest of the first column. In each problem we first give at least the initial steps, and then the final result **E**. The particular sequence of elementary row operations used is not unique; you might find **E** in a quite different way.

1. $\begin{bmatrix} 1 & 2 \\ 3 & 7 \end{bmatrix} \xrightarrow{R2-3R1} \begin{bmatrix} 1 & 2 \\ 0 & 1 \end{bmatrix} \xrightarrow{R1-2R2} \begin{bmatrix} 1 & 0 \\ 0 & 1 \end{bmatrix}$

3. $\begin{bmatrix} 3 & 7 & 15 \\ 2 & 5 & 11 \end{bmatrix} \xrightarrow{R1-R2} \begin{bmatrix} 1 & 2 & 4 \\ 2 & 5 & 11 \end{bmatrix} \xrightarrow{R2-2R1} \begin{bmatrix} 1 & 2 & 4 \\ 0 & 1 & 3 \end{bmatrix} \xrightarrow{R1-2R2} \begin{bmatrix} 1 & 0 & 2 \\ 0 & 1 & 3 \end{bmatrix}$

5. $\begin{bmatrix} 1 & 2 & -11 \\ 2 & 3 & -19 \end{bmatrix} \xrightarrow{R2-2R1} \begin{bmatrix} 1 & 2 & -11 \\ 0 & -1 & 3 \end{bmatrix} \xrightarrow{(-1)R2} \begin{bmatrix} 1 & 2 & -11 \\ 0 & 1 & -3 \end{bmatrix} \xrightarrow{R1-2R2} \begin{bmatrix} 1 & 0 & -5 \\ 0 & 1 & -3 \end{bmatrix}$

7. $\begin{bmatrix} 1 & 2 & 3 \\ 1 & 4 & 1 \\ 2 & 1 & 9 \end{bmatrix} \xrightarrow{R2-R1} \begin{bmatrix} 1 & 2 & 3 \\ 0 & 2 & -2 \\ 2 & 1 & 9 \end{bmatrix} \xrightarrow{R3-2R1} \begin{bmatrix} 1 & 2 & 3 \\ 0 & 2 & -2 \\ 0 & -3 & 3 \end{bmatrix}$

$$\xrightarrow{(1/2)R2} \begin{bmatrix} 1 & 2 & 3 \\ 0 & 1 & -1 \\ 0 & -3 & 3 \end{bmatrix} \xrightarrow{R3+3R2} \begin{bmatrix} 1 & 2 & 3 \\ 0 & 1 & -1 \\ 0 & 0 & 0 \end{bmatrix} \xrightarrow{R1-2R2} \begin{bmatrix} 1 & 0 & 5 \\ 0 & 1 & -1 \\ 0 & 0 & 0 \end{bmatrix}$$

9. $\begin{bmatrix} 5 & 2 & 18 \\ 0 & 1 & 4 \\ 4 & 1 & 12 \end{bmatrix} \xrightarrow{R1-R3} \begin{bmatrix} 1 & 1 & 6 \\ 0 & 1 & 4 \\ 4 & 1 & 12 \end{bmatrix} \xrightarrow{R3-4R1} \begin{bmatrix} 1 & 1 & 6 \\ 0 & 1 & 4 \\ 0 & -3 & -12 \end{bmatrix}$

$\xrightarrow{R3+3R2} \begin{bmatrix} 1 & 1 & 6 \\ 0 & 1 & 4 \\ 0 & 0 & 0 \end{bmatrix} \xrightarrow{R1-R2} \begin{bmatrix} 1 & 0 & 2 \\ 0 & 1 & 4 \\ 0 & 0 & 0 \end{bmatrix}$

11. $\begin{bmatrix} 3 & 9 & 1 \\ 2 & 6 & 7 \\ 1 & 3 & -6 \end{bmatrix} \xrightarrow{SWAP(R1,R3)} \begin{bmatrix} 1 & 3 & -6 \\ 2 & 6 & 7 \\ 3 & 9 & 1 \end{bmatrix} \xrightarrow{R2-2R1} \begin{bmatrix} 1 & 3 & -6 \\ 0 & 0 & 19 \\ 3 & 9 & 1 \end{bmatrix}$

$\xrightarrow{R3-3R1} \begin{bmatrix} 1 & 3 & -6 \\ 0 & 0 & 19 \\ 0 & 0 & 19 \end{bmatrix} \xrightarrow{(1/19)R2} \begin{bmatrix} 1 & 3 & -6 \\ 0 & 0 & 1 \\ 0 & 0 & 19 \end{bmatrix} \xrightarrow{R3-19R2} \cdots \begin{bmatrix} 1 & 3 & 0 \\ 0 & 0 & 1 \\ 0 & 0 & 0 \end{bmatrix}$

13. $\begin{bmatrix} 2 & 7 & 4 & 0 \\ 1 & 3 & 2 & 1 \\ 2 & 6 & 5 & 4 \end{bmatrix} \xrightarrow{SWAP(R1,R2)} \begin{bmatrix} 1 & 3 & 2 & 1 \\ 2 & 7 & 4 & 0 \\ 2 & 6 & 5 & 4 \end{bmatrix} \xrightarrow{R2-2R1} \begin{bmatrix} 1 & 3 & 2 & 1 \\ 0 & 1 & 0 & -2 \\ 2 & 6 & 5 & 4 \end{bmatrix}$

$\xrightarrow{R2-2R1} \begin{bmatrix} 1 & 3 & 2 & 1 \\ 0 & 1 & 0 & -2 \\ 0 & 0 & 1 & 2 \end{bmatrix} \xrightarrow{R1-3R2} \cdots \rightarrow \begin{bmatrix} 1 & 0 & 0 & 3 \\ 0 & 1 & 0 & -2 \\ 0 & 0 & 1 & 2 \end{bmatrix}$

15. $\begin{bmatrix} 2 & 2 & 4 & 2 \\ 1 & -1 & -4 & 3 \\ 2 & 7 & 19 & -3 \end{bmatrix} \xrightarrow{SWAP(R1,R2)} \begin{bmatrix} 1 & -1 & -4 & 3 \\ 2 & 2 & 4 & 2 \\ 2 & 7 & 19 & -3 \end{bmatrix} \xrightarrow{R2-2R1} \begin{bmatrix} 1 & -1 & -4 & 3 \\ 0 & 4 & 12 & -4 \\ 2 & 7 & 19 & -3 \end{bmatrix}$

$\xrightarrow{R3-2R1} \begin{bmatrix} 1 & -1 & -4 & 3 \\ 0 & 4 & 12 & -4 \\ 0 & 9 & 27 & -9 \end{bmatrix} \xrightarrow{(1/4)R2} \begin{bmatrix} 1 & -1 & -4 & 3 \\ 0 & 1 & 3 & -1 \\ 0 & 9 & 27 & -9 \end{bmatrix}$

$$\xrightarrow{R3-9R2} \begin{bmatrix} 1 & -1 & -4 & 3 \\ 0 & 1 & 3 & -1 \\ 0 & 0 & 0 & 0 \end{bmatrix} \xrightarrow{R1+R2} \begin{bmatrix} 1 & 0 & -1 & 2 \\ 0 & 1 & 3 & -1 \\ 0 & 0 & 0 & 0 \end{bmatrix}$$

17. $$\begin{bmatrix} 1 & 1 & 1 & -1 & -4 \\ 1 & -2 & -2 & 8 & -1 \\ 2 & 3 & -1 & 3 & 11 \end{bmatrix} \xrightarrow{R2-R1} \begin{bmatrix} 1 & 1 & 1 & -1 & -4 \\ 0 & -3 & -3 & 9 & 3 \\ 2 & 3 & -1 & 3 & 11 \end{bmatrix} \xrightarrow{R3-2R1} \begin{bmatrix} 1 & 1 & 1 & -1 & -4 \\ 0 & -3 & -3 & 9 & 3 \\ 0 & 1 & -3 & 5 & 19 \end{bmatrix}$$

$$\xrightarrow{(-1/3)R2} \begin{bmatrix} 1 & 1 & 1 & -1 & -4 \\ 0 & 1 & 1 & -3 & -1 \\ 0 & 1 & -3 & 5 & 19 \end{bmatrix} \xrightarrow{R3-R2} \begin{bmatrix} 1 & 1 & 1 & -1 & -4 \\ 0 & 1 & 1 & -3 & -1 \\ 0 & 0 & -4 & 8 & 20 \end{bmatrix}$$

$$\xrightarrow{(-1/4)R3} \begin{bmatrix} 1 & 1 & 1 & -1 & -4 \\ 0 & 1 & 1 & -3 & -1 \\ 0 & 0 & 1 & -2 & -5 \end{bmatrix} \xrightarrow{R1-R2} \cdots \rightarrow \begin{bmatrix} 1 & 0 & 0 & 2 & -3 \\ 0 & 1 & 0 & -1 & 4 \\ 0 & 0 & 1 & -2 & -5 \end{bmatrix}$$

19. $$\begin{bmatrix} 2 & 7 & -10 & -19 & 13 \\ 1 & 3 & -4 & -8 & 6 \\ 1 & 0 & 2 & 1 & 3 \end{bmatrix} \xrightarrow{SWAP(R1,R3)} \begin{bmatrix} 1 & 0 & 2 & 1 & 3 \\ 1 & 3 & -4 & -8 & 6 \\ 2 & 7 & -10 & -19 & 13 \end{bmatrix}$$

$$\xrightarrow{R2-R1} \begin{bmatrix} 1 & 0 & 2 & 1 & 3 \\ 0 & 3 & -6 & -9 & 3 \\ 2 & 7 & -10 & -19 & 13 \end{bmatrix} \xrightarrow{R3-2R1} \begin{bmatrix} 1 & 0 & 2 & 1 & 3 \\ 0 & 3 & -6 & -9 & 3 \\ 0 & 7 & -14 & -21 & 7 \end{bmatrix}$$

$$\xrightarrow{(1/3)R2} \begin{bmatrix} 1 & 0 & 2 & 1 & 3 \\ 0 & 1 & -2 & -3 & 1 \\ 0 & 7 & -14 & -21 & 7 \end{bmatrix} \xrightarrow{R3-7R2} \begin{bmatrix} 1 & 0 & 2 & 1 & 3 \\ 0 & 1 & -2 & -3 & 1 \\ 0 & 0 & 0 & 0 & 0 \end{bmatrix}$$

In each of Problems 21-30, we give just the first two or three steps in the reduction. Then we display the resulting reduced echelon form **E** of the augmented coefficient matrix **A** of the given linear system, and finally list the resulting solution (if any).

21. Begin by interchanging rows 1 and 2 of **A**. Then subtract twice row 1 both from row 2 and from row 3.

$$\mathbf{E} = \begin{bmatrix} 1 & 0 & 0 & 3 \\ 0 & 1 & 0 & -2 \\ 0 & 0 & 1 & 4 \end{bmatrix}; \quad x_1 = 3,\ x_2 = -2,\ x_3 = 4$$

23. Begin by subtracting twice row 1 of **A** both from row 2 and from row 3. Then add row 2 to row 3.

$$\mathbf{E} = \begin{bmatrix} 1 & 0 & -3 & 14 \\ 0 & 1 & 2 & 3 \\ 0 & 0 & 0 & 0 \end{bmatrix}; \quad x_1 = 4+3t,\ x_2 = 3-2t,\ x_3 = t$$

25. Begin by interchanging rows 1 and 2 of **A**. Then subtract three times row 1 from row 2, and five times row 1 from row 3.

$$\mathbf{E} = \begin{bmatrix} 1 & 0 & -2 & 0 \\ 0 & 1 & 3 & 0 \\ 0 & 0 & 0 & 1 \end{bmatrix}. \qquad \text{The system has no solution.}$$

27.
$$\begin{aligned} x_1 - 4x_2 - 3x_3 - 3x_4 &= 4 \\ 2x_1 - 6x_2 - 5x_3 - 5x_4 &= 5 \\ 3x_1 - x_2 - 4x_3 - 5x_4 &= -7 \end{aligned}$$

[The first printing of the textbook had a misprinted 2 (instead of 4) as the right-hand side constant in the first equation.] Begin by subtracting twice row 1 from row 2 of **A**, and three times row 1 from row 3.

$$\mathbf{E} = \begin{bmatrix} 1 & 0 & 0 & 2 & 3 \\ 0 & 1 & 0 & -1 & -4 \\ 0 & 0 & 1 & 3 & 5 \end{bmatrix}; \quad x_1 = 3-2t,\ x_2 = -4+t,\ x_3 = 5-3t,\ x_4 = t$$

29. Begin by interchanging rows 1 and 2 of **A**. Then subtract three times row 1 from row 2, and four times row 1 from row 3.

$$\mathbf{E} = \begin{bmatrix} 1 & 0 & 1 & 1 & 3 \\ 0 & 1 & -2 & 3 & 5 \\ 0 & 0 & 0 & 0 & 0 \end{bmatrix}; \quad x_1 = 3-s-t,\ x_2 = 5+2s-3t,\ x_3 = s,\ x_4 = t$$

31.
$$\begin{bmatrix} 1 & 2 & 3 \\ 0 & 4 & 5 \\ 0 & 0 & 6 \end{bmatrix} \xrightarrow{(1/6)R3} \begin{bmatrix} 1 & 2 & 3 \\ 0 & 4 & 5 \\ 0 & 0 & 1 \end{bmatrix} \xrightarrow{R2-5R3} \begin{bmatrix} 1 & 2 & 3 \\ 0 & 4 & 0 \\ 0 & 0 & 1 \end{bmatrix} \xrightarrow{(1/4)R2} \begin{bmatrix} 1 & 2 & 3 \\ 0 & 1 & 0 \\ 0 & 0 & 1 \end{bmatrix}$$

$$\xrightarrow{R1-2R2} \begin{bmatrix} 1 & 0 & 3 \\ 0 & 1 & 0 \\ 0 & 0 & 1 \end{bmatrix} \xrightarrow{R1-3R3} \begin{bmatrix} 1 & 0 & 0 \\ 0 & 1 & 0 \\ 0 & 0 & 1 \end{bmatrix}$$

33. If the upper left element of a 2×2 reduced echelon matrix is 1, then the possibilities are $\begin{bmatrix} 1 & 0 \\ 0 & 1 \end{bmatrix}$ and $\begin{bmatrix} 1 & * \\ 0 & 0 \end{bmatrix}$, depending on whether there is a nonzero element in the second row. If the upper left element is zero — so both elements of the second row are also 0, then the possibilities are $\begin{bmatrix} 0 & 1 \\ 0 & 0 \end{bmatrix}$ and $\begin{bmatrix} 0 & 0 \\ 0 & 0 \end{bmatrix}$.

35. **(a)** If (x_0, y_0) is a solution, then it follows that

$$a(kx_0) + b(ky_0) = k(ax_0 + by_0) = k \cdot 0 = 0,$$
$$c(kx_0) + d(ky_0) = k(cx_0 + dy_0) = k \cdot 0 = 0$$

so (kx_0, ky_0) is also a solution.

(b) If (x_1, y_1) and (x_2, y_2) are solutions, then it follows that

$$a(x_1 + x_2) + b(y_1 + y_2) = (ax_1 + by_1) + (ax_2 + by_2) = 0 + 0 = 0,$$
$$c(x_1 + x_2) + d(y_1 + y_2) = (cx_1 + dy_1) + (cx_2 + dy_2) = 0 + 0 = 0$$

so $(x_1 + x_2, y_1 + y_2)$ is also a solution.

37. If $ad - bc = 0$ then, much as in Problem 32, we see that the second row of the reduced echelon form of the coefficient matrix is allzero. Hence there is a free variable, and thus the given homogeneous system has a nontrivial solution involving a parameter t.

39. It is given that the augmented coefficient matrix of the homogeneous 3×3 system has the form

$$\begin{bmatrix} a_1 & b_1 & c_1 & 0 \\ a_2 & b_2 & c_2 & 0 \\ pa_1 + qa_2 & pb_1 + qb_2 & pc_1 + qc_2 & 0 \end{bmatrix}.$$

Upon subtracting both p times row 1 and q times row 2 from row 3, we get the matrix

$$\begin{bmatrix} a_1 & b_1 & c_1 & 0 \\ a_2 & b_2 & c_2 & 0 \\ 0 & 0 & 0 & 0 \end{bmatrix}$$

corresponding to two homogeneous linear equations in three unknowns. Hence there is at least one free variable, and thus the system has a nontrivial family of solutions.

SECTION 3.4

MATRIX OPERATIONS

The objective of this section is simple to state. It is not merely knowledge of, but complete mastery of matrix addition and multiplication (particularly the latter). Matrix multiplication must be practiced until it is carried out not only accurately but quickly and with confidence — until you can hardly look at two matrices **A** and **B** without thinking of "pouring" the ith row of **A** down the jth column of **B**.

1. $3\begin{bmatrix} 3 & -5 \\ 2 & 7 \end{bmatrix} + 4\begin{bmatrix} -1 & 0 \\ 3 & -4 \end{bmatrix} = \begin{bmatrix} 9 & -15 \\ 6 & 21 \end{bmatrix} + \begin{bmatrix} -4 & 0 \\ 12 & -16 \end{bmatrix} = \begin{bmatrix} 5 & -15 \\ 18 & 5 \end{bmatrix}$

3. $-2\begin{bmatrix} 5 & 0 \\ 0 & 7 \\ 3 & -1 \end{bmatrix} + 4\begin{bmatrix} -4 & 5 \\ 3 & 2 \\ 7 & 4 \end{bmatrix} = \begin{bmatrix} -10 & 0 \\ 0 & -14 \\ -6 & 2 \end{bmatrix} + \begin{bmatrix} -16 & 20 \\ 12 & 8 \\ 28 & 16 \end{bmatrix} = \begin{bmatrix} -26 & 20 \\ 12 & -6 \\ 22 & 18 \end{bmatrix}$

5. $\begin{bmatrix} 2 & -1 \\ 3 & 2 \end{bmatrix}\begin{bmatrix} -4 & 2 \\ 1 & 3 \end{bmatrix} = \begin{bmatrix} -9 & 1 \\ -10 & 12 \end{bmatrix}$; $\begin{bmatrix} -4 & 2 \\ 1 & 3 \end{bmatrix}\begin{bmatrix} 2 & -1 \\ 3 & 2 \end{bmatrix} = \begin{bmatrix} -2 & 8 \\ 11 & 5 \end{bmatrix}$

7. $\begin{bmatrix} 1 & 2 & 3 \end{bmatrix}\begin{bmatrix} 3 \\ 4 \\ 6 \end{bmatrix} = \begin{bmatrix} 26 \end{bmatrix}$; $\begin{bmatrix} 3 \\ 4 \\ 6 \end{bmatrix}\begin{bmatrix} 1 & 2 & 3 \end{bmatrix} = \begin{bmatrix} 3 & 6 & 9 \\ 4 & 8 & 12 \\ 5 & 10 & 15 \end{bmatrix}$

9. $\begin{bmatrix} 0 & -2 \\ 3 & 1 \\ -4 & 5 \end{bmatrix}\begin{bmatrix} 3 \\ -2 \end{bmatrix} = \begin{bmatrix} 4 \\ 7 \\ -22 \end{bmatrix}$ but the product $\begin{bmatrix} 3 \\ -2 \end{bmatrix}\begin{bmatrix} 0 & -2 \\ 3 & 1 \\ -4 & 5 \end{bmatrix}$ is not defined.

11. $\mathbf{AB} = \begin{bmatrix} 3 & -5 \end{bmatrix}\begin{bmatrix} 2 & 7 & 5 & 6 \\ -1 & 4 & 2 & 3 \end{bmatrix} = \begin{bmatrix} 11 & 1 & 5 & 3 \end{bmatrix}$ but the product **BA** is not defined.

13. $\mathbf{A(BC)} = \begin{bmatrix} 3 & 1 \\ -1 & 4 \end{bmatrix}\left(\begin{bmatrix} 2 & 5 \\ -3 & 1 \end{bmatrix}\begin{bmatrix} 0 & 1 \\ 2 & 3 \end{bmatrix}\right) = \begin{bmatrix} 3 & 1 \\ -1 & 4 \end{bmatrix}\begin{bmatrix} 10 & 17 \\ 2 & 0 \end{bmatrix} = \begin{bmatrix} 32 & 51 \\ -2 & -17 \end{bmatrix}$

$$(\mathbf{AB})\mathbf{C} = \left(\begin{bmatrix} 3 & 1 \\ -1 & 4 \end{bmatrix}\begin{bmatrix} 2 & 5 \\ -3 & 1 \end{bmatrix}\right)\begin{bmatrix} 0 & 1 \\ 2 & 3 \end{bmatrix} = \begin{bmatrix} 3 & 16 \\ -14 & -1 \end{bmatrix}\begin{bmatrix} 0 & 1 \\ 2 & 3 \end{bmatrix} = \begin{bmatrix} 32 & 51 \\ -2 & -17 \end{bmatrix}$$

15. $\mathbf{A}(\mathbf{BC}) = \begin{bmatrix} 3 \\ 2 \end{bmatrix}\left([1 \ -1 \ 2]\begin{bmatrix} 2 & 0 \\ 0 & 3 \\ 1 & 4 \end{bmatrix}\right) = \begin{bmatrix} 3 \\ 2 \end{bmatrix}[4 \ 5] = \begin{bmatrix} 12 & 15 \\ 8 & 10 \end{bmatrix}$

$(\mathbf{AB})\mathbf{C} = \left(\begin{bmatrix} 3 \\ 2 \end{bmatrix}[1 \ -1 \ 2]\right)\begin{bmatrix} 2 & 0 \\ 0 & 3 \\ 1 & 4 \end{bmatrix} = \begin{bmatrix} 3 & -3 & 6 \\ 2 & -2 & 4 \end{bmatrix}\begin{bmatrix} 2 & 0 \\ 0 & 3 \\ 1 & 4 \end{bmatrix} = \begin{bmatrix} 12 & 15 \\ 8 & 10 \end{bmatrix}$

Each of the homogeneous linear systems in Problems 17-22 is already in echelon form, so it remains only to write (by back substitution) the solution, first in parametric form and then in vector form.

17. $x_3 = s, \quad x_4 = t, \quad x_1 = 5s - 4t, \quad x_2 = -2s + 7t$

    $\mathbf{x} = s(5, -2, 1, 0) + t(-4, 7, 0, 1)$

19. $x_4 = s, \quad x_5 = t, \quad x_1 = -3s + t, \quad x_2 = 2s - 6t, \quad x_3 = -s + 8t$

    $\mathbf{x} = s(-3, 2, -1, 1, 0) + t(1, -6, 8, 0, 1)$

21. $x_3 = r, \quad x_4 = s, \quad x_5 = t, \quad x_1 = r - 2s - 7t, \quad x_2 = -2r + 3s - 4t$

    $\mathbf{x} = r(1, -2, 1, 0, 0) + s(-2, 3, 0, 1, 0) + t(-7, -4, 0, 0, 1)$

23. The matrix equation $\begin{bmatrix} 2 & 1 \\ 3 & 2 \end{bmatrix}\begin{bmatrix} a & b \\ c & d \end{bmatrix} = \begin{bmatrix} 1 & 0 \\ 0 & 1 \end{bmatrix}$ entails the four scalar equations

    $2a + c = 1 \qquad 2b + d = 0$
    $3a + 2c = 0 \qquad 3b + 2d = 1$

    that we readily solve for $a = 2, b = -1, c = -3, d = 2$. Hence the apparent inverse matrix of $\mathbf{A}$, such that $\mathbf{AB} = \mathbf{I}$, is $\mathbf{B} = \begin{bmatrix} 2 & -1 \\ -3 & 2 \end{bmatrix}$. Indeed, we find that $\mathbf{BA} = \mathbf{I}$ as well.

25. The matrix equation $\begin{bmatrix} 5 & 7 \\ 2 & 3 \end{bmatrix}\begin{bmatrix} a & b \\ c & d \end{bmatrix} = \begin{bmatrix} 1 & 0 \\ 0 & 1 \end{bmatrix}$ entails the four scalar equations

$$5a+7c = 1 \qquad 5b+7d = 0$$
$$2a+3c = 0 \qquad 2b+3d = 1$$

that we readily solve for $a = 3$, $b = -7$, $c = -2$, $d = 5$. Hence the apparent inverse matrix of $\mathbf{A}$, such that $\mathbf{AB} = \mathbf{I}$, is $\mathbf{B} = \begin{bmatrix} 3 & -7 \\ -2 & 5 \end{bmatrix}$. Indeed, we find that $\mathbf{BA} = \mathbf{I}$ as well.

27. $$\begin{bmatrix} a_1 & 0 & 0 & \cdots & 0 \\ 0 & a_2 & 0 & \cdots & 0 \\ 0 & 0 & a_3 & \cdots & 0 \\ \vdots & \vdots & \vdots & \ddots & \vdots \\ 0 & 0 & 0 & \cdots & a_n \end{bmatrix} \begin{bmatrix} b_1 & 0 & 0 & \cdots & 0 \\ 0 & b_2 & 0 & \cdots & 0 \\ 0 & 0 & b_3 & \cdots & 0 \\ \vdots & \vdots & \vdots & \ddots & \vdots \\ 0 & 0 & 0 & \cdots & b_n \end{bmatrix} = \begin{bmatrix} a_1 b_1 & 0 & 0 & \cdots & 0 \\ 0 & a_2 b_2 & 0 & \cdots & 0 \\ 0 & 0 & a_3 b_3 & \cdots & 0 \\ \vdots & \vdots & \vdots & \ddots & \vdots \\ 0 & 0 & 0 & \cdots & a_n b_n \end{bmatrix}$$

Thus the product of two diagonal matrices of the same size is obtained simply by multiplying corresponding diagonal elements. Then the commutativity of scalar multiplication immediately implies that $\mathbf{AB} = \mathbf{BA}$ for diagonal matrices.

29. $$(a+d)\mathbf{A} - (ad-bc)\mathbf{I} = (a+d) \begin{bmatrix} a & b \\ c & d \end{bmatrix} - (ad-bc) \begin{bmatrix} 1 & 0 \\ 0 & 1 \end{bmatrix}$$
$$= \begin{bmatrix} (a^2+ad)-(ad-bc) & ab+bd \\ ac+cd & (ad+d^2)-(ad-bc) \end{bmatrix} = \begin{bmatrix} a^2+bc & ab+bd \\ ac+cd & bc+d^2 \end{bmatrix}$$
$$= \begin{bmatrix} a & b \\ c & d \end{bmatrix} \begin{bmatrix} a & b \\ c & d \end{bmatrix} = \mathbf{A}^2$$

31. (a) If $\mathbf{A} = \begin{bmatrix} 2 & -1 \\ -4 & 3 \end{bmatrix}$ and $\mathbf{B} = \begin{bmatrix} 1 & 5 \\ 3 & 7 \end{bmatrix}$ then

$$(\mathbf{A}+\mathbf{B})(\mathbf{A}-\mathbf{B}) = \begin{bmatrix} 3 & 4 \\ -1 & 10 \end{bmatrix} \begin{bmatrix} 1 & -6 \\ -7 & -4 \end{bmatrix} = \begin{bmatrix} -25 & -34 \\ -71 & -34 \end{bmatrix}$$

but

$$\mathbf{A}^2 - \mathbf{B}^2 = \begin{bmatrix} 8 & -5 \\ -20 & 13 \end{bmatrix} - \begin{bmatrix} 16 & 40 \\ 24 & 64 \end{bmatrix} = \begin{bmatrix} -8 & -45 \\ -44 & -51 \end{bmatrix}.$$

(b) If $\mathbf{AB} = \mathbf{BA}$ then

$$(\mathbf{A}+\mathbf{B})(\mathbf{A}-\mathbf{B}) = \mathbf{A}(\mathbf{A}-\mathbf{B}) + \mathbf{B}(\mathbf{A}-\mathbf{B})$$
$$= \mathbf{A}^2 - \mathbf{AB} + \mathbf{BA} - \mathbf{B}^2 = \mathbf{A}^2 - \mathbf{B}^2.$$

33. Four different 2×2 matrices $\mathbf{A}$ with $\mathbf{A}^2 = \mathbf{I}$ are

$$\begin{bmatrix} 1 & 0 \\ 0 & 1 \end{bmatrix}, \begin{bmatrix} -1 & 0 \\ 0 & 1 \end{bmatrix}, \begin{bmatrix} 1 & 0 \\ 0 & -1 \end{bmatrix}, \text{ and } \begin{bmatrix} -1 & 0 \\ 0 & -1 \end{bmatrix}.$$

35. If $\mathbf{A} = \begin{bmatrix} 2 & -1 \\ 2 & -1 \end{bmatrix} \neq \mathbf{0}$ then $\mathbf{A}^2 = (1)\mathbf{A} - (0)\mathbf{I} = \mathbf{A}$.

37. If $\mathbf{A} = \begin{bmatrix} 0 & 1 \\ -1 & 0 \end{bmatrix} \neq \mathbf{0}$ then $\mathbf{A}^2 = (0)\mathbf{A} - (1)\mathbf{I} = -\mathbf{I}$.

39. If $\mathbf{A}\mathbf{x}_1 = \mathbf{A}\mathbf{x}_2 = \mathbf{0}$, then

$$\mathbf{A}(c_1\mathbf{x}_1 + c_2\mathbf{x}_2) = c_1(\mathbf{A}\mathbf{x}_1) + c_2(\mathbf{A}\mathbf{x}_2) = c_1(\mathbf{0}) + c_2(\mathbf{0}) = \mathbf{0}.$$

41. If $\mathbf{AB} = \mathbf{BA}$ then

$$\begin{aligned}(\mathbf{A}+\mathbf{B})^3 &= (\mathbf{A}+\mathbf{B})(\mathbf{A}+\mathbf{B})^2 = (\mathbf{A}+\mathbf{B})(\mathbf{A}^2 + 2\mathbf{AB} + \mathbf{B}^2) \\ &= \mathbf{A}(\mathbf{A}^2 + 2\mathbf{AB} + \mathbf{B}^2) + \mathbf{B}(\mathbf{A}^2 + 2\mathbf{AB} + \mathbf{B}^2) \\ &= (\mathbf{A}^3 + 2\mathbf{A}^2\mathbf{B} + \mathbf{AB}^2) + (\mathbf{A}^2\mathbf{B} + 2\mathbf{AB}^2 + \mathbf{B}^3) \\ &= \mathbf{A}^3 + 3\mathbf{A}^2\mathbf{B} + 3\mathbf{AB}^2 + \mathbf{B}^3.\end{aligned}$$

To compute $(\mathbf{A}+\mathbf{B})^4$, write $(\mathbf{A}+\mathbf{B})^4 = (\mathbf{A}+\mathbf{B})(\mathbf{A}+\mathbf{B})^3$ and proceed similarly, substituting the expansion of $(\mathbf{A}+\mathbf{B})^3$ just obtained.

43. First, matrix multiplication gives $\mathbf{A}^2 = \begin{bmatrix} 2 & -1 & -1 \\ -1 & 2 & -1 \\ -1 & -1 & 2 \end{bmatrix} = \begin{bmatrix} 6 & -3 & -3 \\ -3 & 6 & -3 \\ -3 & -3 & 6 \end{bmatrix} = 3\mathbf{A}$. Then

$$\begin{aligned}\mathbf{A}^3 &= \mathbf{A}^2 \cdot \mathbf{A} = 3\mathbf{A} \cdot \mathbf{A} = 3\mathbf{A}^2 = 3 \cdot 3\mathbf{A} = 9\mathbf{A}, \\ \mathbf{A}^4 &= \mathbf{A}^3 \cdot \mathbf{A} = 9\mathbf{A} \cdot \mathbf{A} = 9\mathbf{A}^2 = 9 \cdot 3\mathbf{A} = 27\mathbf{A},\end{aligned}$$

and so forth.

SECTION 3.5

INVERSES OF MATRICES

The computational objective of this section is clearcut — to find the inverse of a given invertible matrix. From a more general viewpoint, Theorem 7 on the properties of nonsingular matrices summarizes most of the basic theory of this chapter.

In Problems 1-8 we first give the inverse matrix $\mathbf{A}^{-1}$ and then calculate the solution vector $\mathbf{x}$.

1. $\mathbf{A}^{-1} = \begin{bmatrix} 3 & -2 \\ -4 & 3 \end{bmatrix}$; $\mathbf{x} = \begin{bmatrix} 3 & -2 \\ -4 & 3 \end{bmatrix} \begin{bmatrix} 5 \\ 6 \end{bmatrix} = \begin{bmatrix} 3 \\ -2 \end{bmatrix}$

3. $\mathbf{A}^{-1} = \begin{bmatrix} 6 & -7 \\ -5 & 6 \end{bmatrix}$; $\mathbf{x} = \begin{bmatrix} 6 & -7 \\ -5 & 6 \end{bmatrix} \begin{bmatrix} 2 \\ -3 \end{bmatrix} = \begin{bmatrix} 33 \\ -28 \end{bmatrix}$

5. $\mathbf{A}^{-1} = \dfrac{1}{2}\begin{bmatrix} 4 & -2 \\ -5 & 3 \end{bmatrix}$; $\mathbf{x} = \dfrac{1}{2}\begin{bmatrix} 4 & -2 \\ -5 & 3 \end{bmatrix}\begin{bmatrix} 5 \\ 6 \end{bmatrix} = \dfrac{1}{2}\begin{bmatrix} 8 \\ -7 \end{bmatrix}$

7. $\mathbf{A}^{-1} = \dfrac{1}{4}\begin{bmatrix} 7 & -9 \\ -5 & 7 \end{bmatrix}$; $\mathbf{x} = \dfrac{1}{4}\begin{bmatrix} 7 & -9 \\ -5 & 7 \end{bmatrix}\begin{bmatrix} 3 \\ 2 \end{bmatrix} = \dfrac{1}{4}\begin{bmatrix} 3 \\ -1 \end{bmatrix}$

In Problems 9-22 we give at least the first few steps in the reduction of the augmented matrix whose right half is the identity matrix of appropriate size. We wind up with its echelon form, whose left half is an identity matrix and whose right half is the desired inverse matrix.

9. $\begin{bmatrix} 5 & 6 & 1 & 0 \\ 4 & 5 & 0 & 1 \end{bmatrix} \xrightarrow{R1-R2} \begin{bmatrix} 1 & 1 & 1 & -1 \\ 4 & 5 & 0 & 1 \end{bmatrix} \xrightarrow{R2-4R1} \begin{bmatrix} 1 & 1 & 1 & -1 \\ 0 & 1 & -4 & 5 \end{bmatrix}$

$\xrightarrow{R1-R2} \begin{bmatrix} 1 & 0 & 5 & -6 \\ 0 & 1 & -4 & 5 \end{bmatrix}$; thus $\mathbf{A}^{-1} = \begin{bmatrix} 5 & -6 \\ -4 & 5 \end{bmatrix}$

11. $\begin{bmatrix} 1 & 5 & 1 & 1 & 0 & 0 \\ 2 & 5 & 0 & 0 & 1 & 0 \\ 2 & 7 & 1 & 0 & 0 & 1 \end{bmatrix} \xrightarrow{R2-2R1} \begin{bmatrix} 1 & 5 & 1 & 1 & 0 & 0 \\ 0 & -5 & -2 & -2 & 1 & 0 \\ 2 & 7 & 1 & 0 & 0 & 1 \end{bmatrix}$

$\xrightarrow{R3-2R1} \begin{bmatrix} 1 & 5 & 1 & 1 & 0 & 0 \\ 0 & -5 & -2 & -2 & 1 & 0 \\ 0 & -3 & -1 & -2 & 0 & 1 \end{bmatrix} \xrightarrow{R1+R3} \begin{bmatrix} 1 & 2 & 0 & -1 & 0 & 1 \\ 0 & -5 & -2 & -2 & 1 & 0 \\ 0 & -3 & -1 & -2 & 0 & 1 \end{bmatrix}$

$\xrightarrow{R2-2R3} \begin{bmatrix} 1 & 2 & 0 & -1 & 0 & 1 \\ 0 & 1 & 0 & 2 & 1 & -2 \\ 0 & -3 & -1 & -2 & 0 & 1 \end{bmatrix} \xrightarrow{R3+3R2} \begin{bmatrix} 1 & 2 & 0 & -1 & 0 & 1 \\ 0 & 1 & 0 & 2 & 1 & -2 \\ 0 & 0 & -1 & 4 & 3 & -5 \end{bmatrix}$

$\xrightarrow{(-1)R3} \cdots \xrightarrow{R1-2R2} \begin{bmatrix} 1 & 0 & 0 & -5 & -2 & 5 \\ 0 & 1 & 0 & 2 & 1 & -2 \\ 0 & 0 & 1 & -4 & -3 & 5 \end{bmatrix}$; thus $\mathbf{A}^{-1} = \begin{bmatrix} -5 & -2 & 5 \\ 2 & 1 & -2 \\ -4 & -3 & 5 \end{bmatrix}$

13. $\begin{bmatrix} 2 & 7 & 3 & 1 & 0 & 0 \\ 1 & 3 & 2 & 0 & 1 & 0 \\ 3 & 7 & 9 & 0 & 0 & 1 \end{bmatrix} \xrightarrow{SWAP(R1,R2)} \begin{bmatrix} 1 & 3 & 2 & 0 & 1 & 0 \\ 2 & 7 & 3 & 1 & 0 & 0 \\ 3 & 7 & 9 & 0 & 0 & 1 \end{bmatrix}$

$\xrightarrow{R2-2R1} \begin{bmatrix} 1 & 3 & 2 & 0 & 1 & 0 \\ 0 & 1 & -1 & 1 & -2 & 0 \\ 3 & 7 & 9 & 0 & 0 & 1 \end{bmatrix} \xrightarrow{R3-3R1} \begin{bmatrix} 1 & 3 & 2 & 0 & 1 & 0 \\ 0 & 1 & -1 & 1 & -2 & 0 \\ 0 & -2 & 3 & 0 & -3 & 1 \end{bmatrix}$

$\xrightarrow{R3+2R2} \cdots \rightarrow \begin{bmatrix} 1 & 0 & 0 & -13 & 42 & -5 \\ 0 & 1 & 0 & 3 & -9 & 1 \\ 0 & 0 & 1 & 2 & -7 & 1 \end{bmatrix}$; thus $\mathbf{A}^{-1} = \begin{bmatrix} -13 & 42 & -5 \\ 3 & -9 & 1 \\ 2 & -7 & 1 \end{bmatrix}$

15. $\begin{bmatrix} 1 & 1 & 5 & 1 & 0 & 0 \\ 1 & 4 & 13 & 0 & 1 & 0 \\ 3 & 2 & 12 & 0 & 0 & 1 \end{bmatrix} \xrightarrow{R2-R1} \begin{bmatrix} 1 & 1 & 5 & 1 & 0 & 0 \\ 0 & 3 & 8 & -1 & 1 & 0 \\ 3 & 2 & 12 & 0 & 0 & 1 \end{bmatrix}$

$\xrightarrow{R3-3R1} \begin{bmatrix} 1 & 1 & 5 & 1 & 0 & 0 \\ 0 & 3 & 8 & -1 & 1 & 0 \\ 0 & -1 & -3 & -3 & 0 & 1 \end{bmatrix} \xrightarrow{R2+2R3} \begin{bmatrix} 1 & 1 & 5 & 1 & 0 & 0 \\ 0 & 1 & 2 & -7 & 1 & 2 \\ 0 & -1 & -3 & -3 & 0 & 1 \end{bmatrix}$

$\xrightarrow{R3+R2} \cdots \rightarrow \begin{bmatrix} 1 & 0 & 0 & -22 & 2 & 7 \\ 0 & 1 & 0 & -27 & 3 & 8 \\ 0 & 0 & 1 & 10 & -1 & -3 \end{bmatrix}$; thus $\mathbf{A}^{-1} = \begin{bmatrix} -22 & 2 & 7 \\ -27 & 3 & 8 \\ 10 & -1 & -3 \end{bmatrix}$

17. $\begin{bmatrix} 1 & -3 & 0 & 1 & 0 & 0 \\ -1 & 2 & -1 & 0 & 1 & 0 \\ 0 & -2 & 2 & 0 & 0 & 1 \end{bmatrix} \xrightarrow{R2+R1} \begin{bmatrix} 1 & -3 & 0 & 1 & 0 & 0 \\ 0 & -1 & -1 & 1 & 1 & 0 \\ 0 & -2 & 2 & 0 & 0 & 1 \end{bmatrix}$

$\xrightarrow{(-1)R2} \begin{bmatrix} 1 & -3 & 0 & 1 & 0 & 0 \\ 0 & 1 & 1 & -1 & -1 & 0 \\ 0 & -2 & 2 & 0 & 0 & 1 \end{bmatrix} \xrightarrow{R3+2R2} \begin{bmatrix} 1 & -3 & 0 & 1 & 0 & 0 \\ 0 & 1 & 1 & -1 & -1 & 0 \\ 0 & 0 & 4 & -2 & -2 & 1 \end{bmatrix}$

$\xrightarrow{(1/4)R3} \cdots \rightarrow \begin{bmatrix} 1 & 0 & 0 & -\frac{1}{2} & -\frac{3}{2} & -\frac{3}{4} \\ 0 & 1 & 0 & -\frac{1}{2} & -\frac{1}{2} & -\frac{1}{4} \\ 0 & 0 & 1 & -\frac{1}{2} & -\frac{1}{2} & \frac{1}{4} \end{bmatrix}$; thus $\mathbf{A}^{-1} = \frac{1}{4}\begin{bmatrix} -2 & -6 & -3 \\ -2 & -2 & -1 \\ -2 & -2 & 1 \end{bmatrix}$

19. $\begin{bmatrix} 1 & 4 & 3 & 1 & 0 & 0 \\ 1 & 4 & 5 & 0 & 1 & 0 \\ 2 & 5 & 1 & 0 & 0 & 1 \end{bmatrix} \xrightarrow{R2-R1} \begin{bmatrix} 1 & 4 & 3 & 1 & 0 & 0 \\ 0 & 0 & 2 & -1 & 1 & 0 \\ 2 & 5 & 1 & 0 & 0 & 1 \end{bmatrix}$

$\xrightarrow{SWAP(R2,R3)} \begin{bmatrix} 1 & 4 & 3 & 1 & 0 & 0 \\ 2 & 5 & 1 & 0 & 0 & 1 \\ 0 & 0 & 2 & -1 & 1 & 0 \end{bmatrix} \xrightarrow{R2-2R1} \begin{bmatrix} 1 & 4 & 3 & 1 & 0 & 0 \\ 0 & -3 & -5 & -2 & 0 & 1 \\ 0 & 0 & 2 & -1 & 1 & 0 \end{bmatrix}$

$\xrightarrow{(-1/3)R2} \begin{bmatrix} 1 & 4 & 3 & 1 & 0 & 0 \\ 0 & 1 & \frac{5}{3} & \frac{2}{3} & 0 & -\frac{1}{3} \\ 0 & 0 & 2 & -1 & 1 & 0 \end{bmatrix} \xrightarrow{(1/2)R3} \cdots \to \begin{bmatrix} 1 & 0 & 0 & -\frac{7}{2} & \frac{11}{6} & \frac{4}{3} \\ 0 & 1 & 0 & \frac{3}{2} & -\frac{5}{6} & -\frac{1}{3} \\ 0 & 0 & 1 & -\frac{1}{2} & \frac{1}{2} & 0 \end{bmatrix};$

thus $\mathbf{A}^{-1} = \frac{1}{6}\begin{bmatrix} -21 & 11 & 8 \\ 9 & -5 & -2 \\ -3 & -3 & 0 \end{bmatrix}$

21. $\begin{bmatrix} 0 & 0 & 1 & 0 & 1 & 0 & 0 & 0 \\ 1 & 0 & 0 & 0 & 0 & 1 & 0 & 0 \\ 0 & 1 & 2 & 0 & 0 & 0 & 1 & 0 \\ 3 & 0 & 0 & 1 & 0 & 0 & 0 & 1 \end{bmatrix} \xrightarrow{SWAP(R1,R2)} \begin{bmatrix} 1 & 0 & 0 & 0 & 0 & 1 & 0 & 0 \\ 0 & 0 & 1 & 0 & 1 & 0 & 0 & 0 \\ 0 & 1 & 2 & 0 & 0 & 0 & 1 & 0 \\ 3 & 0 & 0 & 1 & 0 & 0 & 0 & 1 \end{bmatrix}$

$\xrightarrow{SWAP(R2,R3)} \begin{bmatrix} 1 & 0 & 0 & 0 & 0 & 1 & 0 & 0 \\ 0 & 1 & 2 & 0 & 0 & 0 & 1 & 0 \\ 0 & 0 & 1 & 0 & 1 & 0 & 0 & 0 \\ 3 & 0 & 0 & 1 & 0 & 0 & 0 & 1 \end{bmatrix} \xrightarrow{R4-3R1} \begin{bmatrix} 1 & 0 & 0 & 0 & 0 & 1 & 0 & 0 \\ 0 & 1 & 2 & 0 & 0 & 0 & 1 & 0 \\ 0 & 0 & 1 & 0 & 1 & 0 & 0 & 0 \\ 0 & 0 & 0 & 1 & 0 & -3 & 0 & 1 \end{bmatrix}$

$\xrightarrow{R2-2R3} \begin{bmatrix} 1 & 0 & 0 & 0 & 0 & 1 & 0 & 0 \\ 0 & 1 & 0 & 0 & -2 & 0 & 1 & 0 \\ 0 & 0 & 1 & 0 & 1 & 0 & 0 & 0 \\ 0 & 0 & 0 & 1 & 0 & -3 & 0 & 1 \end{bmatrix};$ thus $\mathbf{A}^{-1} = \begin{bmatrix} 0 & 1 & 0 & 0 \\ -2 & 0 & 1 & 0 \\ 1 & 0 & 0 & 0 \\ 0 & -3 & 0 & 1 \end{bmatrix}$

In Problems 23-28 we first give the inverse matrix $\mathbf{A}^{-1}$ and then calculate the solution matrix $\mathbf{X}$.

23. $\mathbf{A}^{-1} = \begin{bmatrix} 4 & -3 \\ -5 & 4 \end{bmatrix};\quad \mathbf{X} = \begin{bmatrix} 4 & -3 \\ -5 & 4 \end{bmatrix}\begin{bmatrix} 1 & 3 & -5 \\ -1 & -2 & 5 \end{bmatrix} = \begin{bmatrix} 7 & 18 & -35 \\ -9 & -23 & 45 \end{bmatrix}$

Section 3.5

25. $\mathbf{A}^{-1} = \begin{bmatrix} 11 & -9 & 4 \\ -2 & 2 & -1 \\ -2 & 1 & 0 \end{bmatrix}$; $\mathbf{X} = \begin{bmatrix} 11 & -9 & 4 \\ -2 & 2 & -1 \\ -2 & 1 & 0 \end{bmatrix} \begin{bmatrix} 1 & 0 & 3 \\ 0 & 2 & 2 \\ -1 & 1 & 0 \end{bmatrix} = \begin{bmatrix} 7 & -14 & 15 \\ -1 & 3 & -2 \\ -2 & 2 & -4 \end{bmatrix}$

27. $\mathbf{A}^{-1} = \begin{bmatrix} 7 & -20 & 17 \\ 0 & -1 & 1 \\ -2 & 6 & -5 \end{bmatrix}$; $\mathbf{X} = \begin{bmatrix} 7 & -20 & 17 \\ 0 & -1 & 1 \\ -2 & 6 & -5 \end{bmatrix} \begin{bmatrix} 0 & 0 & 1 & 1 \\ 0 & 1 & 0 & 1 \\ 1 & 0 & 1 & 0 \end{bmatrix} = \begin{bmatrix} 17 & -20 & 24 & -13 \\ 1 & -1 & 1 & -1 \\ -5 & 6 & -7 & 4 \end{bmatrix}$

29. (a) The fact that $\mathbf{A}^{-1}$ is the inverse of $\mathbf{A}$ means that $\mathbf{A}\mathbf{A}^{-1} = \mathbf{A}^{-1}\mathbf{A} = \mathbf{I}$. That is, that when $\mathbf{A}^{-1}$ is multiplied either on the right or on the left by $\mathbf{A}$, the result is the identity matrix $\mathbf{I}$. By the same token, this means that $\mathbf{A}$ is the inverse of $\mathbf{A}^{-1}$.

 (b) $\mathbf{A}^n(\mathbf{A}^{-1})^n = \mathbf{A}^{n-1} \cdot \mathbf{A}\mathbf{A}^{-1} \cdot (\mathbf{A}^{-1})^{n-1} = \mathbf{A}^{n-1} \cdot \mathbf{I} \cdot (\mathbf{A}^{-1})^{n-1} = \cdots = \mathbf{I}$. Similarly, $(\mathbf{A}^{-1})^n \mathbf{A}^n = \mathbf{I}$, so it follows that $(\mathbf{A}^{-1})^n$ is the inverse of $\mathbf{A}^n$.

31. Let $p = -r > 0$, $q = -s > 0$, and $\mathbf{B} = \mathbf{A}^{-1}$. Then

$$\mathbf{A}^r \mathbf{A}^s = \mathbf{A}^{-p} \mathbf{A}^{-q} = (\mathbf{A}^{-1})^p (\mathbf{A}^{-1})^q$$
$$= \mathbf{B}^p \mathbf{B}^q = \mathbf{B}^{p+q} \quad \text{(because } p, q > 0\text{)}$$
$$= (\mathbf{A}^{-1})^{p+q} = \mathbf{A}^{-p-q} = \mathbf{A}^{r+s}$$

as desired, and $(\mathbf{A}^r)^s = (\mathbf{A}^{-p})^{-q} = (\mathbf{B}^p)^{-q} = \mathbf{B}^{-pq} = \mathbf{A}^{pq} = \mathbf{A}^{rs}$ similarly.

33. In particular, $\mathbf{A}\mathbf{e}_j = \mathbf{e}_j$ where $\mathbf{e}_j$ denotes the jth column vector of the identity matrix $\mathbf{I}$. Hence it follows from Fact 2 that $\mathbf{AI} = \mathbf{I}$, and therefore $\mathbf{A} = \mathbf{I}^{-1} = \mathbf{I}$.

35. If the jth column of $\mathbf{A}$ is all zeros and $\mathbf{B}$ is any $n \times n$ matrix, then the jth column of $\mathbf{BA}$ is all zeros, so $\mathbf{BA} \ne \mathbf{I}$. Hence $\mathbf{A}$ has no inverse matrix. Similarly, if the ith row of $\mathbf{A}$ is all zeros, then so is the ith row of $\mathbf{AB}$.

37. Direct multiplication shows that $\mathbf{A}\mathbf{A}^{-1} = \mathbf{A}^{-1}\mathbf{A} = \mathbf{I}$.

39. $\mathbf{EA} = \begin{bmatrix} 1 & 0 & 0 \\ 0 & 1 & 0 \\ 2 & 0 & 1 \end{bmatrix} \begin{bmatrix} a_{11} & a_{12} & a_{13} \\ a_{21} & a_{22} & a_{23} \\ a_{31} & a_{32} & a_{33} \end{bmatrix} = \begin{bmatrix} a_{11} & a_{12} & a_{13} \\ a_{21} & a_{22} & a_{23} \\ a_{31} + 2a_{11} & a_{32} + a_{12} & a_{33} + a_{13} \end{bmatrix}$

41. This follows immediately from the fact that the ijth element of $\mathbf{AB}$ is the product of the ith row of $\mathbf{A}$ and the jth column of $\mathbf{B}$.

43. Let $\mathbf{E}_1, \mathbf{E}_2, \cdots, \mathbf{E}_k$ be the elementary matrices corresponding to the elementary row operations that reduce $\mathbf{A}$ to $\mathbf{B}$. Then Theorem 5 gives $\mathbf{B} = \mathbf{E}_k \mathbf{E}_{k-1} \cdots \mathbf{E}_2 \mathbf{E}_1 \mathbf{A} = \mathbf{GA}$ where $\mathbf{G} = \mathbf{E}_k \mathbf{E}_{k-1} \cdots \mathbf{E}_2 \mathbf{E}_1$.

45. One can simply photocopy the portion of the proof of Theorem 7 that follows Equation (20). Starting only with the assumption that $\mathbf{A}$ and $\mathbf{B}$ are square matrices with $\mathbf{AB} = \mathbf{I}$, it is proved there that $\mathbf{A}$ and $\mathbf{B}$ are then invertible.

SECTION 3.6

DETERMINANTS

1. $\begin{vmatrix} 0 & 0 & 3 \\ 4 & 0 & 0 \\ 0 & 5 & 0 \end{vmatrix} = +(3) \begin{vmatrix} 4 & 0 \\ 0 & 5 \end{vmatrix} = 3 \cdot 4 \cdot 5 = 60$

3. $\begin{vmatrix} 1 & 0 & 0 & 0 \\ 2 & 0 & 5 & 0 \\ 3 & 6 & 9 & 8 \\ 4 & 0 & 10 & 7 \end{vmatrix} = +(1) \begin{vmatrix} 0 & 5 & 0 \\ 6 & 9 & 8 \\ 0 & 10 & 7 \end{vmatrix} = -(5) \begin{vmatrix} 6 & 8 \\ 0 & 7 \end{vmatrix} = -5(42-0) = -210$

5. $\begin{vmatrix} 0 & 0 & 1 & 0 & 0 \\ 2 & 0 & 0 & 0 & 0 \\ 0 & 0 & 0 & 3 & 0 \\ 0 & 0 & 0 & 0 & 4 \\ 0 & 5 & 0 & 0 & 0 \end{vmatrix} = +1 \begin{vmatrix} 2 & 0 & 0 & 0 \\ 0 & 0 & 3 & 0 \\ 0 & 0 & 0 & 4 \\ 0 & 5 & 0 & 0 \end{vmatrix} = +2 \begin{vmatrix} 0 & 3 & 0 \\ 0 & 0 & 4 \\ 5 & 0 & 0 \end{vmatrix} = 2(+5) \begin{vmatrix} 3 & 0 \\ 0 & 4 \end{vmatrix} = 2 \cdot 5 \cdot 3 \cdot 4 = 120$

7. $\begin{vmatrix} 1 & 1 & 1 \\ 2 & 2 & 2 \\ 3 & 3 & 3 \end{vmatrix} \overset{R2-2R1}{=} \begin{vmatrix} 1 & 1 & 1 \\ 0 & 0 & 0 \\ 3 & 3 & 3 \end{vmatrix} = 0$

9. $\begin{vmatrix} 3 & -2 & 5 \\ 0 & 5 & 17 \\ 6 & -4 & 12 \end{vmatrix} \overset{R3-2R1}{=} \begin{vmatrix} 3 & -2 & 5 \\ 0 & 5 & 17 \\ 0 & 0 & 2 \end{vmatrix} = +2 \begin{vmatrix} 3 & 5 \\ 0 & 2 \end{vmatrix} = 5(6-0) = 30$

11. $\begin{vmatrix} 1 & 2 & 3 & 4 \\ 0 & 5 & 6 & 7 \\ 0 & 0 & 8 & 9 \\ 2 & 4 & 6 & 9 \end{vmatrix} \overset{R4-2R1}{=} \begin{vmatrix} 1 & 2 & 3 & 4 \\ 0 & 5 & 6 & 7 \\ 0 & 0 & 8 & 9 \\ 0 & 0 & 0 & 1 \end{vmatrix} = +1 \begin{vmatrix} 5 & 6 & 7 \\ 0 & 8 & 9 \\ 0 & 0 & 1 \end{vmatrix} = +5 \begin{vmatrix} 8 & 9 \\ 0 & 1 \end{vmatrix} = 5 \cdot 8 = 40$

13. $\begin{vmatrix} -4 & 4 & -1 \\ -1 & -2 & 2 \\ 1 & 4 & 3 \end{vmatrix} \overset{R2+R3}{=} \begin{vmatrix} -4 & 4 & -1 \\ 0 & 2 & 5 \\ 1 & 4 & 3 \end{vmatrix} \overset{R1+4R3}{=} \begin{vmatrix} 0 & 20 & 11 \\ 0 & 2 & 5 \\ 1 & 4 & 3 \end{vmatrix} = +1 \begin{vmatrix} 20 & 11 \\ 2 & 5 \end{vmatrix} = 100 - 22 = 78$

15. $\begin{vmatrix} -2 & 5 & 4 \\ 5 & 3 & 1 \\ 1 & 4 & 5 \end{vmatrix} \overset{R1+2R3}{=} \begin{vmatrix} 0 & 13 & 14 \\ 5 & 3 & 1 \\ 1 & 4 & 5 \end{vmatrix} \overset{R2-5R3}{=} \begin{vmatrix} 0 & 13 & 14 \\ 0 & -17 & -24 \\ 1 & 4 & 5 \end{vmatrix} = +1 \begin{vmatrix} 13 & 14 \\ -17 & -24 \end{vmatrix} = -74$

17. $\begin{vmatrix} 2 & 3 & 3 & 1 \\ 0 & 4 & 3 & -3 \\ 2 & -1 & -1 & -3 \\ 0 & -4 & -3 & 2 \end{vmatrix} \overset{R3-R1}{=} \begin{vmatrix} 2 & 3 & 3 & 1 \\ 0 & 4 & 3 & -3 \\ 0 & -4 & -4 & -4 \\ 0 & -4 & -3 & 2 \end{vmatrix} = 2 \begin{vmatrix} 4 & 3 & -3 \\ -4 & -4 & -4 \\ -4 & -3 & 2 \end{vmatrix} \overset{\substack{R2+R1 \\ R3+R1}}{=} 2 \begin{vmatrix} 4 & 3 & -3 \\ 0 & -1 & -7 \\ 0 & 0 & -1 \end{vmatrix} = 8$

19. $\begin{vmatrix} 1 & 0 & 0 & 3 \\ 0 & 1 & -2 & 0 \\ -2 & 3 & -2 & 3 \\ 0 & -3 & 3 & 3 \end{vmatrix} \overset{R3+2R1}{=} \begin{vmatrix} 1 & 0 & 0 & 3 \\ 0 & 1 & -2 & 0 \\ 0 & 3 & -2 & 9 \\ 0 & -3 & 3 & 3 \end{vmatrix} = 1 \begin{vmatrix} 1 & -2 & 0 \\ 3 & -2 & 9 \\ -3 & 3 & 3 \end{vmatrix} \overset{C2+2C1}{=} \begin{vmatrix} 1 & 0 & 0 \\ 3 & 4 & 9 \\ -3 & -3 & 3 \end{vmatrix} = 39$

21. $\Delta = \begin{vmatrix} 3 & 4 \\ 5 & 7 \end{vmatrix} = 1; \quad x = \frac{1}{\Delta} \begin{vmatrix} 2 & 4 \\ 1 & 7 \end{vmatrix} = 10, \quad y = \frac{1}{\Delta} \begin{vmatrix} 3 & 2 \\ 5 & 1 \end{vmatrix} = -7$

23. $\Delta = \begin{vmatrix} 17 & 7 \\ 12 & 5 \end{vmatrix} = 1; \quad x = \frac{1}{\Delta} \begin{vmatrix} 6 & 7 \\ 4 & 5 \end{vmatrix} = 2, \quad y = \frac{1}{\Delta} \begin{vmatrix} 17 & 6 \\ 12 & 4 \end{vmatrix} = -4$

25. $\Delta = \begin{vmatrix} 5 & 6 \\ 3 & 4 \end{vmatrix} = 2; \quad x = \frac{1}{\Delta} \begin{vmatrix} 12 & 6 \\ 6 & 4 \end{vmatrix} = 6, \quad y = \frac{1}{\Delta} \begin{vmatrix} 5 & 12 \\ 3 & 6 \end{vmatrix} = -3$

27. $\Delta = \begin{vmatrix} 5 & 2 & -2 \\ 1 & 5 & -3 \\ 5 & -3 & 5 \end{vmatrix} = 96; \quad x_1 = \frac{1}{\Delta} \begin{vmatrix} 1 & 2 & -2 \\ -2 & 5 & -3 \\ 2 & -3 & 5 \end{vmatrix} = \frac{1}{3},$

$$x_2 = \frac{1}{\Delta}\begin{vmatrix} 5 & 1 & -2 \\ 1 & -2 & -3 \\ 5 & 2 & 5 \end{vmatrix} = -\frac{2}{3}, \qquad x_3 = \frac{1}{\Delta}\begin{vmatrix} 5 & 2 & 1 \\ 1 & 5 & -2 \\ 5 & -3 & 2 \end{vmatrix} = -\frac{1}{3}$$

29. $\Delta = \begin{vmatrix} 3 & -1 & -5 \\ 4 & -4 & -3 \\ 1 & 0 & -5 \end{vmatrix} = 23; \qquad x_1 = \frac{1}{\Delta}\begin{vmatrix} 3 & -1 & -5 \\ -4 & -4 & -3 \\ 2 & 0 & -5 \end{vmatrix} = 2,$

$$x_2 = \frac{1}{\Delta}\begin{vmatrix} 3 & 3 & -5 \\ 4 & -4 & -3 \\ 1 & 2 & -5 \end{vmatrix} = 3, \qquad x_3 = \frac{1}{\Delta}\begin{vmatrix} 3 & -1 & 3 \\ 4 & -4 & -4 \\ 1 & 0 & 2 \end{vmatrix} = 0$$

31. $\Delta = \begin{vmatrix} 2 & 0 & -5 \\ 4 & -5 & 3 \\ -2 & 1 & 1 \end{vmatrix} = 14; \qquad x_1 = \frac{1}{\Delta}\begin{vmatrix} -3 & 0 & -5 \\ 3 & -5 & 3 \\ 1 & 1 & 1 \end{vmatrix} = -\frac{8}{7},$

$$x_2 = \frac{1}{\Delta}\begin{vmatrix} 2 & -3 & -5 \\ 4 & 3 & 3 \\ -2 & 1 & 1 \end{vmatrix} = -\frac{10}{7}, \qquad x_3 = \frac{1}{\Delta}\begin{vmatrix} 2 & 0 & -3 \\ 4 & -5 & 3 \\ -2 & 1 & 1 \end{vmatrix} = \frac{1}{7}$$

33. $\det \mathbf{A} = -4, \qquad \mathbf{A}^{-1} = \frac{1}{4}\begin{bmatrix} 4 & 4 & 4 \\ 16 & 15 & 13 \\ 28 & 25 & 23 \end{bmatrix}$

35. $\det \mathbf{A} = 35, \qquad \mathbf{A}^{-1} = \frac{1}{35}\begin{bmatrix} -15 & 25 & -26 \\ 10 & -5 & 8 \\ 15 & -25 & 19 \end{bmatrix}$

37. $\det \mathbf{A} = 29, \qquad \mathbf{A}^{-1} = \frac{1}{29}\begin{bmatrix} 11 & -14 & -15 \\ -17 & 19 & 10 \\ 18 & -15 & -14 \end{bmatrix}$

39. $\det \mathbf{A} = 37, \qquad \mathbf{A}^{-1} = \frac{1}{37}\begin{bmatrix} -21 & -1 & -13 \\ 4 & 9 & 6 \\ -6 & 5 & -9 \end{bmatrix}$

41. If $\mathbf{A} = \begin{bmatrix} \mathbf{a}_1 \\ \mathbf{a}_2 \end{bmatrix}$ and $\mathbf{B} = [\mathbf{b}_1 \ \ \mathbf{b}_2]$ in terms of the two row vectors of $\mathbf{A}$ and the two column vectors of $\mathbf{B}$, then $\mathbf{AB} = \begin{bmatrix} \mathbf{a}_1\mathbf{b}_1 & \mathbf{a}_1\mathbf{b}_2 \\ \mathbf{a}_2\mathbf{b}_1 & \mathbf{a}_2\mathbf{b}_2 \end{bmatrix}$, so

$$(\mathbf{AB})^T = \begin{bmatrix} \mathbf{a}_1\mathbf{b}_1 & \mathbf{a}_2\mathbf{b}_1 \\ \mathbf{a}_1\mathbf{b}_2 & \mathbf{a}_2\mathbf{b}_2 \end{bmatrix} = \begin{bmatrix} \mathbf{b}_1^T \\ \mathbf{b}_2^T \end{bmatrix} \begin{bmatrix} \mathbf{a}_1^T & \mathbf{a}_1^T \end{bmatrix} = \mathbf{B}^T\mathbf{A}^T,$$

because the rows of $\mathbf{A}$ are the columns of $\mathbf{A}^T$ and the columns of $\mathbf{B}$ are the rows of $\mathbf{B}^T$.

43. We expand the left-hand determinant along its first column:

$$\begin{vmatrix} ka_{11} & a_{12} & a_{13} \\ ka_{21} & a_{22} & a_{23} \\ ka_{31} & a_{32} & a_{33} \end{vmatrix}$$

$$= ka_{11}(a_{12}a_{23} - a_{22}a_{13}) - ka_{21}(a_{12}a_{33} - a_{32}a_{13}) + ka_{31}(a_{12}a_{23} - a_{22}a_{13})$$

$$= k\left[a_{11}(a_{12}a_{23} - a_{22}a_{13}) - a_{21}(a_{12}a_{33} - a_{32}a_{13}) + a_{31}(a_{12}a_{23} - a_{22}a_{13})\right]$$

$$= k\begin{vmatrix} a_{11} & a_{12} & a_{13} \\ a_{21} & a_{22} & a_{23} \\ a_{31} & a_{32} & a_{33} \end{vmatrix}$$

45. We expand the left-hand determinant along its third column:

$$\begin{vmatrix} a_1 & b_1 & c_1+d_1 \\ a_2 & b_2 & c_2+d_2 \\ a_3 & b_3 & c_3+d_3 \end{vmatrix}$$

$$= (c_1+d_1)(a_2b_3 - a_3b_2) - (c_2+d_2)(a_1b_3 - a_3b_1) + (c_3+d_3)(a_1b_2 - a_2b_1)$$

$$= c_1(a_2b_3 - a_3b_2) - c_2(a_1b_3 - a_3b_1) + c_3(a_1b_2 - a_2b_1)$$

$$+ d_1(a_2b_3 - a_3b_2) - d_2(a_1b_3 - a_3b_1) + d_3(a_1b_2 - a_2b_1)$$

$$= \begin{vmatrix} a_1 & b_1 & c_1 \\ a_2 & b_2 & c_2 \\ a_3 & b_3 & c_3 \end{vmatrix} + \begin{vmatrix} a_1 & b_1 & d_1 \\ a_2 & b_2 & d_2 \\ a_3 & b_3 & d_3 \end{vmatrix}$$

47. We illustrate these properties with 2×2 matrices $\mathbf{A} = [a_{ij}]$ and $\mathbf{B} = [b_{ij}]$.

(a) $\left(\mathbf{A}^T\right)^T = \begin{bmatrix} a_{11} & a_{21} \\ a_{12} & a_{22} \end{bmatrix}^T = \begin{bmatrix} a_{11} & a_{12} \\ a_{22} & a_{22} \end{bmatrix} = \mathbf{A}$

(b) $(c\mathbf{A})^T = \begin{bmatrix} ca_{11} & ca_{12} \\ ca_{21} & ca_{22} \end{bmatrix}^T = \begin{bmatrix} ca_{11} & ca_{21} \\ ca_{12} & ca_{22} \end{bmatrix} = c\begin{bmatrix} a_{11} & a_{21} \\ a_{12} & a_{22} \end{bmatrix} = c\mathbf{A}^T$

(c) $(\mathbf{A}+\mathbf{B})^T = \begin{bmatrix} a_{11}+b_{11} & a_{12}+b_{12} \\ a_{21}+b_{21} & a_{22}+b_{22} \end{bmatrix}^T = \begin{bmatrix} a_{11}+b_{11} & a_{21}+b_{21} \\ a_{12}+b_{12} & a_{22}+b_{22} \end{bmatrix}$

$= \begin{bmatrix} a_{11} & a_{21} \\ a_{12} & a_{22} \end{bmatrix} + \begin{bmatrix} b_{11} & b_{21} \\ b_{12} & b_{22} \end{bmatrix} = \mathbf{A}^T + \mathbf{B}^T$

49. If we write $\mathbf{A} = \begin{vmatrix} a_1 & b_1 & c_1 \\ a_2 & b_2 & c_2 \\ a_3 & b_3 & c_3 \end{vmatrix}$ and $\mathbf{A}^T = \begin{vmatrix} a_1 & a_2 & a_3 \\ b_1 & b_2 & b_3 \\ c_1 & c_2 & c_3 \end{vmatrix}$, then expansion of $|\mathbf{A}|$ along its first row and of $|\mathbf{A}^T|$ along its first column both give the result

$a_1(b_2 c_3 - b_3 c_2) + b_1(a_2 c_3 - a_3 c_2) + c_1(a_2 b_3 - a_3 b_2)$.

51. If $\mathbf{A}^n = \mathbf{0}$ then $|\mathbf{A}|^n = \mathbf{0}$, so it follows immediately that $|\mathbf{A}| = 0$.

53. If $\mathbf{A} = \mathbf{P}^{-1}\mathbf{B}\mathbf{P}$ then $|\mathbf{A}| = |\mathbf{P}^{-1}\mathbf{B}\mathbf{P}| = |\mathbf{P}^{-1}||\mathbf{B}||\mathbf{P}| = |\mathbf{P}|^{-1}|\mathbf{B}||\mathbf{P}| = |\mathbf{B}|$.

55. If either $\mathbf{AB} = \mathbf{I}$ or $\mathbf{BA} = \mathbf{I}$ is given, then it follows from Problem 54 that $\mathbf{A}$ and $\mathbf{B}$ are both invertible because their product (one way or the other) is invertible. Hence $\mathbf{A}^{-1}$ exists. So if (for instance) it is $\mathbf{AB} = \mathbf{I}$ that is given, then multiplication by $\mathbf{A}^{-1}$ on the right yields $\mathbf{B} = \mathbf{A}^{-1}$.

57. If $\mathbf{A} = \begin{bmatrix} a & d & f \\ 0 & b & e \\ 0 & 0 & b \end{bmatrix}$ then $\mathbf{A}^{-1} = \frac{1}{abc}\begin{bmatrix} bc & -cd & de-bf \\ 0 & ac & -ae \\ 0 & 0 & ab \end{bmatrix}$.

59. These are almost immediate computations.

61. Subtraction of the first row from both the second and the third row gives

$\begin{vmatrix} 1 & a & a^2 \\ 1 & b & b^2 \\ 1 & c & c^2 \end{vmatrix} = \begin{vmatrix} 1 & a & a^2 \\ 0 & b-a & b^2-a^2 \\ 0 & c-a & c^2-a^2 \end{vmatrix} = (b-a)(c^2-a^2) - (c-a)(b^2-a^2)$

Section 3.6

$$= (b-a)(c-a)(c+a) - (c-a)(b-a)(b+a)$$
$$= (b-a)(c-a)[(c+a)-(b+a)] = (b-a)(c-a)(c-b).$$

62. Expansion of the 4×4 determinant defining $P(y)$ along its 4th row yields

$$P(y) = y^3 \begin{vmatrix} 1 & x_1 & x_1^2 \\ 1 & x_2 & x_2^2 \\ 1 & x_3 & x_3^2 \end{vmatrix} + \cdots = y^3 V(x_1, x_2, x_3) + \text{lower-degree terms in } y.$$

Because it is clear from the determinant definition of $P(y)$ that $P(x_1) = P(x_2) = P(x_3) = 0$, the three roots of the cubic polynomial $P(y)$ are x_1, x_2, x_3. The factor theorem therefore says that $P(y) = k(y - x_1)(y - x_2)(y - x_3)$ for some constant k, and the calculation above implies that

$$k = V(x_1, x_2, x_3) = (x_3 - x_1)(x_3 - x_2)(x_2 - x_1).$$

Finally we see that

$$V(x_1, x_2, x_3, x_4) = P(x_4) = V(x_1, x_2, x_3) \cdot (x_4 - x_1)(x_4 - x_2)(x_4 - x_1)$$
$$= (x_4 - x_1)(x_4 - x_2)(x_4 - x_1)(x_3 - x_1)(x_3 - x_2)(x_2 - x_1),$$

which is the desired formula for $V(x_1, x_2, x_3, x_4)$.

63. The same argument as in Problem 62 yields

$$P(y) = V(x_1, x_2, \cdots, x_{n-1}) \cdot (y - x_1)(y - x_2) \cdots (y - x_{n-1}).$$

Therefore

$$V(x_1, x_2, \cdots, x_n) = (x_n - x_1)(x_n - x_2) \cdots (x_n - x_{n-1}) V(x_1, x_2, \cdots, x_{n-1})$$
$$= (x_n - x_1)(x_n - x_2) \cdots (x_n - x_{n-1}) \prod_{i>j}^{n-1} (x_i - x_j)$$
$$= \prod_{i>j}^{n} (x_i - x_j).$$

SECTION 3.7

LINEAR EQUATIONS AND CURVE FITTING

In Problems 1-10 we first set up the linear system in the coefficients $a, b, \ldots$ that we get by substituting each given point (x_i, y_i) into the desired interpolating polynomial equation $y = a + bx + \cdots$. Then we give the polynomial that results from solution of this linear system.

1. $y(x) = a + bx$

$$\begin{bmatrix} 1 & 1 \\ 1 & 3 \end{bmatrix} \begin{bmatrix} a \\ b \end{bmatrix} = \begin{bmatrix} 1 \\ 7 \end{bmatrix} \Rightarrow a = -2, \; b = 3 \quad \text{so} \quad y(x) = -2 + 3x$$

3. $y(x) = a + bx + cx^2$

$$\begin{bmatrix} 1 & 0 & 0 \\ 1 & 1 & 1 \\ 1 & 2 & 4 \end{bmatrix} \begin{bmatrix} a \\ b \\ c \end{bmatrix} = \begin{bmatrix} 3 \\ 1 \\ -5 \end{bmatrix} \Rightarrow a = 3, \; b = 0, \; c = -2 \quad \text{so} \quad y(x) = 3 - 2x^2$$

5. $y(x) = a + bx + cx^2$

$$\begin{bmatrix} 1 & 1 & 1 \\ 1 & 2 & 4 \\ 1 & 3 & 9 \end{bmatrix} \begin{bmatrix} a \\ b \\ c \end{bmatrix} = \begin{bmatrix} 3 \\ 3 \\ 5 \end{bmatrix} \Rightarrow a = 5, \; b = -3, \; c = 1 \quad \text{so} \quad y(x) = 5 - 3x + x^2$$

7. $y(x) = a + bx + cx^2 + dx^3$

$$\begin{bmatrix} 1 & -1 & 1 & -1 \\ 1 & 0 & 0 & 0 \\ 1 & 1 & 1 & 1 \\ 1 & 2 & 4 & 8 \end{bmatrix} \begin{bmatrix} a \\ b \\ c \\ d \end{bmatrix} = \begin{bmatrix} 1 \\ 0 \\ 1 \\ -4 \end{bmatrix}$$

$\Rightarrow a = 0, \; b = \dfrac{4}{3}, \; c = 1, \; d = -\dfrac{4}{3} \quad \text{so} \quad y(x) = \dfrac{1}{3}\left(4x + 3x^2 - 4x^3\right)$

9. $y(x) = a + bx + cx^2 + dx^3$

$$\begin{bmatrix} 1 & -2 & 4 & -8 \\ 1 & -1 & 1 & -1 \\ 1 & 1 & 1 & 1 \\ 1 & 2 & 4 & 8 \end{bmatrix} \begin{bmatrix} a \\ b \\ c \\ d \end{bmatrix} = \begin{bmatrix} -2 \\ 2 \\ 10 \\ 26 \end{bmatrix}$$

$\Rightarrow a = 4, \; b = 3, \; c = 2, \; d = 1 \quad \text{so} \quad y(x) = 4 + 3x + 2x^2 + x^3$

In Problems 11-14 we first set up the linear system in the coefficients A, B, C that we get by substituting each given point (x_i, y_i) into the circle equation $Ax + By + C = -x^2 - y^2$ (see

Eq. (9) in the text). Then we give the circle that results from solution of this linear system.

11. $Ax + By + C = -x^2 - y^2$

$$\begin{bmatrix} -1 & -1 & 1 \\ 6 & 6 & 1 \\ 7 & 5 & 1 \end{bmatrix} \begin{bmatrix} A \\ B \\ C \end{bmatrix} = \begin{bmatrix} -2 \\ -72 \\ -74 \end{bmatrix} \quad \Rightarrow \quad A = -6, \ B = -4, \ C = -12$$

$x^2 + y^2 - 6x - 4y - 12 = 0$

$(x-3)^2 + (y-2)^2 = 25$ center $(3, 2)$ and radius 5

13. $Ax + By + C = -x^2 - y^2$

$$\begin{bmatrix} 1 & 0 & 1 \\ 0 & -5 & 1 \\ -5 & -4 & 1 \end{bmatrix} \begin{bmatrix} A \\ B \\ C \end{bmatrix} = \begin{bmatrix} -1 \\ -25 \\ -41 \end{bmatrix} \quad \Rightarrow \quad A = 4, \ B = 4, \ C = -5$$

$x^2 + y^2 + 4x + 4y - 5 = 0$

$(x+2)^2 + (y+2)^2 = 13$ center $(-3, -2)$ and radius $\sqrt{13}$

In Problems 15-18 we first set up the linear system in the coefficients A, B, C that we get by substituting each given point (x_i, y_i) into the central conic equation $Ax^2 + Bxy + Cy^2 = 1$ (see Eq. (10) in the text). Then we give the equation that results from solution of this linear system.

15. $Ax^2 + Bxy + Cy^2 = 1$

$$\begin{bmatrix} 0 & 0 & 25 \\ 25 & 0 & 0 \\ 25 & 25 & 25 \end{bmatrix} \begin{bmatrix} A \\ B \\ C \end{bmatrix} = \begin{bmatrix} 1 \\ 1 \\ 1 \end{bmatrix} \quad \Rightarrow \quad A = \frac{1}{25}, \ B = -\frac{1}{25}, \ C = \frac{1}{25}$$

$x^2 - xy + y^2 = 25$

17. $Ax^2 + Bxy + Cy^2 = 1$

$$\begin{bmatrix} 0 & 0 & 1 \\ 1 & 0 & 0 \\ 100 & 100 & 100 \end{bmatrix} \begin{bmatrix} A \\ B \\ C \end{bmatrix} = \begin{bmatrix} 1 \\ 1 \\ 1 \end{bmatrix} \quad \Rightarrow \quad A = 1, \ B = -\frac{199}{100}, \ C = 1$$

$100x^2 - 199xy + 100y^2 = 100$

19. We substitute each of the two given points into the equation $y = A + \dfrac{B}{x}$.

$$\begin{bmatrix} 1 & 1 \\ 1 & \frac{1}{2} \end{bmatrix} \begin{bmatrix} A \\ B \end{bmatrix} = \begin{bmatrix} 5 \\ 4 \end{bmatrix} \quad \Rightarrow \quad A = 3, \ B = 2 \ \text{ so } \ y = 3 + \dfrac{2}{x}$$

In Problems 21 and 22 we fit the sphere equation $(x-h)^2 + (y-k)^2 + (z-l)^2 = r^2$ in the expanded form $Ax + By + Cz + D = -x^2 - y^2 - z^2$ that is analogous to Eq. (9) in the text (for a circle).

21. $Ax + By + Cz + D = -x^2 - y^2 - z^2$

$$\begin{bmatrix} 4 & 6 & 15 & 1 \\ 13 & 5 & 7 & 1 \\ 5 & 14 & 6 & 1 \\ 5 & 5 & -9 & 1 \end{bmatrix} \begin{bmatrix} A \\ B \\ C \\ D \end{bmatrix} = \begin{bmatrix} -277 \\ -243 \\ -257 \\ -131 \end{bmatrix} \quad \Rightarrow \quad A = -2, \ B = -4, \ C = -6, \ D = -155$$

$x^2 + y^2 + z^2 - 2x - 4y - 6z - 155 = 0$

$(x-1)^2 + (y-2)^2 + (z-3)^2 = 169 \quad$ center $(1, 2, 3)$ and radius 13

In Problems 23-26 we first take $t = 0$ in 1970 to fit a quadratic polynomial $P(t) = a + bt + ct^2$. Then we write the quadratic polynomial $Q(T) = P(T - 1970)$ that expresses the predicted population in terms of the actual calendar year T.

23. $P(t) = a + bt + ct^2$

$$\begin{bmatrix} 1 & 0 & 0 \\ 1 & 10 & 100 \\ 1 & 20 & 400 \end{bmatrix} \begin{bmatrix} a \\ b \\ c \end{bmatrix} = \begin{bmatrix} 49.061 \\ 49.137 \\ 50.809 \end{bmatrix}$$

$P(t) = 49.061 - 0.0722t + 0.00798t^2$

$Q(T) = 31160.9 - 31.5134T + 0.00798T^2$

25. $P(t) = a + bt + ct^2$

$$\begin{bmatrix} 1 & 0 & 0 \\ 1 & 10 & 100 \\ 1 & 20 & 400 \end{bmatrix} \begin{bmatrix} a \\ b \\ c \end{bmatrix} = \begin{bmatrix} 62.813 \\ 75.367 \\ 85.446 \end{bmatrix}$$

$$P(t) = 62.813 + 1.37915t - 0.012375t^2$$

$$Q(T) = -50680.3 + 50.1367T - 0.012375T^2$$

In Problems 27-30 we first take $t = 0$ in 1960 to fit a cubic polynomial $P(t) = a + bt + ct^2 + dt^3$. Then we write the cubic polynomial $Q(T) = P(T - 1960)$ that expresses the predicted population in terms of the actual calendar year T.

27. $\quad P(t) = a + bt + ct^2 + dt^3$

$$\begin{bmatrix} 1 & 0 & 0 & 0 \\ 1 & 10 & 100 & 1000 \\ 1 & 20 & 400 & 8000 \\ 1 & 30 & 900 & 27000 \end{bmatrix} \begin{bmatrix} a \\ b \\ c \\ d \end{bmatrix} = \begin{bmatrix} 44.678 \\ 49.061 \\ 49.137 \\ 50.809 \end{bmatrix}$$

$$P(t) = 44.678 + 0.850417t - 0.05105t^2 + 0.000983833t^3$$

$$Q(T) = -7.60554 \times 10^6 + 11539.4T - 5.83599T^2 + 0.000983833T^3$$

29. $\quad P(t) = a + bt + ct^2 + dt^3$

$$\begin{bmatrix} 1 & 0 & 0 & 0 \\ 1 & 10 & 100 & 1000 \\ 1 & 20 & 400 & 8000 \\ 1 & 30 & 900 & 27000 \end{bmatrix} \begin{bmatrix} a \\ b \\ c \\ d \end{bmatrix} = \begin{bmatrix} 54.973 \\ 62.813 \\ 75.367 \\ 85.446 \end{bmatrix}$$

$$P(t) = 54.973 + 0.308667t + 0.059515t^2 - 0.00119817t^3$$

$$Q(T) = 9.24972 \times 10^6 - 14041.6T + 7.10474T^2 - 0.00119817T^3$$

In Problems 31-34 we take $t = 0$ in 1950 to fit a quartic polynomial $P(t) = a + bt + ct^2 + dt^3 + et^4$. Then we write the quartic polynomial $Q(T) = P(T - 1950)$ that expresses the predicted population in terms of the actual calendar year T.

31. $\quad P(t) = a + bt + ct^2 + dt^3 + et^4$.

$$\begin{bmatrix} 1 & 0 & 0 & 0 & 0 \\ 1 & 10 & 100 & 1000 & 10000 \\ 1 & 20 & 400 & 8000 & 160000 \\ 1 & 30 & 900 & 27000 & 810000 \\ 1 & 40 & 1600 & 64000 & 2560000 \end{bmatrix} \begin{bmatrix} a \\ b \\ c \\ d \\ e \end{bmatrix} = \begin{bmatrix} 39.478 \\ 44.678 \\ 49.061 \\ 49.137 \\ 50.809 \end{bmatrix}$$

$$P(t) = 39.478 + 0.209692t + 0.0564163t^2 - 0.00292992t^3 + 0.0000391375t^4$$

$$Q(T) = 5.87828 \times 10^8 - 1.19444 \times 10^6 T + 910.118T^2 - 0.308202T^3 + 0.0000391375T^4$$

33. $P(t) = a + bt + ct^2 + dt^3 + et^4$.

$$\begin{bmatrix} 1 & 0 & 0 & 0 & 0 \\ 1 & 10 & 100 & 1000 & 10000 \\ 1 & 20 & 400 & 8000 & 160000 \\ 1 & 30 & 900 & 27000 & 810000 \\ 1 & 40 & 1600 & 64000 & 2560000 \end{bmatrix} \begin{bmatrix} a \\ b \\ c \\ d \\ e \end{bmatrix} = \begin{bmatrix} 47.197 \\ 54.973 \\ 62.813 \\ 75.367 \\ 85.446 \end{bmatrix}$$

$$P(t) = 47.197 + 1.22537t - 0.0771921t^2 + 0.00373475t^3 - 0.0000493292t^4$$

$$Q(T) = -7.41239 \times 10^8 + 1.50598 \times 10^6 T - 1147.37T^2 + 0.388502T^3 - 0.0000493292T^4$$

35. Expansion of the determinant along the first row gives an equation of the form $ay + bx^2 + cx + d = 0$ that can be solved for $y = Ax^2 + Bx + C$. If the coordinates of any one of the three given points $(x_1, y_1), (x_2, y_2), (x_3, y_3)$ are substituted in the first row, then the determinant has two identical rows and therefore vanishes.

37. Expansion of the determinant along the first row gives an equation of the form $a(x^2 + y^2) + bx + cy + d = 0$, and we get the desired form of the equation of a circle upon division by a. If the coordinates of any one of the three given points $(x_1, y_1), (x_2, y_2)$, and (x_3, y_3) are substituted in the first row, then the determinant has two identical rows and therefore vanishes.

39. Expansion of the determinant along the first row gives an equation of the form $ax^2 + bxy + cy^2 + d = 0$, which can be written in the central conic form $Ax^2 + Bxy + Cy^2 = 1$ upon division by $-d$. If the coordinates of any one of the three given points $(x_1, y_1), (x_2, y_2)$, and (x_3, y_3) are substituted in the first row, then the determinant has two identical rows and therefore vanishes.

$$= x^2 \begin{vmatrix} 0 & 16 & 1 \\ 0 & 0 & 1 \\ 25 & 25 & 1 \end{vmatrix} - xy \begin{vmatrix} 0 & 16 & 1 \\ 9 & 0 & 1 \\ 25 & 25 & 1 \end{vmatrix} + y \begin{vmatrix} 0 & 0 & 1 \\ 9 & 0 & 1 \\ 25 & 25 & 1 \end{vmatrix} - \begin{vmatrix} 0 & 0 & 16 \\ 9 & 0 & 0 \\ 25 & 25 & 25 \end{vmatrix}$$

$$= 400x^2 - 481xy + 225y^2 - 3600 = 0.$$

CHAPTER 4

VECTOR SPACES

The treatment of vector spaces in this chapter is very concrete. Prior to the final section of the chapter, almost all of the vector spaces appearing in examples and problems are subspaces of Cartesian coordinate spaces of n-tuples of real numbers. The main motivation throughout is the fact that the solution space of a homogeneous linear system $\mathbf{Ax} = \mathbf{0}$ is precisely such a "concrete" vector space.

SECTION 4.1

THE VECTOR SPACE $\mathbf{R}^3$

Here the fundamental concepts of vectors, linear independence, and vector spaces are introduced in the context of the familiar 2-dimensional coordinate plane $\mathbf{R}^2$ and 3-space $\mathbf{R}^3$. The concept of a subspace of a vector space is illustrated, the proper nontrivial subspaces of $\mathbf{R}^3$ being simply lines and planes through the origin.

1. $|\mathbf{a}-\mathbf{b}| = |(2,5,-4)-(1,-2,-3)| = |(1,7,-1)| = \sqrt{51}$
    $2\mathbf{a}+\mathbf{b} = 2(2,5,-4)+(1,-2,-3) = (4,10,-8)+(1,-2,-3) = (5,8,-11)$
    $3\mathbf{a}-4\mathbf{b} = 3(2,5,-4)-4(1,-2,-3) = (6,15,-12)-(4,-8,-12) = (2,23,0)$

3. $|\mathbf{a}-\mathbf{b}| = |(2\mathbf{i}-3\mathbf{j}+5\mathbf{k})-(5\mathbf{i}+3\mathbf{j}-7\mathbf{k})| = |-3\mathbf{i}-6\mathbf{j}+12\mathbf{k}| = \sqrt{189} = 3\sqrt{21}$
    $2\mathbf{a}+\mathbf{b} = 2(2\mathbf{i}-3\mathbf{j}+5\mathbf{k})+(5\mathbf{i}+3\mathbf{j}-7\mathbf{k})$
    $\phantom{2\mathbf{a}+\mathbf{b}} = (4\mathbf{i}-6\mathbf{j}+10\mathbf{k})+(5\mathbf{i}+3\mathbf{j}-7\mathbf{k}) = 9\mathbf{i}-3\mathbf{j}+3\mathbf{k}$
    $3\mathbf{a}-4\mathbf{b} = 3(2\mathbf{i}-3\mathbf{j}+5\mathbf{k})-4(5\mathbf{i}+3\mathbf{j}-7\mathbf{k})$
    $\phantom{3\mathbf{a}-4\mathbf{b}} = (6\mathbf{i}-9\mathbf{j}+15\mathbf{k})-(20\mathbf{i}+12\mathbf{j}-28\mathbf{k}) = -14\mathbf{i}-21\mathbf{j}+43\mathbf{k}$

5. $\mathbf{v} = \tfrac{3}{2}\mathbf{u}$, so the vectors $\mathbf{u}$ and $\mathbf{v}$ are linearly dependent.

7. $a\mathbf{u}+b\mathbf{v} = a(2,2)+b(2,-2) = (2a+2b, 2a-2b) = \mathbf{0}$ implies $a = b = 0$, so the vectors $\mathbf{u}$ and $\mathbf{v}$ are linearly independent.

In each of Problems 9-14, we set up and solve (as in Example 2 of this section) the system

$$a\mathbf{u} + b\mathbf{v} = \begin{bmatrix} u_1 & v_1 \\ u_2 & v_2 \end{bmatrix} \begin{bmatrix} a \\ b \end{bmatrix} = \begin{bmatrix} w_1 \\ w_2 \end{bmatrix} = \mathbf{w}$$

to find the coefficient values a and b such that $\mathbf{w} = a\mathbf{u} + b\mathbf{v}$,

9. $\begin{bmatrix} 1 & -1 \\ -2 & 3 \end{bmatrix} \begin{bmatrix} a \\ b \end{bmatrix} = \begin{bmatrix} 1 \\ 0 \end{bmatrix}$ $\Rightarrow$ $a = 3$, $b = 2$ so $\mathbf{w} = 3\mathbf{u} + 2\mathbf{v}$

11. $\begin{bmatrix} 5 & 2 \\ 7 & 3 \end{bmatrix} \begin{bmatrix} a \\ b \end{bmatrix} = \begin{bmatrix} 1 \\ 1 \end{bmatrix}$ $\Rightarrow$ $a = 1$, $b = -2$ so $\mathbf{w} = \mathbf{u} - 2\mathbf{v}$

13. $\begin{bmatrix} 7 & 3 \\ 5 & 4 \end{bmatrix} \begin{bmatrix} a \\ b \end{bmatrix} = \begin{bmatrix} 5 \\ -2 \end{bmatrix}$ $\Rightarrow$ $a = 2$, $b = -2$ so $\mathbf{w} = 2\mathbf{u} - 3\mathbf{v}$

In Problems 15-18, we calculate the determinant $|\mathbf{u} \ \mathbf{v} \ \mathbf{w}|$ so as to determine (using Theorem 4) whether the three vectors $\mathbf{u}$, $\mathbf{v}$, and $\mathbf{w}$ are linearly dependent (det = 0) or linearly independent (det $\neq$ 0).

15. $\begin{vmatrix} 3 & 5 & 8 \\ -1 & 4 & 3 \\ 2 & -6 & -4 \end{vmatrix} = 0$ so the three vectors are linearly dependent.

17. $\begin{vmatrix} 1 & 3 & 1 \\ -1 & 0 & -2 \\ 2 & 1 & 2 \end{vmatrix} = -5 \neq 0$ so the three vectors are linearly independent.

In Problems 19-24, we attempt to solve the homogeneous system $\mathbf{Ax} = \mathbf{0}$ by reducing the coefficient matrix $\mathbf{A} = \begin{bmatrix} \mathbf{u} & \mathbf{v} & \mathbf{w} \end{bmatrix}$ to echelon form $\mathbf{E}$. If we find that the system has only the trivial solution $a = b = c = 0$, this means that the vectors $\mathbf{u}$, $\mathbf{v}$, and $\mathbf{w}$ are linearly independent. Otherwise, a nontrivial solution $\mathbf{x} = \begin{bmatrix} a & b & c \end{bmatrix}^T \neq \mathbf{0}$ provides us with a nontrivial linear combination $a\mathbf{u} + b\mathbf{v} + c\mathbf{w} \neq \mathbf{0}$ that shows the three vectors are linearly dependent.

19. $\mathbf{A} = \begin{bmatrix} 2 & -3 & 0 \\ 0 & 1 & -2 \\ 1 & -1 & -1 \end{bmatrix} \rightarrow \begin{bmatrix} 1 & 0 & -3 \\ 0 & 1 & -2 \\ 0 & 0 & 0 \end{bmatrix} = \mathbf{E}$

The nontrivial solution $a = 3$, $b = 2$, $c = 1$ gives $3\mathbf{u} + 2\mathbf{v} + \mathbf{w} = \mathbf{0}$, so the three vectors are linearly dependent.

Section 4.1

79

21. $\mathbf{A} = \begin{bmatrix} 1 & -2 & 3 \\ 1 & -1 & 7 \\ -2 & 6 & 2 \end{bmatrix} \rightarrow \begin{bmatrix} 1 & 0 & 11 \\ 0 & 1 & 4 \\ 0 & 0 & 0 \end{bmatrix} = \mathbf{E}$

The nontrivial solution $a = 11$, $b = 4$, $c = -1$ gives $11\mathbf{u} + 4\mathbf{v} - \mathbf{w} = \mathbf{0}$, so the three vectors are linearly dependent.

23. $\mathbf{A} = \begin{bmatrix} 2 & 5 & 2 \\ 0 & 4 & -1 \\ 3 & -2 & 1 \end{bmatrix} \rightarrow \begin{bmatrix} 1 & 0 & 0 \\ 0 & 1 & 0 \\ 0 & 0 & 1 \end{bmatrix} = \mathbf{E}$

The system $\mathbf{Ax} = \mathbf{0}$ has only the trivial solution $a = b = c = 0$, so the vectors $\mathbf{u}$, $\mathbf{v}$, and $\mathbf{w}$ are linearly independent.

In Problems 25-28, we solve the nonhomogeneous system $\mathbf{Ax} = \mathbf{t}$ by reducing the augmented coefficient matrix $\mathbf{A} = \begin{bmatrix} \mathbf{u} & \mathbf{v} & \mathbf{w} & \mathbf{t} \end{bmatrix}$ to echelon form $\mathbf{E}$. The solution vector $\mathbf{x} = \begin{bmatrix} a & b & c \end{bmatrix}^T$ appears as the final column of $\mathbf{E}$, and provides us with the desired linear combination $\mathbf{t} = a\mathbf{u} + b\mathbf{v} + c\mathbf{w}$.

25. $\mathbf{A} = \begin{bmatrix} 1 & 3 & 1 & 2 \\ -2 & 0 & -1 & -7 \\ 2 & 1 & 2 & 9 \end{bmatrix} \rightarrow \begin{bmatrix} 1 & 0 & 0 & 2 \\ 0 & 1 & 0 & -1 \\ 0 & 0 & 1 & 3 \end{bmatrix} = \mathbf{E}$

Thus $a = 2$, $b = -1$, $c = 3$ so $\mathbf{t} = 2\mathbf{u} - \mathbf{v} + 3\mathbf{w}$.

27. $\mathbf{A} = \begin{bmatrix} 1 & -1 & 4 & 0 \\ 4 & -2 & 4 & 0 \\ 3 & 2 & 1 & 19 \end{bmatrix} \rightarrow \begin{bmatrix} 1 & 0 & 0 & 2 \\ 0 & 1 & 0 & 6 \\ 0 & 0 & 1 & 1 \end{bmatrix} = \mathbf{E}$

Thus $a = 2$, $b = 6$, $c = 1$ so $\mathbf{t} = 2\mathbf{u} + 6\mathbf{v} + \mathbf{w}$.

29. Given vectors $(0, y, z)$ and $(0, v, w)$ in V, we see that their sum $(0, y+v, z+w)$ and the scalar multiple $c(0, y, z) = (0, cy, cz)$ both have first component 0, and therefore are elements of V.

31. If (x, y, z) and (u, v, w) are in V, then

$$2(x+u) = (2x) + (2u) = (3y) + (3v) = 3(y+v),$$

so their sum $(x+u, y+v, z+w)$ is in V. Similarly,

$$2(cx) = c(2x) = c(3y) = 3(cy),$$

so the scalar multiple (cx, cy, cz) is in V.

33. $(0,1,0)$ is in V but the sum $(0,1,0) + (0,1,0) = (0,2,0)$ is not in V; thus V is not closed under addition. Alternatively, $2(0,1,0) = (0,2,0)$ is not in V, so V is not closed under multiplication by scalars.

35. Evidently V is closed under addition of vectors. However, $(0,0,1)$ is in V but $(-1)(0,0,1) = (0,0,-1)$ is not, so V is not closed under multiplication by scalars.

37. Pick a fixed element $\mathbf{u}$ in the (nonempty) vector space V. Then, with $c = 0$, the scalar multiple $c\mathbf{u} = 0\mathbf{u} = \mathbf{0}$ must be in V. Thus V necessarily contains the zero vector $\mathbf{0}$.

39. It suffices to show that every vector $\mathbf{v}$ in V is a scalar multiple of the given nonzero vector $\mathbf{u}$ in V. If $\mathbf{u}$ and $\mathbf{v}$ were linearly independent, then — as illustrated in Example 2 of this section — every vector in $\mathbf{R}^2$ could be expressed as a linear combination of $\mathbf{u}$ and $\mathbf{v}$. In this case it would follow that V is all of $\mathbf{R}^2$ (since, by Problem 38, V is closed under taking linear combinations). But we are given that V is a proper subspace of $\mathbf{R}^2$, so we must conclude that $\mathbf{u}$ and $\mathbf{v}$ are linearly dependent vectors. Since $\mathbf{u} \neq \mathbf{0}$, it follows that the arbitrary vector $\mathbf{v}$ in V is a scalar multiple of $\mathbf{u}$, and thus V is precisely the set of all scalar multiples of $\mathbf{u}$. In geometric language, the subspace V is then the straight line through the origin determined by the nonzero vector $\mathbf{u}$.

41. If the vectors $\mathbf{u}$ and $\mathbf{v}$ are in the intersection V of the subspaces V_1 and V_2, then their sum $\mathbf{u} + \mathbf{v}$ is in V_1 because both vectors are in V_1, and $\mathbf{u} + \mathbf{v}$ is in V_2 because both are in V_2. Therefore $\mathbf{u} + \mathbf{v}$ is in V, and thus V is closed under addition of vectors. Similarly, the intersection V is closed under multiplication by scalars, and is therefore itself a subspace.

SECTION 4.2

THE VECTOR SPACE $\mathbf{R}^n$ AND SUBSPACES

The main objective in this section is for the student to understand what types of subsets of the vector space $\mathbf{R}^n$ of n-tuples of real numbers are subspaces — playing the role in $\mathbf{R}^n$ of lines and planes through the origin in $\mathbf{R}^3$. Our first reason for studying subspaces is the fact that the *solution space* of any homogeneous linear system $\mathbf{Ax} = \mathbf{0}$ is a subspace of $\mathbf{R}^n$.

1. If $\mathbf{x} = (x_1, x_2, 0)$ and $\mathbf{y} = (y_1, y_2, 0)$ are vectors in W, then their sum

$$\mathbf{x} + \mathbf{y} = (x_1, x_2, 0) + (y_1, y_2, 0) = (x_1 + y_1, x_2 + y_2, 0)$$

and the scalar multiple $c\mathbf{x} = (cx_1, cx_2, 0)$ both have third coordinate zero, and therefore are also elements of W. Hence W is a subspace of $\mathbf{R}^3$.

3. The typical vector in W is of the form $\mathbf{x} = (x_1, 1, x_3)$ with second coordinate 1. But the particular scalar multiple $2\mathbf{x} = (2x_1, 2, 2x_3)$ of such a vector has second coordinate $2 \neq 1$, and thus is not in W. Hence W is not closed under multiplication by scalars, and therefore is not a subspace of $\mathbf{R}^3$. (Since $2\mathbf{x} = \mathbf{x} + \mathbf{x}$, W is not closed under vector addition either.)

5. Suppose $\mathbf{x} = (x_1, x_2, x_3, x_4)$ and $\mathbf{y} = (y_1, y_2, y_3, y_4)$ are vectors in W, so

$$x_1 + 2x_2 + 3x_3 + 4x_4 = 0 \quad \text{and} \quad y_1 + 2y_2 + 3y_3 + 4y_4 = 0.$$

Then their sum $\mathbf{s} = \mathbf{x} + \mathbf{y} = (x_1 + y_1, x_2 + y_2, x_3 + y_3, x_4 + y_4) = (s_1, s_2, s_3, s_4)$ satisfies the same condition

$$\begin{aligned} s_1 + 2s_2 + 3s_3 + 4s_4 &= (x_1 + y_1) + 2(x_2 + y_2) + 3(x_3 + y_3) + 4(x_4 + y_4) \\ &= (x_1 + 2x_2 + 3x_3 + 4x_4) + (y_1 + 2y_2 + 3y_3 + 4y_4) = 0 + 0 = 0, \end{aligned}$$

and thus is an element of W. Similarly, the scalar multiple $\mathbf{m} = c\mathbf{x} = (cx_1, cx_2, cx_3, cx_4) = (m_1, m_2, m_3, m_4)$ satisfies the condition

$$m_1 + 2m_2 + 3m_3 + 4m_4 = cx_1 + 2cx_2 + 3cx_3 + 4cx_4 = c(x_1 + 2x_2 + 3x_3 + 4x_4) = 0,$$

and hence is also an element of W. Therefore W is a subspace of $\mathbf{R}^3$.

7. The vectors $\mathbf{x} = (1, 1)$ and $\mathbf{y} = (1, -1)$ are in W, but their sum $\mathbf{x} + \mathbf{y} = (2, 0)$ is not, because $|2| \neq |0|$. Hence W is not a subspace of $\mathbf{R}^2$.

9. The vector $\mathbf{x} = (1, 0)$ is in W, but its scalar multiple $2\mathbf{x} = (2, 0)$ is not, because $(2)^2 + (0)^2 = 4 \neq 1$. Hence W is not a subspace of $\mathbf{R}^2$.

11. Suppose $\mathbf{x} = (x_1, x_2, x_3, x_4)$ and $\mathbf{y} = (y_1, y_2, y_3, y_4)$ are vectors in W, so

$$x_1 + x_2 = x_3 + x_4 \quad \text{and} \quad y_1 + y_2 = y_3 + y_4.$$

Then their sum $\mathbf{s} = \mathbf{x} + \mathbf{y} = (x_1 + y_1, x_2 + y_2, x_3 + y_3, x_4 + y_4) = (s_1, s_2, s_3, s_4)$ satisfies the same condition

$$\begin{aligned} s_1 + s_2 &= (x_1 + y_1) + (x_2 + y_2) = (x_1 + x_2) + (y_1 + y_2) \\ &= (x_3 + x_4) + (y_3 + y_4) = (x_3 + y_3) + (x_4 + y_4) = s_3 + s_4 \end{aligned}$$

and thus is an element of W. Similarly, the scalar multiple $\mathbf{m} = c\mathbf{x} = (cx_1, cx_2, cx_3, cx_4) = (m_1, m_2, m_3, m_4)$ satisfies the condition

$$m_1 + m_2 = cx_1 + cx_2 = c(x_1 + x_2) = c(x_3 + x_4) = cx_3 + cx_4 = m_3 + m_4,$$

and hence is also an element of W. Therefore W is a subspace of $\mathbf{R}^3$.

13. The vectors $\mathbf{x} = (1, 0, 1, 0)$ and $\mathbf{y} = (0, 1, 0, 1)$ are in W (because the product of the 4 components is 0 in each case) but their sum $\mathbf{s} = \mathbf{x} + \mathbf{y} = (1, 1, 1, 1)$ is not, because $s_1 s_2 s_3 s_4 = 1 \neq 0$. Hence W is not a subspace of $\mathbf{R}^2$.

In Problems 15-22, we first reduce the coefficient matrix $\mathbf{A}$ to echelon form $\mathbf{E}$ in order to solve the given homogeneous system $\mathbf{Ax = 0}$.

15. $\mathbf{A} = \begin{bmatrix} 1 & -4 & 1 & -4 \\ 1 & 2 & 1 & 8 \\ 1 & 1 & 1 & 6 \end{bmatrix} \rightarrow \begin{bmatrix} 1 & 0 & 1 & 4 \\ 0 & 1 & 0 & 2 \\ 0 & 0 & 0 & 0 \end{bmatrix} = \mathbf{E}$

Thus $x_3 = s$ and $x_4 = t$ are free variables. We solve for $x_1 = -s - 4t$ and $x_2 = -2t$, so

$$\mathbf{x} = (x_1, x_2, x_3, x_4) = (-s - 4t, -2t, s, t)$$
$$= (-s, 0, s, 0) + (-4t, -2t, 0, t) = s\mathbf{u} + t\mathbf{v}$$

where $\mathbf{u} = (-1, 0, 1, 0)$ and $\mathbf{v} = (-4, -2, 0, 1)$.

17. $\mathbf{A} = \begin{bmatrix} 1 & 3 & 8 & -1 \\ 1 & -3 & -10 & 5 \\ 1 & 4 & 11 & -2 \end{bmatrix} \rightarrow \begin{bmatrix} 1 & 0 & -1 & 2 \\ 0 & 1 & 3 & -1 \\ 0 & 0 & 0 & 0 \end{bmatrix} = \mathbf{E}$

Thus $x_3 = s$ and $x_4 = t$ are free variables. We solve for $x_1 = s - 2t$ and $x_2 = -3s + t$, so

$$\mathbf{x} = (x_1, x_2, x_3, x_4) = (-s - 5t, -s - 3t, s, t)$$
$$= (s, -3s, s, 0) + (-2t, t, 0, t) = s\mathbf{u} + t\mathbf{v}$$

where $\mathbf{u} = (1, -3, 1, 0)$ and $\mathbf{v} = (-2, 1, 0, 1)$.

19. $A = \begin{bmatrix} 1 & -3 & -5 & -6 \\ 2 & 1 & 4 & -4 \\ 1 & 3 & 7 & 1 \end{bmatrix} \to \begin{bmatrix} 1 & 0 & 1 & 0 \\ 0 & 1 & 2 & 0 \\ 0 & 0 & 0 & 1 \end{bmatrix} = E$

Thus $x_3 = t$ is a free variable and $x_4 = 0$. We solve for $x_1 = -t$ and $x_2 = -2t$, so

$$\mathbf{x} = (x_1, x_2, x_3, x_4) = (-t, -2t, t, 0) = t\mathbf{u}$$

where $\mathbf{u} = (-1, -2, 1, 0)$.

21. $A = \begin{bmatrix} 1 & 7 & 2 & -3 \\ 2 & 7 & 1 & -4 \\ 3 & 5 & -1 & -5 \end{bmatrix} \to \begin{bmatrix} 1 & 0 & 0 & 3 \\ 0 & 1 & 0 & -2 \\ 0 & 0 & 1 & 4 \end{bmatrix} = E$

Thus $x_4 = t$ is a free variable. We solve for $x_1 = -3t$, $x_2 = 2t$, and $x_4 = -4t$. so

$$\mathbf{x} = (x_1, x_2, x_3, x_4) = (-3t, 2t, -4t, t) = t\mathbf{u}$$

where $\mathbf{u} = (-3, 2, -4, 1)$.

23. Let $\mathbf{u}$ be a vector in W. Then $0\mathbf{u}$ is also in W. But $0\mathbf{u} = (0+0)\mathbf{u} = 0\mathbf{u} + 0\mathbf{u}$, so upon subtracting $0\mathbf{u}$ from each side, we see that $0\mathbf{u} = \mathbf{0}$, the zero vector.

25. If W is a subspace, then it contains the scalar multiples $a\mathbf{u}$ and $b\mathbf{v}$, and hence contains their sum $a\mathbf{u} + b\mathbf{v}$. Conversely, if the subset W is closed under taking linear combinations of pairs of vectors, then it contains $(1)\mathbf{u} + (1)\mathbf{v} = \mathbf{u} + \mathbf{v}$ and $(c)\mathbf{u} + (0)\mathbf{v} = c\mathbf{u}$, and hence is a subspace.

27. Let $a_1\mathbf{u} + b_1\mathbf{v}$ and $a_2\mathbf{u} + b_2\mathbf{v}$ be two vectors in $W = \{a\mathbf{u} + b\mathbf{v}\}$. Then the sum

$$(a_1\mathbf{u} + b_1\mathbf{v}) + (a_2\mathbf{u} + b_2\mathbf{v}) = (a_1 + a_2)\mathbf{u} + (b_1 + b_2)\mathbf{v}$$

and the scalar multiple $c(a_1\mathbf{u} + b_1\mathbf{v}) = (ca_1)\mathbf{u} + (cb_1)\mathbf{v}$ are again scalar multiples of $\mathbf{u}$ and $\mathbf{v}$, and hence are themselves elements of W. Hence W is a subspace.

29. If $A\mathbf{x}_0 = \mathbf{b}$ and $\mathbf{y} = \mathbf{x} - \mathbf{x}_0$, then

$$A\mathbf{y} = A(\mathbf{x} - \mathbf{x}_0) = A\mathbf{x} - A\mathbf{x}_0 = A\mathbf{x} - \mathbf{b}.$$

Hence it is clear that $A\mathbf{y} = \mathbf{0}$ if and only if $A\mathbf{x} = \mathbf{b}$.

31. Let $\mathbf{w}_1$ and $\mathbf{w}_2$ be two vectors in the sum $U + V$. Then $\mathbf{w}_i = \mathbf{u}_i + \mathbf{v}_i$ where $\mathbf{u}_i$ is in U and $\mathbf{v}_i$ is in V ($i = 1, 2$). Then the linear combination

$$aw_1 + bw_2 = a(u_1 + v_1) + b(u_2 + v_2) = (au_1 + bu_2) + (av_1 + bv_2)$$

is the sum of the vectors $au_1 + bu_2$ in U and $av_1 + bv_2$ in U, and therefore is an element of $U + V$. Thus $U + V$ is a subspace. If U and V are noncollinear lines through the origin in $\mathbf{R}^3$, then $U + V$ is a plane through the origin.

SECTION 4.3

LINEAR COMBINATIONS AND INDEPENDENCE OF VECTORS

In this section we use two types of computational problems as aids in understanding linear independence and dependence. The first of these problems is that of expressing a vector $\mathbf{w}$ as a linear combination of k given vectors $\mathbf{v}_1, \mathbf{v}_2, \cdots, \mathbf{v}_k$ (if possible). The second is that of determining whether k given vectors $\mathbf{v}_1, \mathbf{v}_2, \cdots, \mathbf{v}_k$ are linearly independent. For vectors in $\mathbf{R}^n$, each of these problems reduces to solving a linear system of n equations in k unknowns. Thus an abstract question of linear independence or dependence becomes a concrete question of whether or not a given linear system has a nontrivial solution.

1. $\mathbf{v}_2 = \tfrac{3}{2}\mathbf{v}_1$, so the two vectors $\mathbf{v}_1$ and $\mathbf{v}_2$ are linearly dependent.

3. The three vectors $\mathbf{v}_1$, $\mathbf{v}_2$, and $\mathbf{v}_3$ are linearly dependent, as are any 3 vectors in $\mathbf{R}^2$. The reason is that the vector equation $c_1\mathbf{v}_1 + c_2\mathbf{v}_2 + c_3\mathbf{v}_3 = \mathbf{0}$ reduces to a homogeneous linear system of 2 equations in the 3 unknowns c_1, c_2, and c_3, and any such system has a nontrivial solution.

5. The equation $c_1\mathbf{v}_1 + c_2\mathbf{v}_2 + c_3\mathbf{v}_3 = \mathbf{0}$ yields

$$c_1(1,0,0) + c_2(0,-2,0) + c_3(0,0,3) = (c_1,-2c_2,3c_3) = (0,0,0),$$

and therefore implies immediately that $c_1 = c_2 = c_3 = 0$. Hence the given vectors $\mathbf{v}_1$, $\mathbf{v}_2$, and $\mathbf{v}_3$ are linearly independent.

7. The equation $c_1\mathbf{v}_1 + c_2\mathbf{v}_2 + c_3\mathbf{v}_3 = \mathbf{0}$ yields

$$c_1(2,1,0,0) + c_2(3,0,1,0) + c_3(4,0,0,1) = (2c_1 + 3c_2, c_1, c_2, c_3) = (0,0,0,0).$$

Obviously it follows immediately that $c_1 = c_2 = c_3 = 0$. Hence the given vectors $\mathbf{v}_1$, $\mathbf{v}_2$, and $\mathbf{v}_3$ are linearly independent.

In Problems 9-16 we first set up the linear system to be solved for the linear combination coefficients $\{c_i\}$, and then show the reduction of its augmented coefficient matrix $\mathbf{A}$ to reduced echelon form $\mathbf{E}$.

9.
$$c_1\mathbf{v}_1 + c_2\mathbf{v}_2 = \mathbf{w}$$

$$\mathbf{A} = \begin{bmatrix} 5 & 3 & 1 \\ 3 & 2 & 0 \\ 4 & 5 & -7 \end{bmatrix} \rightarrow \begin{bmatrix} 1 & 0 & 2 \\ 0 & 1 & -3 \\ 0 & 0 & 0 \end{bmatrix} = \mathbf{E}$$

We see that the system of 3 equations in 2 unknowns has the unique solution $c_1 = 2, c_2 = -3$, so $\mathbf{w} = 2\mathbf{v}_1 - 3\mathbf{v}_2$.

11.
$$c_1\mathbf{v}_1 + c_2\mathbf{v}_2 = \mathbf{w}$$

$$\mathbf{A} = \begin{bmatrix} 7 & 3 & 1 \\ -6 & -3 & 0 \\ 4 & 2 & 0 \\ 5 & 3 & -1 \end{bmatrix} \rightarrow \begin{bmatrix} 1 & 0 & 1 \\ 0 & 1 & -2 \\ 0 & 0 & 0 \\ 0 & 0 & 0 \end{bmatrix} = \mathbf{E}$$

We see that the system of 4 equations in 2 unknowns has the unique solution $c_1 = 1, c_2 = -2$, so $\mathbf{w} = \mathbf{v}_1 - 2\mathbf{v}_2$.

13.
$$c_1\mathbf{v}_1 + c_2\mathbf{v}_2 = \mathbf{w}$$

$$\mathbf{A} = \begin{bmatrix} 1 & 5 & 5 \\ 5 & -3 & 2 \\ -3 & 4 & -2 \end{bmatrix} \rightarrow \begin{bmatrix} 1 & 0 & 0 \\ 0 & 1 & 0 \\ 0 & 0 & 1 \end{bmatrix} = \mathbf{E}$$

The last row of $\mathbf{E}$ corresponds to the scalar equation $0c_1 + 0c_2 = 1$, so the system of 3 equations in 2 unknowns is inconsistent. This means that $\mathbf{w}$ cannot be expressed as a linear combination of $\mathbf{v}_1$ and $\mathbf{v}_2$.

15.
$$c_1\mathbf{v}_1 + c_2\mathbf{v}_2 + c_3\mathbf{v}_3 = \mathbf{w}$$

$$\mathbf{A} = \begin{bmatrix} 2 & 3 & 1 & 4 \\ -1 & 0 & 2 & 5 \\ 4 & 1 & -1 & 6 \end{bmatrix} \rightarrow \begin{bmatrix} 1 & 0 & 0 & 3 \\ 0 & 1 & 0 & -2 \\ 0 & 0 & 1 & 4 \end{bmatrix} = \mathbf{E}$$

We see that the system of 3 equations in 3 unknowns has the unique solution $c_1 = 3, c_2 = -2, c_3 = 4$, so $\mathbf{w} = 3\mathbf{v}_1 - 2\mathbf{v}_2 + 4\mathbf{v}_3$.

In Problems 17-22, $\mathbf{A} = [\mathbf{v}_1 \ \mathbf{v}_2 \ \mathbf{v}_3]$ is the coefficient matrix of the homogeneous linear system corresponding to the vector equation $c_1\mathbf{v}_1 + c_2\mathbf{v}_2 + c_3\mathbf{v}_3 = \mathbf{0}$. Inspection of the indicated reduced echelon form $\mathbf{E}$ of $\mathbf{A}$ then reveals whether or not a nontrivial solution exists.

17. $\mathbf{A} = \begin{bmatrix} 1 & 2 & 3 \\ 0 & -3 & 5 \\ 1 & 4 & 2 \end{bmatrix} \rightarrow \begin{bmatrix} 1 & 0 & 0 \\ 0 & 1 & 0 \\ 0 & 0 & 1 \end{bmatrix} = \mathbf{E}$

We see that the system of 3 equations in 3 unknowns has the unique solution $c_1 = c_2 = c_3 = 0$, so the vectors $\mathbf{v}_1, \mathbf{v}_2, \mathbf{v}_3$ are linearly independent.

19. $\mathbf{A} = \begin{bmatrix} 2 & 5 & 2 \\ 0 & 4 & -1 \\ 3 & -2 & 1 \\ 0 & 1 & -1 \end{bmatrix} \rightarrow \begin{bmatrix} 1 & 0 & 0 \\ 0 & 1 & 0 \\ 0 & 0 & 1 \\ 0 & 0 & 0 \end{bmatrix} = \mathbf{E}$

We see that the system of 4 equations in 3 unknowns has the unique solution $c_1 = c_2 = c_3 = 0$, so the vectors $\mathbf{v}_1, \mathbf{v}_2, \mathbf{v}_3$ are linearly independent.

21. $\mathbf{A} = \begin{bmatrix} 3 & 1 & 1 \\ 0 & -1 & 2 \\ 1 & 0 & 1 \\ 2 & 1 & 0 \end{bmatrix} \rightarrow \begin{bmatrix} 1 & 0 & 1 \\ 0 & 1 & -2 \\ 0 & 0 & 0 \\ 0 & 0 & 0 \end{bmatrix} = \mathbf{E}$

We see that the system of 4 equations in 3 unknowns has a 1-dimensional solution space. If we choose $c_3 = -1$ then $c_1 = 1$ and $c_2 = -2$. Therefore $\mathbf{v}_1 - 2\mathbf{v}_2 - \mathbf{v}_3 = \mathbf{0}$.

23. Because $\mathbf{v}_1$ and $\mathbf{v}_2$ are linearly independent, the vector equation

$$c_1\mathbf{u}_1 + c_2\mathbf{u}_2 = c_1(\mathbf{v}_1 + \mathbf{v}_2) + c_2(\mathbf{v}_1 - \mathbf{v}_2) = \mathbf{0}$$

yields the homogeneous linear system

$$c_1 + c_2 = 0$$
$$c_1 - c_2 = 0.$$

It follows readily that $c_1 = c_2 = 0$, and therefore that the vectors $\mathbf{u}_1$ and $\mathbf{u}_2$ are linearly independent.

25. Because the vectors $\mathbf{v}_1, \mathbf{v}_2, \mathbf{v}_3$ are linearly independent, the vector equation

$$c_1\mathbf{u}_1+c_2\mathbf{u}_2+c_3\mathbf{u}_3 \;=\; c_1(\mathbf{v}_1)+c_2(\mathbf{v}_1+2\mathbf{v}_2)+c_3(\mathbf{v}_1+2\mathbf{v}_2+3\mathbf{v}_3) \;=\; \mathbf{0}$$

yields the homogeneous linear system

$$\begin{aligned} c_1 + c_2 + c_3 &= 0 \\ 2c_2 + 2c_3 &= 0 \\ 3c_3 &= 0. \end{aligned}$$

It follows by back-substitution that $c_1 = c_2 = c_3 = 0$, and therefore that the vectors $\mathbf{u}_1, \mathbf{u}_2, \mathbf{u}_3$ are linearly independent.

27. If the elements of S are $\mathbf{v}_1, \mathbf{v}_2, \cdots, \mathbf{v}_k$ with $\mathbf{v}_1 = \mathbf{0}$, then we can take $c_1 = 1$ and $c_2 = \cdots = c_k = 0$. This choice gives coefficients $c_1, c_2, \cdots, c_k$ not all zero such that $c_1\mathbf{v}_1 + c_2\mathbf{v}_2 + \cdots + c_k\mathbf{v}_k = \mathbf{0}$. This means that the vectors $\mathbf{v}_1, \mathbf{v}_2, \cdots, \mathbf{v}_k$ are linearly dependent.

29. If some subset of S were linearly dependent, then Problem 28 would imply immediately that S itself is linearly dependent (contrary to hypothesis).

31. If S is contained in span(T), then every vector in S is a linear combination of vectors in T. Hence every vector in span(S) is a linear combination of linear combinations of vectors in T. Therefore every vector in span(S) is a linear combination of vectors in T, and therefore is itself in span(T). Thus span(S) is a subset of span(T).

33. The determinant of the $k \times k$ identity matrix is nonzero, so it follows immediately from Theorem 3 in this section that the vectors $\mathbf{v}_1, \mathbf{v}_2, \cdots, \mathbf{v}_k$ are linearly independent.

35. Because the vectors $\mathbf{v}_1, \mathbf{v}_2, \cdots, \mathbf{v}_k$ are linearly independent, Theorem 3 implies that some $k \times k$ submatrix $\mathbf{A}_0$ of $\mathbf{A}$ has nonzero determinant. Let $\mathbf{A}_0$ consist of the rows $i_1, i_2, \cdots, i_k$ of the matrix $\mathbf{A}$, and let $\mathbf{C}_0$ denote the $k \times k$ submatrix consisting of the same rows of the product matrix $\mathbf{C} = \mathbf{AB}$. Then $\mathbf{C}_0 = \mathbf{A}_0\mathbf{B}$, so $|\mathbf{C}_0| = |\mathbf{A}_0||\mathbf{B}| \neq 0$ because (by hypothesis) the $k \times k$ matrix $\mathbf{B}$ is also nonsingular. Therefore Theorem 3 implies that the column vectors of $\mathbf{AB}$ are linearly independent.

SECTION 4.4

BASES AND DIMENSION FOR VECTOR SPACES

A basis $\{\mathbf{v}_1, \mathbf{v}_2, \cdots, \mathbf{v}_k\}$ for a subspace W of $\mathbf{R}^n$ enables up to visualize W as a k-dimensional plane (or "hyperplane") through the origin in $\mathbf{R}^n$. In case W is the solution space of a homogeneous linear system, a basis for W is a maximal linearly independent set of solutions of the system, and every other solution is a linear combination of these particular solutions.

1. The vectors $\mathbf{v}_1$ and $\mathbf{v}_2$ are linearly independent (because neither is a scalar multiple of the other) and therefore form a basis for $\mathbf{R}^2$.

3. Any four vectors in $\mathbf{R}^3$ are linearly dependent, so the given vectors do not form a basis for $\mathbf{R}^3$.

5. The three given vectors $\mathbf{v}_1, \mathbf{v}_2, \mathbf{v}_3$ all lie in the 2-dimensional subspace $x_1 = 0$ of $\mathbf{R}^3$. Therefore they are linearly dependent, and hence do not form a basis for $\mathbf{R}^3$.

7. $\text{Det}([\mathbf{v}_1 \ \mathbf{v}_2 \ \mathbf{v}_3]) = 1 \neq 0$, so the three vectors are linearly independent, and hence do form a basis for $\mathbf{R}^3$.

9. The single equation $x - 2y + 5z = 0$ is already a system in reduced echelon form, with free variables y and z. With $y = s, z = t, x = 2s - 5t$ we get the solution vector

$$(x, y, z) = (2s - 5t, s, t) = s(2,1,0) + t(-5,0,1).$$

Hence the plane $x - 2y + 5z = 0$ is a 2-dimensional subspace of $\mathbf{R}^3$ with basis consisting of the vectors $\mathbf{v}_1 = (2,1,0)$ and $\mathbf{v}_2 = (-5,0,1)$.

11. The line of intersection of the planes in Problems 9 and 11 is the solution space of the system
$$x - 2y + 5z = 0$$
$$y - z = 0.$$

This system is in echelon form with free variable $z = t$. With $y = t$ and $x = -3t$ we have the solution vector $(-3t, t, t) = t(-3,1,1)$. Thus the line is a 1-dimensional subspace of $\mathbf{R}^3$ with basis consisting of the vector $\mathbf{v} = (-3,1,1)$.

13. The typical vector in $\mathbf{R}^4$ of the form (a,b,c,d) with $a = 3c$ and $b = 4d$ can be written as
$$\mathbf{v} = (3c, 4d, c, d) = c(3,0,1,0) + d(0,4,0,1).$$

Hence the subspace consisting of all such vectors is 2-dimensional with basis consisting of the vectors $\mathbf{v}_1 = (3,0,1,0)$ and $\mathbf{v}_2 = (0,4,0,1)$.

In Problems 15-26, we show first the reduction of the coefficient matrix **A** to echelon form **E**. Then we write the typical solution vector as a linear combination of basis vectors for the subspace of the given system.

15. $\mathbf{A} = \begin{bmatrix} 1 & -2 & 3 \\ 2 & -3 & 1 \end{bmatrix} \rightarrow \begin{bmatrix} 1 & 0 & -11 \\ 0 & 1 & -7 \end{bmatrix} = \mathbf{E}$

With free variable $x_3 = t$ and $x_1 = 11t$, $x_2 = 7t$ we get the solution vector $\mathbf{x} = (11t, 7t, t) = t(11, 7, 1)$. Thus the solution space of the given system is 1-dimensional with basis consisting of the vector $\mathbf{v}_1 = (11, 7, 1)$.

17. $\mathbf{A} = \begin{bmatrix} 1 & -3 & 2 & -4 \\ 2 & -5 & 7 & -3 \end{bmatrix} \rightarrow \begin{bmatrix} 1 & 0 & 11 & 11 \\ 0 & 1 & 3 & 5 \end{bmatrix} = \mathbf{E}$

With free variables $x_3 = s$, $x_4 = t$ and with $x_1 = -11s - 11t$, $x_2 = -3s - 5t$ we get the solution vector

$$\mathbf{x} = (-11s - 11t, -3s - 5t, s, t) = s(-11, -3, 1, 0) + t(-11, -5, 0, 1).$$

Thus the solution space of the given system is 2-dimensional with basis consisting of the vectors $\mathbf{v}_1 = (-11, -3, 1, 0)$ and $\mathbf{v}_2 = (-11, -5, 0, 1)$.

19. $\mathbf{A} = \begin{bmatrix} 1 & -3 & -8 & -5 \\ 2 & 1 & -4 & 11 \\ 1 & 3 & 3 & 13 \end{bmatrix} \rightarrow \begin{bmatrix} 1 & 0 & -3 & 4 \\ 0 & 1 & 2 & 3 \\ 0 & 0 & 0 & 0 \end{bmatrix} = \mathbf{E}$

With free variables $x_3 = s$, $x_4 = t$ and with $x_1 = 3s - 4t$, $x_2 = -2s - 3t$ we get the solution vector

$$\mathbf{x} = (3s - 4t, -2s - 3t, s, t) = s(3, -2, 1, 0) + t(-4, -3, 0, 1).$$

Thus the solution space of the given system is 2-dimensional with basis consisting of the vectors $\mathbf{v}_1 = (3, -2, 1, 0)$ and $\mathbf{v}_2 = (-4, -3, 0, 1)$.

21. $\mathbf{A} = \begin{bmatrix} 1 & -4 & -3 & -7 \\ 2 & -1 & 1 & 7 \\ 1 & 2 & 3 & 11 \end{bmatrix} \rightarrow \begin{bmatrix} 1 & 0 & 1 & 5 \\ 0 & 1 & 1 & 3 \\ 0 & 0 & 0 & 0 \end{bmatrix} = \mathbf{E}$

With free variables $x_3 = s$, $x_4 = t$ and with $x_1 = -s - 5t$, $x_2 = -s - 3t$ we get the solution vector

$$\mathbf{x} = (-s - 5t, -s - 3t, s, t) = s(-1, -1, 1, 0) + t(-5, -3, 0, 1).$$

Thus the solution space of the given system is 2-dimensional with basis consisting of the vectors $\mathbf{v}_1 = (-1,-1,1,0)$ and $\mathbf{v}_2 = (-5,-3,0,1)$.

23. $\mathbf{A} = \begin{bmatrix} 1 & 5 & 13 & 14 \\ 2 & 5 & 11 & 12 \\ 2 & 7 & 17 & 19 \end{bmatrix} \rightarrow \begin{bmatrix} 1 & 0 & -2 & 0 \\ 0 & 1 & 3 & 0 \\ 0 & 0 & 0 & 1 \end{bmatrix} = \mathbf{E}$

With free variable $x_3 = s$ and with $x_1 = 2s$, $x_2 = -3s$, $x_4 = 0$ we get the solution vector $\mathbf{x} = (2s, -3s, s, 0) = s(2, -3, 1, 0)$. Thus the solution space of the given system is 1-dimensional with basis consisting of the vector $\mathbf{v}_1 = (2, -3, 1, 0)$.

25. $\mathbf{A} = \begin{bmatrix} 1 & 2 & 7 & -9 & 31 \\ 2 & 4 & 7 & -11 & 34 \\ 3 & 6 & 5 & -11 & 29 \end{bmatrix} \rightarrow \begin{bmatrix} 1 & 2 & 0 & -2 & 3 \\ 0 & 0 & 1 & -1 & 4 \\ 0 & 0 & 0 & 0 & 0 \end{bmatrix} = \mathbf{E}$

With free variables $x_2 = r$, $x_4 = s$, $x_5 = t$ and with $x_1 = -2r + 2s - 3t$, $x_3 = s - 4t$ we get the solution vector

$$\mathbf{x} = (-2r + 2s - 3t, r, s - 4t, s, t) = r(-2, 1, 0, 0, 0) + s(2, 0, 1, 1, 0) + t(-3, 0, -4, 0, 1).$$

Thus the solution space of the given system is 3-dimensional with basis consisting of the vectors $\mathbf{v}_1 = (-2, 1, 0, 0, 0)$, $\mathbf{v}_2 = (2, 0, 1, 1, 0)$, and $\mathbf{v}_3 = (-3, 0, -4, 0, 1)$.

27. If the vectors $\mathbf{v}_1, \mathbf{v}_2, \cdots, \mathbf{v}_n$ are linearly independent, and $\mathbf{w}$ is another vector in V, then the vectors $\mathbf{w}, \mathbf{v}_1, \mathbf{v}_2, \cdots, \mathbf{v}_n$ are linearly dependent (because no $n+1$ vectors in the n-dimensional vector space V are linearly independent). Hence there exist scalars $c, c_1, c_2, \cdots, c_n$ not all zero such that

$$c\mathbf{w} + c_1\mathbf{v}_1 + c_2\mathbf{v}_2 + \cdots + c_n\mathbf{v}_n = \mathbf{0}.$$

If $c = 0$ then the coefficients $c_1, c_2, \cdots, c_n$ would not all be zero, and hence this equation would say (contrary to hypothesis) that the vectors $\mathbf{v}_1, \mathbf{v}_2, \cdots, \mathbf{v}_n$ are linearly dependent. Therefore $c \ne 0$, so we can solve for $\mathbf{w}$ as a linear combination of the vectors $\mathbf{v}_1, \mathbf{v}_2, \cdots, \mathbf{v}_n$. Thus the linearly independent vectors $\mathbf{v}_1, \mathbf{v}_2, \cdots, \mathbf{v}_n$ span V, and therefore form a basis for V.

29. Suppose $c\mathbf{v} + c_1\mathbf{v}_1 + c_2\mathbf{v}_2 + \cdots + c_k\mathbf{v}_k = \mathbf{0}$. Then $c = 0$ because, otherwise, we could solve for $\mathbf{v}$ as a linear combination of the vectors $\mathbf{v}_1, \mathbf{v}_2, \cdots, \mathbf{v}_k$. But this is impossible, because $\mathbf{v}$ is not in the subspace W spanned by $\mathbf{v}_1, \mathbf{v}_2, \cdots, \mathbf{v}_k$. It follows that $c_1\mathbf{v}_1 + c_2\mathbf{v}_2 + \cdots + c_k\mathbf{v}_k = \mathbf{0}$, which implies that $c_1 = c_2 = \cdots = c_k = 0$ also, because the vectors $\mathbf{v}_1, \mathbf{v}_2, \cdots, \mathbf{v}_k$ are linearly independent. Hence we have shown that the $k+1$ vectors $\mathbf{v}, \mathbf{v}_1, \mathbf{v}_2, \cdots, \mathbf{v}_k$ are linearly independent.

31. If $\mathbf{v}_{k+1}$ is a linear combination of the vectors $\mathbf{v}_1, \mathbf{v}_2, \cdots, \mathbf{v}_k$, then obviously every linear combination of the vectors $\mathbf{v}_1, \mathbf{v}_2, \cdots, \mathbf{v}_k, \mathbf{v}_{k+1}$ is also a linear combination of $\mathbf{v}_1, \mathbf{v}_2, \cdots, \mathbf{v}_k$. But the former set of $k+1$ vectors spans V, so the latter set of k vectors also spans V.

33. If S is a maximal linearly independent set in V, the we see immediately that every other vector in V is a linear combination of the vectors in S. Thus S also spans V, and is therefore a basis for V.

35. Let $S = \{\mathbf{v}_1, \mathbf{v}_2, \cdots, \mathbf{v}_n\}$ be a uniquely spanning set for V. Then the fact, that

$$\mathbf{0} = 0\,\mathbf{v}_1 + 0\,\mathbf{v}_2 + \cdots + 0\,\mathbf{v}_n$$

is the unique expression of the zero vector $\mathbf{0}$ as a linear combination of the vectors in S, means that S is a linearly independent set of vectors. Hence S is a basis for V.

SECTION 4.5

GENERAL VECTOR SPACES

In each of Problems 1-12, a certain subset of a vector space is described. This subset is a subspace of the vector space if and only if it is closed under the formation of linear combinations of its elements. Recall also that every subspace of a vector space must contain the zero vector.

1. It is a subspace of M_{33}, because any linear combination of diagonal 3×3 matrices — with only zeros off the principal diagonal — obviously is again a diagonal matrix.

3. The set of all nonsingular 3×3 matrices does not contain the zero matrix, so it is not a subspace.

5. The set of all functions $f : \mathbf{R} \to \mathbf{R}$ with $f(0) = 0$ is a vector space, because if $f(0) = g(0) = 0$ then $(af + bg)(0) = af(0) + bg(0) = a \cdot 0 + b \cdot 0 = 0$.

7. The set of all functions $f : \mathbf{R} \to \mathbf{R}$ with $f(0) = 0$ and $f(1) = 1$ is not a vector space. For instance, if $g = 2f$ then $g(1) = 2f(1) = 2 \cdot 1 = 2 \neq 1$, so g is not such a function. Also, this set does not contain the zero function.

For Problems 9-12, let us call a polynomial of the form $a_0 + a_1 x + a_2 x^2 + a_3 x^3$ a "degree at most 3" polynomial.

9. The set of all degree at most 3 polynomials with nonzero leading coefficient $a_3 \neq 0$ is not a vector space, because it does not contain the zero polynomial (with all coefficients zero).

11. The set of all degree at most 3 polynomials with coefficient sum zero is a vector space, because any linear combination of such polynomials obviously is such a polynomial.

13. The functions $\sin x$ and $\cos x$ are linearly independent, because neither is a scalar multiple of the other. (This follows, for instance, from the facts that $\sin(0) = 0$, $\cos(0) = 1$ and $\sin(\pi/2) = 1$, $\cos(\pi/2) = 0$, noting that any scalar multiple of a function with a zero value must have the value 0 at the same point.)

15. If
$$c_1(1+x) + c_2(1-x) + c_3(1-x^2) = (c_1 + c_2 + c_3) + (c_1 - c_2)x - c_3 x^2 = 0,$$
then
$$c_1 + c_2 + c_3 = c_1 - c_2 = c_3 = 0.$$

It follows easily that $c_1 = c_2 = c_3 = 0$, so we conclude that the functions $(1+x)$, $(1-x)$, and $(1-x^2)$ are linearly independent.

17. $\cos 2x = \cos^2 x - \sin^2 x$ according to a well-known trigonometric identity. Thus these three trigonometric functions are linearly dependent.

19. Multiplication by $(x-2)(x-3)$ yields

$$x - 5 = A(x-3) + B(x-2) = (A+B)x - (3A+2B).$$

Hence $A + B = 1$ and $3A + 2B = 5$, and it follows readily that $A = 3$ and $B = -2$.

21. Multiplication by $x(x^2 + 4)$ yields

$$8 = A(x^2 + 4) + Bx^2 + Cx = 4A + Cx + (A+B)x^2.$$

Hence $4A = 8$, $C = 0$ and $A + B = 0$. It follows readily that $A = 2$ and $B = -2$.

23. If $y'''(x) = 0$ then

$$y''(x) = \int y'''(x)\,dx = \int (0)\,dx = A,$$
$$y'(x) = \int y''(x)\,dx = \int A\,dx = Ax + B, \text{ and}$$
$$y(x) = \int y'(x)\,dx = \int (Ax+B)\,dx = \tfrac{1}{2}Ax^2 + Bx + C,$$

where A, B, and C are arbitrary constants of integration. It follows that the function $y(x)$ is a solution of the differential equation $y'''(x) = 0$ if and only if it is a quadratic (at most 2nd degree) polynomial. Thus the solution space is 3-dimensional with basis $\{1, x, x^2\}$.

25. If $y(x)$ is any solution of the second-order differential equation $y'' - 5y' = 0$ and $v(x) = y'(x)$, then $v(x)$ is a solution of the first-order differential equation $v'(x) = 5v(x)$ with the familiar exponential solution $v(x) = Ce^{5x}$. Therefore

$$y(x) = \int y'(x)\,dx = \int v(x)\,dx = \int Ce^{5x}\,dx = \tfrac{1}{5}Ce^{5x} + D.$$

We therefore see that the solution space of the equation $y'' - 5y' = 0$ is 2-dimensional with basis $\{1, e^{5x}\}$.

27. If we take the positive sign in Eq. (20) of the text, then we have $v^2 = y^2 + a^2$ where $v(x) = y'(x)$. Then

$$\left(\frac{dy}{dx}\right)^2 = y^2 + a^2, \quad \text{so} \quad \frac{dx}{dy} = \frac{1}{\sqrt{y^2 + a^2}}$$

(taking the positive square root as in the text). Then

$$x = \int \frac{dy}{\sqrt{y^2 + a^2}} = \int \frac{a\,du}{\sqrt{a^2 u^2 + a^2}} \qquad (y = au)$$

$$= \int \frac{du}{\sqrt{u^2 + 1}} = \sinh^{-1} u + b = \sinh^{-1}\frac{y}{a} + b.$$

It follows that

$$y(x) = a\sinh(x - b) = a\left(\sinh x \cosh b - \cosh x \sinh b\right)$$
$$= A\cosh x + B\sinh x.$$

29. (a) The verification in a component-wise manner that V is a vector space is the same as the verification that $\mathbf{R}^n$ is a vector space, except with vectors having infinitely many components rather than finitely many components. It boils down to the fact that a linear combination of infinite sequences of real numbers is itself such a sequence,

$$a \cdot \{x_n\}_1^\infty + b \cdot \{y_n\}_1^\infty = \{ax_n + by_n\}_1^\infty.$$

(b) If $e_n = \{0, \cdots, 0, 1, 0, 0, \cdots\}$ is the indicated infinite sequence with 1 in the nth position, then the fact that

$$c_1 e_1 + c_2 e_2 + \cdots + c_k e_k = \{c_1, c_2, \cdots, c_k, 0, 0, \cdots\}$$

evidently implies that any finite set $e_1, e_2, \cdots, e_k$ of these vectors is linearly independent. Thus V contains "arbitrarily large" sets of linearly independent vectors, and therefore is infinite-dimensional.

31. **(a)** If $z_1 = a_1 + ib_1$ and $z_2 = a_2 + ib_2$, then direct computation shows that

$$T(c_1 z_1 + c_2 z_2) = c_1 T(z_1) + c_2 T(z_2) = \begin{bmatrix} c_1 a_1 + c_2 a_2 & -c_1 b_1 - b_2 a_2 \\ c_1 b_1 + b_2 a_2 & c_1 a_1 + c_2 a_2 \end{bmatrix}.$$

(b) If $z_1 = a_1 + ib_1$ and $z_2 = a_2 + ib_2$, then $z_1 z_2 = (a_1 a_2 - b_1 b_2) + i(a_1 b_2 + a_2 b_1)$ and direct computation shows that

$$T(z_1 z_2) = T(z_1) T(z_2) = \begin{bmatrix} a_1 a_2 - b_1 b_2 & -a_1 b_2 - a_2 b_1 \\ a_1 b_2 + a_2 b_1 & a_1 a_2 - b_1 b_2 \end{bmatrix}.$$

(b) If $z = a + ib$ then

$$\frac{1}{z} = \frac{1}{a+bi} \cdot \frac{a-bi}{a-bi} = \frac{a-bi}{a^2+b^2}.$$

Therefore

$$T(z^{-1}) = \frac{1}{a^2+b^2} \begin{bmatrix} a & b \\ -b & a \end{bmatrix} = \begin{bmatrix} a & -b \\ b & a \end{bmatrix}^{-1} = T(z)^{-1}.$$

CHAPTER 5

LINEAR EQUATIONS OF HIGHER ORDER

SECTION 5.1

INTRODUCTION: SECOND-ORDER LINEAR EQUATIONS

In this section the central ideas of the theory of linear differential equations are introduced and illustrated concretely in the context of **second-order** equations. These key concepts include superposition of solutions (Theorem 1), existence and uniqueness of solutions (Theorem 2), linear independence, the Wronskian (Theorem 3), and general solutions (Theorem 4). This discussion of second-order equations serves as preparation for the treatment of nth order linear equations in Section 5.2. Although the concepts in this section may seem somewhat abstract to students, the problems set is quite tangible and largely computational.

In each of Problems 1-16 the verification that y_1 and y_2 satisfy the given differential equation is a routine matter. As in Example 2, we then impose the given initial conditions on the general solution $y = c_1 y_1 + c_2 y_2$. This yields two linear equations that determine the values of the constants c_1 and c_2.

1. Imposition of the initial conditions $y(0) = 0$, $y'(0) = 5$ on the general solution $y(x) = c_1 e^x + c_2 e^{-x}$ yields the two equations $c_1 + c_2 = 0$, $c_1 - c_2 = 0$ with solution $c_1 = 5/2$, $c_2 = -5/2$. Hence the desired particular solution is $y(x) = 5(e^x - e^{-x})/2$.

3. Imposition of the initial conditions $y(0) = 3$, $y'(0) = 8$ on the general solution $y(x) = c_1 \cos 2x + c_2 \sin 2x$ yields the two equations $c_1 = 3$, $2c_2 = 8$ with solution $c_1 = 3$, $c_2 = 4$. Hence the desired particular solution is $y(x) = 3 \cos 2x + 4 \sin 2x$.

5. Imposition of the initial conditions $y(0) = 1$, $y'(0) = 0$ on the general solution $y(x) = c_1 e^x + c_2 e^{2x}$ yields the two equations $c_1 + c_2 = 1$, $c_1 + 2c_2 = 0$ with solution $c_1 = 2$, $c_2 = -1$. Hence the desired particular solution is $y(x) = 2e^x - e^{2x}$.

7. Imposition of the initial conditions $y(0) = -2$, $y'(0) = 8$ on the general solution $y(x) = c_1 + c_2 e^{-x}$ yields the two equations $c_1 + c_2 = -2$, $-c_2 = 8$ with solution $c_1 = 6$, $c_2 = -8$. Hence the desired particular solution is $y(x) = 6 - 8e^{-x}$.

9. Imposition of the initial conditions $y(0) = 2$, $y'(0) = -1$ on the general solution $y(x) = c_1 e^{-x} + c_2 x e^{-x}$ yields the two equations $c_1 = 2$, $-c_1 + c_2 = -1$ with solution $c_1 = 2$, $c_2 = 1$. Hence the desired particular solution is $y(x) = 2e^{-x} + xe^{-x}$.

11. Imposition of the initial conditions $y(0) = 0$, $y'(0) = 5$ on the general solution $y(x) = c_1 e^x \cos x + c_2 e^x \sin x$ yields the two equations $c_1 = 0$, $c_1 + c_2 = 5$ with solution $c_1 = 0$, $c_2 = 5$. Hence the desired particular solution is $y(x) = 5e^x \sin x$.

13. Imposition of the initial conditions $y(1) = 3$, $y'(1) = 1$ on the general solution $y(x) = c_1 x + c_2 x^2$ yields the two equations $c_1 + c_2 = 3$, $c_1 + 2c_2 = 1$ with solution $c_1 = 5$, $c_2 = -2$. Hence the desired particular solution is $y(x) = 5x - 2x^2$.

15. Imposition of the initial conditions $y(1) = 7$, $y'(1) = 2$ on the general solution $y(x) = c_1 x + c_2 x \ln x$ yields the two equations $c_1 = 7$, $c_1 + c_2 = 2$ with solution $c_1 = 7$, $c_2 = -5$. Hence the desired particular solution is $y(x) = 7x - 5x \ln x$.

17. If $y = c/x$ then $y' + y^2 = -c/x^2 + c^2/x^2 = c(c-1)/x^2 \ne 0$ unless either $c = 0$ or $c = 1$.

19. If $y = 1 + \sqrt{x}$ then $yy'' + (y')^2 = (1+\sqrt{x})(-x^{-3/2}/4) + (x^{-1/2}/2)^2 = -x^{-3/2}/4 \ne 0$.

21. Linearly independent, because $x^3 = +x^2|x|$ if $x > 0$, whereas $x^3 = -x^2|x|$ if $x < 0$.

23. Linearly independent, because $f(x) = +g(x)$ if $x > 0$, whereas $f(x) = -g(x)$ if $x < 0$.

25. $f(x) = e^x \sin x$ and $g(x) = e^x \cos x$ are linearly independent, because $f(x) = k g(x)$ would imply that $\sin x = k \cos x$, whereas $\sin x$ and $\cos x$ are linearly independent.

27. Let $L[y] = y'' + py' + qy$. Then $L[y_c] = 0$ and $L[y_p] = f$, so
$$L[y_c + y_p] = L[y_c] + L[y_p] = 0 + f = f.$$

29. There is no contradiction because if the given differential equation is divided by x^2 to get the form in Equation (8) in the text, then the resulting functions $p(x) = -4/x$ and $q(x) = 6/x^2$ are not continuous at $x = 0$.

31. $W(y_1, y_2) = -2x$ vanishes at $x = 0$, whereas if y_1 and y_2 were (linearly independent) solutions of an equation $y'' + py' + qy = 0$ with p and q both continuous on an open interval I containing $x = 0$, then Theorem 3 would imply that $W \ne 0$ on I.

In Problems 33-42 we give the characteristic equation, its roots, and the corresponding general solution.

Section 5.1

97

33. $r^2 - 3r + 2 = 0;$ $r = 1, 2;$ $y(x) = c_1 e^x + c_2 e^{2x}$

35. $r^2 + 5r = 0;$ $r = 0, -5;$ $y(x) = c_1 + c_2 e^{-5x}$

37. $2r^2 - r - 2 = 0;$ $r = 1, -1/2;$ $y(x) = c_1 e^{-x/2} + c_2 e^x$

39. $4r^2 + 4r + 1 = 0;$ $r = -1/2, -1/2;$ $y(x) = (c_1 + c_2 x)e^{-x/2}$

41. $6r^2 - 7r - 20 = 0;$ $r = -4/3, 5/2;$ $y(x) = c_1 e^{-4x/3} + c_2 e^{5x/2}$

In Problems 43-48 we first write and simplify the equation with the indicated characteristic roots, and then write the corresponding differential equation.

43. $(r-0)(r+10) = r^2 + 10r = 0;$ $y'' + 10y' = 0$

45. $(r+10)(r+10) = r^2 + 20r + 100 = 0;$ $y'' + 20y' + 100y = 0$

47. $(r-0)(r-0) = r^2 = 0;$ $y'' = 0$

49. The solution curve with $y(0)=1$, $y'(0)=6$ is $y(x) = 8e^{-x} - 7e^{-2x}$. We find that $y'(x) = 0$ when $x = \ln(7/4)$ so $e^{-x} = 4/7$ and $e^{-2x} = 16/49$. It follows that $y(\ln(7/4)) = 16/7$, so the high point on the curve is $(\ln(7/4)), 16/7) \approx (0.56, 2.29)$, which looks consistent with Fig. 5.1.6.

SECTION 5.2

GENERAL SOLUTIONS OF LINEAR EQUATIONS

Students should check each of Theorems 1 through 4 in this section to see that, in the case $n = 2$, it reduces to the corresponding theorem in Section 5.1. Similarly, the computational problems for this section largely parallel those for the previous section. By the end of Section 5.2 students should understand that, although we do not prove the existence-uniqueness theorem now, it provides the basis for everything we do with linear differential equations.

The linear combinations listed in Problems 1-6 were discovered "by inspection" — that is, by trial and error.

1. $(5/2)(2x) + (-8/3)(3x^2) + (-1)(5x - 8x^2) = 0$

3. $(1)(0) + (0)(\sin x) + (0)(e^x) = 0$

5. $(1)(17) + (-34)(\cos^2 x) + (17)(\cos 2x) = 0$, because $2\cos^2 x = 1 + \cos 2x$.

7. $W = \begin{vmatrix} 1 & x & x^2 \\ 0 & 1 & 2x \\ 0 & 0 & 2 \end{vmatrix} = 2$ is nonzero everywhere.

9. $W = e^x(\cos^2 x + \sin^2 x) = e^x \neq 0$

11. $W = x^3 e^{2x}$ is nonzero if $x \neq 0$.

In each of Problems 13-20 we first form the general solution

$$y(x) = c_1 y_1(x) + c_2 y_2(x) + c_3 y_3(x),$$

then calculate $y'(x)$ and $y''(x)$, and finally impose the given initial conditions to determine the values of the coefficients c_1, c_2, c_3.

13. Imposition of the initial conditions $y(0) = 1$, $y'(0) = 2$, $y''(0) = 0$ on the general solution $y(x) = c_1 e^x + c_2 e^{-x} + c_3 e^{-2x}$ yields the three equations

$$c_1 + c_2 + c_3 = 1, \quad c_1 - c_2 - 2c_3 = 2, \quad c_1 + c_2 + 4c_3 = 0$$

with solution $c_1 = 4/3$, $c_2 = 0$, $c_3 = -1/3$. Hence the desired particular solution is given by $y(x) = (4e^x - e^{-2x})/3$.

15. Imposition of the initial conditions $y(0) = 2$, $y'(0) = 0$, $y''(0) = 0$ on the general solution $y(x) = c_1 e^x + c_2 x e^x + c_3 x^2 e^{3x}$ yields the three equations

$$c_1 = 2, \quad c_1 + c_2 = 0, \quad c_1 + 2c_2 + 2c_3 = 0$$

with solution $c_1 = 2$, $c_2 = -2$, $c_3 = 1$. Hence the desired particular solution is given by $y(x) = (2 - 2x + x^2)e^x$.

17. Imposition of the initial conditions $y(0) = 3$, $y'(0) = -1$, $y''(0) = 2$ on the general solution $y(x) = c_1 + c_2 \cos 3x + c_3 \sin 3x$ yields the three equations

$$c_1 + c_2 = 3, \quad 3c_3 = -1, \quad -9c_2 = 2$$

with solution $c_1 = 29/9$, $c_2 = -2/9$, $c_3 = -1/3$. Hence the desired particular solution is given by $y(x) = (29 - 2\cos 3x - 3\sin 3x)/9$.

19. Imposition of the initial conditions $y(1) = 6$, $y'(1) = 14$, $y''(1) = 22$ on the general solution $y(x) = c_1 x + c_2 x^2 + c_3 x^3$ yields the three equations

$$c_1 + c_2 + c_3 = 6, \quad c_1 + 2c_2 + 3c_3 = 14, \quad 2c_2 + 6c_3 = 22$$

with solution $c_1 = 1$, $c_2 = 2$, $c_3 = 3$. Hence the desired particular solution is given by $y(x) = x + 2x^2 + 3x^3$.

In each of Problems 21–24 we first form the general solution

$$y(x) = y_c(x) + y_p(x) = c_1 y_1(x) + c_2 y_2(x) + y_p(x),$$

then calculate $y'(x)$, and finally impose the given initial conditions to determine the values of the coefficients c_1 and c_2.

21. Imposition of the initial conditions $y(0) = 2$, $y'(0) = -2$ on the general solution $y(x) = c_1 \cos x + c_2 \sin x + 3x$ yields the two equations $c_1 = 2$, $c_2 + 3 = -2$ with solution $c_1 = 2$, $c_2 = -5$. Hence the desired particular solution is given by $y(x) = 2 \cos x - 5 \sin x + 3x$.

23. Imposition of the initial conditions $y(0) = 3$, $y'(0) = 11$ on the general solution $y(x) = c_1 e^{-x} + c_2 e^{3x} - 2$ yields the two equations $c_1 + c_2 - 2 = 3$, $-c_1 + 3c_2 = 11$ with solution $c_1 = 1$, $c_2 = 4$. Hence the desired particular solution is given by $y(x) = e^{-x} + 4e^{3x} - 2$.

25. $L[y] = L[y_1 + y_2] = L[y_1] + L[y_2] = f + g$

27. The equations

$$c_1 + c_2 x + c_3 x^2 = 0, \quad c_2 + 2c_3 x + 0, \quad 2c_3 = 0$$

(the latter two obtained by successive differentiation of the first one) evidently imply — by substituting $x = 0$ — that $c_1 = c_2 = c_3 = 0$.

29. If $c_0 e^{rx} + c_1 x e^{rx} + \cdots + c_n x^n e^{rx} = 0$, then division by e^{rx} yields

$$c_0 + c_1 x + \cdots + c_n x^n = 0,$$

so the result of Problem 28 applies.

31. **(a)** Substitution of $x = a$ in the differential equation gives $y''(a) = -p\, y'(a) - q(a)$.

(b) If $y(0) = 1$ and $y'(0) = 0$, then the equation $y'' - 2y' - 5y = 0$ implies that $y''(0) = 2y'(0) + 5y(0) = 5$.

33. This follows from the fact that

$$\begin{vmatrix} 1 & 1 & 1 \\ a & b & c \\ a^2 & b^2 & c^2 \end{vmatrix} = (b-a)(c-b)(c-a).$$

SECTION 5.3

HOMOGENEOUS EQUATIONS WITH CONSTANT COEFFICIENTS

This is a purely computational section devoted to the single most widely applicable type of higher order differential equations — linear ones with constant coefficients. In Problems 1-20, we write first the characteristic equation and its list of roots, then the corresponding general solution of the given differential equation. Explanatory comments are included only when the solution of the characteristic equation is not routine.

1. $r^2 - 4 = (r-2)(r+2) = 0;\quad r = -2, 2;\quad y(x) = c_1 e^{2x} + c_2 e^{-2x}$

3. $r^2 + 3r - 10 = (r+5)(r-2) = 0;\quad r = -5, 2;\quad y(x) = c_1 e^{2x} + c_2 e^{-5x}$

5. $r^2 + 6r + 9 = (r+3)^2 = 0;\quad r = -3, -3;\quad y(x) = c_1 e^{-3x} + c_2 x e^{-3x}$

7. $4r^2 - 12r + 9 = (2r-3)^2 = 0;\quad r = -3/2, -3/2;\quad y(x) = c_1 e^{3x/2} + c_2 x e^{3x/2}$

9. $r^2 + 8r + 25 = 0;\quad r = \left(-8 \pm \sqrt{-36}\right)/2 = -4 \pm 3i;\quad y(x) = e^{-4x}(c_1 \cos 3x + c_2 \sin 3x)$

11. $r^4 - 8r^3 + 16r^2 = r^2(r-4)^2 = 0;\quad r = 0, 0, 4, 4;\quad y(x) = c_1 + c_2 x + c_3 e^{4x} + c_4 x e^{4x}$

13. $9r^3 + 12r^2 + 4r = r(3r+2)^2 = 0;\quad r = 0, -2/3, -2/3$
$y(x) = c_1 + c_2 e^{-2x/3} + c_3 x e^{-2x/3}$

15. $4r^4 - 8r^2 + 16 = (r^2-4)^2 = (r-2)^2(r+2)^2 = 0;\quad r = 2, 2, -2, -2$
$y(x) = c_1 e^{2x} + c_2 x e^{2x} + c_3 e^{-2x} + c_4 x e^{-2x}$

17. $6r^4 + 11r^2 + 4 = (2r^2+1)(3r^2+4) = 0;\quad r = \pm i/\sqrt{2}, \pm 2i/\sqrt{3},$

$$y(x) = c_1\cos(x/\sqrt{2}) + c_2\sin(x/\sqrt{2}) + c_3\cos(2x/\sqrt{3}) + c_4\sin(2x/\sqrt{3})$$

19. $r^3 + r^2 - r - 1 = r(r^2 - 1) + (r^2 - 1) = (r-1)(r+1)^2 = 0;\quad r = 1, -1, -1;$

$$y(x) = c_1 e^x + c_2 e^{-x} + c_3 x e^{-x}$$

21. Imposition of the initial conditions $y(0) = 7$, $y'(0) = 11$ on the general solution $y(x) = c_1 e^x + c_2 e^{3x}$ yields the two equations $c_1 + c_2 = 7$, $c_1 + 3c_2 = 11$ with solution $c_1 = 5$, $c_2 = 2$. Hence the desired particular solution is $y(x) = 5e^x + 2e^{3x}$.

23. Imposition of the initial conditions $y(0) = 3$, $y'(0) = 1$ on the general solution $y(x) = e^{3x}(c_1 \cos 4x + c_2 \sin 4x)$ yields the two equations $c_1 = 3$, $3c_1 + 4c_2 = 1$ with solution $c_1 = 3$, $c_2 = -2$. Hence the desired particular solution is $y(x) = e^{3x}(3\cos 4x - 2\sin 4x)$.

25. Imposition of the initial conditions $y(0) = -1$, $y'(0) = 0$, $y''(0) = 1$ on the general solution $y(x) = c_1 + c_2 x + c_3 e^{-2x/3}$ yields the three equations

$$c_1 + c_3 = -1,\quad c_2 - 2c_3/3 = 0,\quad 4c_3/9 = 1$$

with solution $c_1 = -13/4$, $c_2 = 3/2$, $c_3 = 9/4$. Hence the desired particular solution is $y(x) = (-13 + 6x + 9e^{-2x/3})/4$.

27. First we spot the root $r = 1$. Then long division of the polynomial $r^3 + 3r^2 - 4$ by $r - 1$ yields the quadratic factor $r^2 + 4r + 4 = (r+2)^2$ with roots $r = -2, -2$. Hence the general solution is $y(x) = c_1 e^x + c_2 e^{-2x} + c_3 x e^{-2x}$.

29. First we spot the root $r = -3$. Then long division of the polynomial $r^3 + 27$ by $r + 3$ yields the quadratic factor $r^2 - 3r + 9$ with roots $r = 3(1 \pm i\sqrt{3})/2$. Hence the general solution is $y(x) = c_1 e^{-3x} + e^{3x/2}[c_2\cos(3x\sqrt{3}/2) + c_3 \sin(3x\sqrt{3}/2)]$.

31. The characteristic equation $r^3 + 3r^2 + 4r - 8 = 0$ has the evident root $r = 1$, and long division then yields the quadratic factor $r^2 + 4r + 8 = (r+2)^2 + 4$ corresponding to the complex conjugate roots $-2 \pm 2i$. Hence the general solution is

$$y(x) = c_1 e^x + e^{-2x}(c_2\cos 2x + c_3\sin 2x).$$

33. Knowing that $y = e^{3x}$ is one solution, we divide the characteristic polynomial $r^3 + 3r^2 - 54$ by $r - 3$ and get the quadratic factor

$$r^2 + 6r + 18 = (r+3)^2 + 9.$$

Hence the general solution is $y(x) = c_1 e^{3x} + e^{-3x}(c_2 \cos 3x + c_3 \sin 3x)$.

35. The fact that $y = \cos 2x$ is one solution tells us that $r^2 + 4$ is a factor of the characteristic polynomial

$$6r^4 + 5r^3 + 25r^2 + 20r + 4.$$

Then long division yields the quadratic factor $6r^2 + 5r + 1 = (3r+1)(2r+1)$ with roots $r = -1/2, -1/3$. Hence the general solution is

$$y(x) = c_1 e^{-x/2} + c_2 e^{-x/3} + c_3 \cos 2x + c_4 \sin 2x$$

37. The characteristic equation is $r^4 - r^3 = r^3(r-1) = 0$, so the general solution is $y(x) = A + Bx + Cx^2 + De^x$. Imposition of the given initial conditions yields the equations

$$A + D = 18, \quad B + D = 12, \quad 2C + D = 13, \quad D = 7$$

with solution $A = 11, B = 5, C = 3, D = 7$. Hence the desired particular solution is

$$y(x) = 11 + 5x + 3x^2 + 7e^x.$$

39. $(r-2)^3 = r^3 - 6r^2 + 12r - 8$, so the differential equation is

$$y''' - 6y'' + 12y' - 8y = 0.$$

41. $(r^2 + 4)(r^2 - 4) = r^4 - 16$, so the differential equation is $y^{(4)} - 16y = 0$.

45. The characteristic polynomial is the quadratic polynomial of Problem 44(b). Hence the general solution is

$$y(x) = c_1 e^{-ix} + c_2 e^{3ix} = c_1(\cos x - i \sin x) + c_2(\cos 3x + i \sin 3x).$$

47. The characteristic roots are $r = \pm\sqrt{-2 + 2i\sqrt{3}} = \pm(1 + i\sqrt{3})$ so the general solution is

$$y(x) = c_1 e^{(1+i\sqrt{3})x} + c_2 e^{-(1+i\sqrt{3})x} = c_1 e^x \left(\cos\sqrt{3}\,x + i\sin\sqrt{3}\,x\right) + c_2 e^{-x}\left(\cos\sqrt{3}\,x - i\sin\sqrt{3}\,x\right)$$

49. The general solution is $y = Ae^{2x} + Be^{-x} + C\cos x + D\sin x$. Imposition of the given initial conditions yields the equations

$$A + B + C = 0$$
$$2A - B + D = 0$$
$$4A + B - C = 0$$
$$8A - B - D = 30$$

that we solve for $A = 2$, $B = -5$, $C = 3$, and $D = -9$. Thus

$$y(x) = 2e^{2x} - 5e^{-x} + 3\cos x - 9\sin x.$$

SECTION 5.4

Mechanical Vibrations

In this section we discuss four types of free motion of a mass on a spring — undamped, underdamped, critically damped, and overdamped. However, the undamped and underdamped cases — in which actual oscillations occur — are emphasized because they are both the most interesting and the most important cases for applications.

1. Frequency: $\omega_0 = \sqrt{k/m} = \sqrt{16/4} = 2$ rad/sec $= 1/\pi$ Hz
   Period: $P = 2\pi/\omega_0 = 2\pi/2 = \pi$ sec

3. The spring constant is $k = 15$ N/0.20 m $= 75$ N/m. The solution of $3x'' + 75x = 0$ with $x(0) = 0$ and $x'(0) = -10$ is $x(t) = -2\sin 5t$. Thus the amplitude is 2 m; the frequency is $\omega_0 = \sqrt{k/m} = \sqrt{75/3} = 5$ rad/sec $= 2.5/\pi$ Hz; and the period is $2\pi/5$ sec.

5. The gravitational acceleration at distance R from the center of the earth is $g = GM/R^2$. According to Equation (6) in the text the (circular) frequency ω of a pendulum is given by $\omega^2 = g/L = GM/R^2L$, so its period is $p = 2\pi/\omega = 2\pi R\sqrt{L/GM}$.

7. The period equation $p = 3960\sqrt{100.10} = (3960 + x)\sqrt{100}$ yields $x \approx 1.9795$ mi $\approx$ 10,450 ft for the altitude of the mountain.

11. The fact that the buoy weighs 100 lb means that $mg = 100$ so $m = 100/32$ slugs. The weight of water is 62.4 lb/ft^3, so the $F = ma$ equation of Problem 10 is

    $$(100/32)x'' = 100 - 62.4\pi r^2 x.$$

    It follows that the buoy's circular frequency ω is given by

    $$\omega^2 = (32)(62.4\pi)r^2/100.$$

But the fact that the buoy's period is $p = 2.5$ sec means that $\omega = 2\pi/2.5$. Equating these two results yields $r \approx 0.3173$ ft ≈ 3.8 in.

13. **(a)** The characteristic equation $10r^2 + 9r + 2 = (5r+2)(2r+1) = 0$ has roots $r = -2/5, -1/2$. When we impose the initial conditions $x(0) = 0$, $x'(0) = 5$ on the general solution $x(t) = c_1 e^{-2t/5} + c_2 e^{-t/2}$ we get the particular solution
$$x(t) = 50\left(e^{-2t/5} - e^{-t/2}\right).$$

   (b) The derivative $x'(t) = 25e^{-t/2} - 20e^{-2t/5} = 5e^{-2t/5}\left(5e^{-t/10} - 4\right) = 0$ when $t = 10\ln(5/4) \approx 2.23144$. Hence the mass's farthest distance to the right is given by $x(10\ln(5/4)) = 512/125 = 4.096$.

15. The characteristic equation $(1/2)r^2 + 3r + 4 = 0$ has roots $r = -2, -4$. When we impose the initial conditions $x(0) = 2$, $x'(0) = 0$ on the general solution $x(t) = c_1 e^{-2t} + c_2 e^{-4t}$ we get the particular solution $x(t) = 4e^{-2t} - 2e^{-4t}$ that describes overdamped motion.

17. The characteristic equation $r^2 + 8r + 16 = 0$ has roots $r = -4, -4$. When we impose the initial conditions $x(0) = 5$, $x'(0) = -10$ on the general solution $x(t) = (c_1 + c_2 t)e^{-4t}$ we get the particular solution $x(t) = (5 + 10t)e^{-4t}$ that describes critically damped motion.

19. The characteristic equation $4r^2 + 20r + 169 = 0$ has roots $r = -5/2 \pm 6i$. When we impose the initial conditions $x(0) = 4$, $x'(0) = 16$ on the general solution $x(t) = e^{-5t/2}(A\cos 6t + B\sin 6t)$ we get the particular solution
$$x(t) = e^{-5t/2}[4\cos 6t + (13/3)\sin 2t] \approx (1/3)\sqrt{313}\, e^{-5t/2}\cos(6t - 0.8254)$$
that describes underdamped motion.

21. The characteristic equation $r^2 + 10r + 125 = 0$ has roots $r = -5 \pm 10i$. When we impose the initial conditions $x(0) = 6$, $x'(0) = 50$ on the general solution $x(t) = e^{-5t}(A\cos 10t + B\sin 10t)$ we get the particular solution
$$x(t) = e^{-5t}(6\cos 10t + 8\sin 10t) \approx 10\, e^{-5t}\cos(10t - 0.9273)$$
that describes underdamped motion.

23. **(a)** With $m = 100$ slugs we get $\omega = \sqrt{k/100}$. But we are given that
$$\omega = (80\text{ cycles/min})(2\pi)(1\text{ min}/60\text{ sec}) = 8\pi/3,$$

and equating the two values yields $k \approx 7018$ lb/ft.

(b) With $\omega_1 = 2\pi(78/60)$ sec^{-1}, Equation (21) in the text yields $c \approx 372.31$ lb/(ft/sec). Hence $p = c/2m \approx 1.8615$. Finally $e^{-pt} = 0.01$ gives $t \approx 2.47$ sec.

31. The binomial series

$$(1+x)^\alpha = 1 + \alpha x + \frac{\alpha(\alpha-1)}{2!}x^2 + \frac{\alpha(\alpha-1)(\alpha-2)}{3!}x^3 + \cdots$$

converges if $|x|<1$. (See, for instance, Section 11.8 of Edwards and Penney, *Calculus with Analytic Geometry*, 5th edition, Prentice Hall, 1998.) With $\alpha = 1/2$ and $x = -c^2/4mk$ in Eq. (21) of Section 5.4 in this text, the binomial series gives

$$\omega_1 = \sqrt{\omega_0^2 - p^2} = \sqrt{\frac{k}{m} - \frac{c^2}{4m^2}} = \sqrt{\frac{k}{m}}\sqrt{1 - \frac{c^2}{4mk}}$$

$$= \sqrt{\frac{k}{m}}\left(1 - \frac{c^2}{8mk} - \frac{c^4}{128m^2k^2} - \cdots\right) \approx \omega_0\left(1 - \frac{c^2}{8mk}\right).$$

33. If $x_1 = x(t_1)$ and $x_2 = x(t_2)$ are two successive local maxima, then $\omega_1 t_2 = \omega_1 t_1 + 2\pi$ so

$$x_1 = C\exp(-pt_1)\cos(\omega_1 t_1 - \alpha),$$

$$x_2 = C\exp(-pt_2)\cos(\omega_1 t_2 - \alpha) = C\exp(-pt_2)\cos(\omega_1 t_1 - \alpha).$$

Hence $x_1/x_2 = \exp[-p(t_1 - t_2)]$, and therefore

$$\ln(x_1/x_2) = -p(t_1 - t_2) = 2\pi p/\omega_1.$$

35. The characteristic equation $r^2 + 2r + 1 = 0$ has roots $r = -1, -1$. When we impose the initial conditions $x(0) = 0$, $x'(1) = 0$ on the general solution $x(t) = (c_1 + c_2 t)e^{-t}$ we get the particular solution $x_1(t) = te^{-t}$.

37. The characteristic equation $r^2 + 2r + (1+10^{-2n}) = 0$ has roots $r = -1 \pm 10^{-n}i$. When we impose the initial conditions $x(0) = 0$, $x'(1) = 0$ on the general solution

$$x(t) = e^{-t}\left[A\cos(10^{-n}t) + B\sin(10^{-n}t)\right]$$

we get the equations $c_1 = 0$, $-c_1 + 10^{-n}c_2 = 1$ with solution $c_1 = 0$, $c_2 = 10^n$. This gives the particular solution $x_3(t) = 10^n e^{-t}\sin(10^{-n}t)$.

SECTION 5.5

UNDETERMINED COEFFICIENTS AND VARIATION OF PARAMETERS

The method of undetermined coefficients is based on "educated guessing". If we can guess correctly the **form** of a particular solution of a nonhomogeneous linear equation with constant coefficients, then we can determine the particular solution explicitly by substitution in the given differential equation. It is pointed out at the end of Section 5.5 that this simple approach is not always successful — in which case the method of variation of parameters is available if a complementary function is known. However, undetermined coefficients *does* turn out to work well with a surprisingly large number of the nonhomogeneous linear differential equations that arise in elementary scientific applications.

In each of Problems 1-20 we give first the form of the trial solution y_{trial}, then the equations in the coefficients we get when we substitute y_{trial} into the differential equation and collect like terms, and finally the resulting particular solution y_p.

1. $y_{trial} = Ae^{3x}$; $25A = 1$; $y_p = (1/25)e^{3x}$

3. $y_{trial} = A\cos 3x + B\sin 3x$; $-15A - 3B = 0$, $3A - 15B = 2$;
$y_p = (\cos 3x - 5\sin 3x)/39$

5. First we substitute $\sin^2 x = (1 - \cos 2x)/2$ on the right-hand side of the differential equation. Then:
$y_{trial} = A + B\cos 2x + C\sin 2x$; $A = 1/2$, $-3B + 2C = -1/2$, $-2B - 3C = 0$;
$y_p = (13 + 3\cos 2x - 2\sin 2x)/26$

7. First we substitute $\sinh x = (e^x - e^{-x})/2$ on the right-hand side of the differential equation. Then:
$y_{trial} = Ae^x + Be^{-x}$; $-3A = 1/2$, $-3B = -1/2$; $y_p = (e^{-x} - e^x)/6 = -(1/3)\sinh x$

9. First we note that e^x is part of the complementary function $y_c = c_1 e^x + c_2 e^{-3x}$. Then:
$y_{trial} = A + x(B + Cx)e^x$; $-3A = 0$, $4B + 2C = 0$, $8C = 1$;
$y_p = -(1/3) + (2x^2 - x)e^x/16$.

11. First we note the duplication with the complementary function
$y_c = c_1 x + c_2 \cos 2x + c_3 \sin 2x$. Then:
$y_{trial} = x(A + Bx)$; $4A = -1$, $8B = 3$; $y_p = (3x^2 - 2x)/8$

13. $y_{trial} = e^x(A\cos x + B\sin x);\quad 7A+4B=0,\ -4A+7B=1;$
$y_p = e^x(7\sin x - 4\cos x)/65$

15. This is something of a trick problem. We cannot solve the characteristic equation $r^5 + 5r^4 - 1 = 0$ to find the complementary function, but we can see that it contains no constant term (why?). Hence the trial solution $y_{trial} = A$ leads immediately to the particular solution $y_p = -17$.

17. First we note the duplication with the complementary function $y_c = c_1\cos x + c_2\sin x$. Then:
$$y_{trial} = x[(A+Bx)\cos x + (C+Dx)\sin x];$$
$$2B+2C=0,\ 4D=1,\ -2A+2D=1,\ -4B=0;$$
$$y_p = (x^2\sin x - x\cos x)/4$$

19. First we note the duplication with the part $c_1 + c_2 x$ of the complementary function (which corresponds to the factor r^2 of the characteristic polynomial). Then:
$$y_{trial} = x^2(A + Bx + Cx^2);\quad 4A+12B=-1,\ 12B+48C=0,\ 24C=3;$$
$$y_p = (10x^2 - 4x^3 + x^4)/8$$

In Problems 21-30 we list first the complementary function y_c, then the initially proposed trial function y_i, and finally the actual trial function y_p in which duplication with the complementary function has been eliminated.

21. $y_c = e^x(c_1\cos x + c_2\sin x);\quad y_i = e^x(A\cos x + B\sin x)$
$y_p = x\cdot e^x(A\cos x + B\sin x)$

23. $y_c = c_1\cos x + c_2\sin x;\quad y_i = (A+Bx)\cos 2x + (C+Dx)\sin 2x$
$y_p = x\cdot[(A+Bx)\cos 2x + (C+Dx)\sin 2x]$

25. $y_c = c_1 e^{-x} + c_2 e^{-2x};\quad y_i = (A+Bx)e^{-x} + (C+Dx)e^{-2x}$
$y_p = x\cdot(A+Bx)e^{-x} + x\cdot(C+Dx)e^{-2x}$

27. $y_c = (c_1\cos x + c_2\sin x) + (c_3\cos 2x + c_3\sin 2x)$
$y_i = (A\cos x + B\sin x) + (C\cos 2x + D\sin 2x)$
$y_p = x\cdot[(A\cos x + B\sin x) + (C\cos 2x + D\sin 2x)]$

29. $y_c = (c_1 + c_2 x + c_3 x^2)e^x + c_4 e^{2x} + c_5 e^{-2x};\quad y_i = (A+Bx)e^x + Ce^{2x} + De^{-2x}$

$$y_p = x^3 \cdot (A+Bx)e^x + x \cdot (Ce^{2x}) + x \cdot (De^{-2x})$$

In Problems 31-40 we list first the complementary function y_c, the trial solution y_{tr} for the method of undetermined coefficients, and the corresponding general solution $y_g = y_c + y_p$ where y_p results from determining the coefficients in y_{tr} so as to satisfy the given nonhomogeneous differential equation. Then we list the linear equations obtained by imposing the given initial conditions, and finally the resulting particular solution $y(x)$.

31. $y_c = c_1 \cos 2x + c_2 \sin 2x;$ $y_{tr} = A + Bx;$ $y_g = c_1 \cos 2x + c_2 \sin 2x + x/2;$

$c_1 = 1,\ 2c_2 + 1/2 = 2;$ $y(x) = \cos 2x + (3/4)\sin 2x + x/2$

33. $y_c = c_1 \cos 3x + c_2 \sin 3x;$ $y_{tr} = A\cos 2x + B\sin 2x$

$y_g = c_1 \cos 3x + c_2 \sin 3x + (1/5)\sin 2x$

$c_1 = 1,\ 3c_2 + 2/5 = 0;$ $y(x) = (15\cos 3x - 2\sin 3x + 3\sin 2x)/15$

35. $y_c = e^x(c_1 \cos x + c_2 \sin x);$ $y_{tr} = A + Bx$

$y_g = e^x(c_1 \cos x + c_2 \sin x) + 1 + x/2$

$c_1 + 1 = 3,\ c_1 + c_2 + 1/2 = 0;$ $y(x) = e^x(4\cos x - 5\sin x)/2 + 1 + x/2$

37. $y_c = c_1 + c_2 e^x + c_3 x e^x;$ $y_{tr} = x \cdot (A) + x^2 \cdot (B + Cx)e^x$

$y_g = c_1 + c_2 e^x + c_3 x e^x + x - x^2 e^x/2 + x^3 e^x/6$

$c_1 + c_2 = 0,\ c_2 + c_3 + 1 = 0,\ c_2 + 2c_3 - 1 = 1$

$y(x) = 4 + x + e^x(-24 + 18x - 3x^2 + x^3)/6$

39. $y_c = c_1 + c_2 x + c_3 e^{-x};$ $y_{tr} = x^2 \cdot (A + Bx) + x \cdot (Ce^{-x})$

$y_g = c_1 + c_2 x + c_3 e^{-x} - x^2/2 + x^3/6 + xe^{-x}$

$c_1 + c_3 = 1,\ c_2 - c_3 + 1 = 0,\ c_3 - 3 = 1$

$y(x) = (-18 + 18x - 3x^2 + x^3)/6 + (4+x)e^{-x}$

41. The trial solution $y_{tr} = A + Bx + Cx^2 + Dx^3 + Ex^4 + Fx^5$ leads to the equations

$$2A - B - 2C - 6D + 24E = 0$$
$$-2B - 2C - 6D - 24E + 120F = 0$$
$$-2C - 3D - 12E - 60F = 0$$
$$-2D - 4E - 20F = 0$$
$$-2E - 5F = 0$$
$$-2F = 8$$

that are readily solve by back-substitution. The resulting particular solution is

$$y(x) = -255 - 450x + 30x^2 + 20x^3 + 10x^4 - 4x^5$$

43. **(a)** $\cos 3x + i \sin 3x = (\cos x + i \sin x)^3$
$$= \cos^3 x + 3i\cos^2 x \sin x - 3\cos x \sin^2 x - i \sin^3 x$$

When we equate real parts we get the equation

$$\cos^3 x - 3(\cos x)(1 - \cos^2 x) = 4\cos^3 x - 3\cos x$$

and readily solve for $\cos^3 x = \frac{3}{4}\cos x + \frac{1}{4}\cos 3x$. The formula for $\sin^3 x$ is derived similarly by equating imaginary parts in the first equation above.

(b) Upon substituting the trial solution $y_p = A\cos x + B\sin x + C\cos 3x + D\sin 3x$ in the differential equation $y'' + 4y = \frac{3}{4}\cos x + \frac{1}{4}\cos 3x$, we find that $A = 1/4$, $B = 0$, $C = -1/20$, $D = 0$. The resulting general solution is

$$y(x) = c_1 \cos 2x + c_2 \sin 2x + (1/4)\cos x - (1/20)\cos 3x.$$

45. We substitute

$$\sin^4 x = (1 - \cos 2x)^2/4$$
$$= (1 - 2\cos 2x + \cos^2 2x)/4 = (3 - 4\cos 2x + \cos 4x)/8$$

on the right-hand side of the differential equation, and then substitute the trial solution $y_p = A\cos 2x + B\sin 2x + C\cos 4x + D\sin 4x + E$. We find that $A = -1/10$, $B = 0$, $C = -1/56$, $D = 0$, $E = 1/24$. The resulting general solution is

$$y = c_1 \cos 3x + c_2 \sin 3x + 1/24 - (1/10)\cos 2x - (1/56)\cos 4x.$$

In Problems 47-49 we list the independent solutions y_1 and y_2 of the associated homogeneous equation, their Wronskian $W = W(y_1, y_2)$, the coefficient functions

$$u_1(x) = -\int \frac{y_2(x) f(x)}{W(x)} dx \quad \text{and} \quad u_2(x) = \int \frac{y_1(x) f(x)}{W(x)} dx$$

in the particular solution $y_p = u_1 y_1 + u_2 y_2$ of Eq. (32) in the text, and finally y_p itself.

47. $y_1 = e^{-2x}$, $\quad y_2 = e^{-x}$, $\quad W = e^{-3x}$

$u_1 = -(4/3)e^{3x}$, $\quad u_2 = 2e^{2x}$,

$y_p = (2/3)e^x$

49. $y_1 = e^{2x}$, $\quad y_2 = xe^{2x}$, $\quad W = e^{4x}$

$u_1 = -x^2$, $\quad u_2 = 2x$,

$y_p = x^2 e^{2x}$

51. $y_1 = \cos 2x$, $\quad y_2 = \sin 2x$, $\quad W = 2$

Liberal use of trigonometric sum and product identities yields

$u_1 = (\cos 5x - 5\cos x)/20$, $\quad u_1 = (\sin 5x - 5\sin x)/20$

$y_p = -(1/4)(\cos 2x \cos x - \sin 2x \sin x) + (1/20)(\cos 5x \cos 2x + \sin 5x \sin 2x)$

$\quad = -(1/5)\cos 3x \quad (!)$

53. $y_1 = \cos 3x$, $\quad y_2 = \sin 3x$, $\quad W = 3$

$u_1' = -(2/3)\tan 3x$, $\quad u_2' = 2/3$

$y_p = (2/9)[3x \sin 3x + (\cos 3x) \ln|\cos 3x|]$

55. $y_1 = \cos 2x$, $\quad y_2 = \sin 2x$, $\quad W = 2$

$u_1' = -(1/2)\sin^2 x \sin 2x = -(1/4)(1 - \cos 2x)\sin 2x$

$u_2' = (1/2)\sin^2 x \cos 2x = (1/4)(1 - \cos 2x)\cos 2x$

$y_p = (1 - x \sin 2x)/8$

57. With $y_1 = x$, $y_2 = x^{-1}$, and $f(x) = 72x^3$, Equations (31) in the text take the form

$$xu_1' + x^{-1} u_2' = 0,$$
$$u_1' - x^{-2} u_2' = 72x^3.$$

Upon multiplying the second equation by x and then adding, we readily solve first for

$$u_1' = 36x^3, \quad \text{so} \quad u_1 = 9x^4$$

and then

$$u_2' = -x^2 u_1' = -36x^5, \quad \text{so} \quad u_2 = -6x^6.$$

Then it follows that

$$y_p = y_1 u_1 + y_2 u_2 = (x)(9x^4) + (x^{-1})(-6x^6) = 3x^5.$$

59. $y_1 = x^2,$ $\quad y_2 = x^2 \ln x,$
$W = x^3,$ $\quad f(x) = x^2$
$u_1' = -x \ln x,$ $\quad u_2' = x$
$y_p = x^4/4$

61. $y_1 = \cos(\ln x),$ $\quad y_2 = \sin(\ln x),$ $\quad W = 1/x,$
$f(x) = (\ln x)/x^2$
$u_1 = (\ln x)\cos(\ln x) - \sin(\ln x)$
$u_2 = (\ln x)\sin(\ln x) + \cos(\ln x)$
$y_p = \ln x \quad (!)$

63. This is simply a matter of solving the equations in (31) for the derivatives

$$u_1' = -\frac{y_2(x)f(x)}{W(x)} \quad \text{and} \quad u_2' = \frac{y_1(x)f(x)}{W(x)},$$

integrating each, and then substituting the results in (32).

SECTION 5.6

FORCED OSCILLATIONS AND RESONANCE

1. Trial of $x = A \cos 2t$ yields the particular solution $x_p = 2 \cos 2t$. (Can you see that — because the differential equation contains no first-derivative term — there is no need to include a $\sin 2t$ term in the trial solution?) Hence the general solution is

$$x(t) = c_1 \cos 3t + c_2 \sin 3t + 2 \cos 2t.$$

The initial conditions imply that $c_1 = -2$ and $c_2 = 0$, so $x(t) = 2 \cos 2t - 2 \cos 3t$.

3. First we apply the method of undetermined coefficients — with trial solution

$x = A\cos 5t + B\sin 5t$ — to find the particular solution

$$x_p = (3/15)\cos 5t + (4/15)\sin 5t$$
$$= \frac{1}{3}\left[\frac{3}{5}\cos 5t + \frac{4}{5}\sin 5t\right] = \frac{1}{3}\cos(5t - \beta)$$

where $\beta = \tan^{-1}(4/3) \approx 0.9273$. Hence the general solution is

$$x(t) = c_1\cos 10t + c_2\sin 10t + (3/15)\cos 5t + (4/15)\sin 5t.$$

The initial conditions $x(0) = 25$, $x'(0) = 0$ now yield $c_1 = 372/15$ and $c_2 = -2/15$, so the part of the solution with frequency $\omega = 10$ is

$$x_c = (1/15)(372\cos 10t - 2\sin 10t) =$$
$$= \frac{\sqrt{138388}}{15}\left[\frac{372}{\sqrt{138388}}\cos 10t - \frac{2}{\sqrt{138388}}\sin 10t\right]$$
$$= \frac{\sqrt{138388}}{15}\cos(10t - \alpha)$$

where $\alpha = 2\pi - \tan^{-1}(1/186) \approx 6.2778$ is a fourth-quadrant angle.

5. Substitution of the trial solution $x = C\cos\omega t$ gives $C = F_0/(k - m\omega^2)$. Then imposition of the initial conditions $x(0) = x_0$, $x'(0) = 0$ on the general solution

$$x(t) = c_1\cos\omega_0 t + c_2\sin\omega_0 t + C\cos\omega t \quad (\text{where } \omega_0 = \sqrt{k/m})$$

gives the particular solution $x(t) = (x_0 - C)\cos\omega_0 t + C\cos\omega t$.

In Problems 7-10 we give first the trial solution x_p involving undetermined coefficients A and B, then the equations that determine these coefficients, and finally the resulting steady periodic solution x_{sp}.

7. $x_p = A\cos 3t + B\sin 3t$; $-5A + 12B = 0$, $12A + 5B = 0$

$$x_{sp}(t) = -\frac{50}{169}\cos 3t + \frac{120}{169}\sin 3t = \frac{10}{13}\left(-\frac{5}{13}\cos 3t + \frac{12}{13}\sin 3t\right) = \frac{10}{13}\cos(3t - \alpha)$$

$\alpha = \pi - \tan^{-1}(12/5) \approx 1.9656$ (2nd quadrant angle)

9. $x_p = A\cos 10t + 10\sin 5t$; $-199A + 20B = 0$, $-20A - 199B = 3$

Section 5.6

$$x_{sp}(t) = -\frac{60}{40001}\cos 10t - \frac{597}{40001}\sin 10t$$

$$= \frac{3}{\sqrt{40001}}\left(-\frac{20}{\sqrt{40001}}\cos 10t - \frac{199}{\sqrt{40001}}\sin 10t\right) = \frac{3}{\sqrt{40001}}\cos(10t - \alpha)$$

$\alpha = \pi + \tan^{-1}(199/20) \approx 4.6122$ (3rd quadrant angle)

Each solution in Problems 11-14 has two parts. For the first part, we give first the trial solution x_p involving undetermined coefficients A and B, then the equations that determine these coefficients, and finally the resulting steady periodic solution x_{sp}. For the second part, we give first the general solution $x(t)$ involving the coefficients c_1 and c_2 in the transient solution, then the equations that determine these coefficients, and finally the resulting transient solution x_{tr}.

11. $x_p = A\cos 3t + B\sin 3t;$ $-4A + 12B = 10,$ $12A + 4B = 0$

$$x_{sp}(t) = -\frac{1}{4}\cos 3t + \frac{3}{4}\sin 3t = \frac{\sqrt{10}}{4}\left(-\frac{1}{\sqrt{10}}\cos 3t + \frac{3}{\sqrt{10}}\sin 3t\right) = \frac{\sqrt{10}}{4}\cos(3t - \alpha)$$

$\alpha = \pi - \tan^{-1}(3) \approx 1.8925$ (2nd quadrant angle)

$x(t) = e^{-2t}(c_1\cos t + c_2\sin t) + x_{sp}(t);$ $c_1 - 1/4 = 0,$ $-2c_1 + c_2 + 9/4 = 0$

$$x_{tr}(t) = e^{-2t}\left(\frac{1}{4}\cos t - \frac{7}{4}\sin t\right) = \frac{\sqrt{50}}{4}\left(\frac{1}{\sqrt{50}}\cos t - \frac{7}{\sqrt{50}}\sin t\right) = \frac{\sqrt{50}}{4}\cos(t - \beta)$$

$\beta = 2\pi - \tan^{-1}(7) \approx 4.8543$ (4th quadrant angle)

13. $x_p = A\cos 10t + B\sin 10t;$ $-94A + 20B = 3,$ $20A + 94B = 0$

$$x_{sp}(t) = -\frac{141}{4618}\cos 10t + \frac{30}{4618}\sin 10t$$

$$= \frac{3\sqrt{2309}}{4618}\left(-\frac{47}{\sqrt{2309}}\cos 10t + \frac{10}{\sqrt{2309}}\sin 10t\right) = \frac{3}{2\sqrt{2309}}\cos(10t - \alpha)$$

$\alpha = \pi - \tan^{-1}(10/47) \approx 2.9320$ (2nd quadrant angle)

$x(t) = e^{-t}\left(c_1\cos t\sqrt{5} + c_2\sin t\sqrt{5}\right) + x_{sp}(t);$

$c_1 - 141/4618 = 10,$ $-c_1 + \sqrt{5}\,c_2 + 150/2309 = 0$

$$x_{tr}(t) = \frac{e^{-t}}{4618\sqrt{5}}\left(46321\sqrt{5}\cos t\sqrt{5} + 46021\sin t\sqrt{5}\right)$$

Chapter 5

$$= 3\sqrt{\frac{309083}{23090}}\, e^{-t} \cos\left(t\sqrt{5}-\beta\right) \approx 10.9761\, e^{-t} \cos\left(t\sqrt{5}-\beta\right)$$

$$\beta = \tan^{-1}(46321\sqrt{5}/46021) \approx 1.1527 \quad \text{(1st quadrant angle)}$$

In Problems 15-18 we substitute $x(t) = A(\omega)\cos\omega t + B(\omega)\sin\omega t$ into the differential equation $mx'' + cx' + kx = F_0 \cos\omega t$ with the given numerical values of m, c, k, and F_0. We give first the equations in A and B that result upon collection of coefficients of $\cos\omega t$ and $\sin\omega t$, and then the values of $A(\omega)$ and $B(\omega)$ that we get by solving these equations. Finally, $C = \sqrt{A^2 + B^2}$ gives the amplitude of the resulting forced oscillations as a function of the forcing frequency ω.

15. $(2-\omega^2)A + 2\omega B = 2, \quad -2\omega A + (2-\omega^2)B = 0$

$A = 2(2-\omega^2)/(4+\omega^4), \quad B = 4\omega/(4+\omega^4)$

$C(\omega) = 2/\sqrt{4+\omega^4}$ begins with $C(0) = 1$ and steadily decreases as ω increases. Hence there is no practical resonance frequency.

17. $(45-\omega^2)A + 6\omega B = 50, \quad -6\omega A + (45-\omega^2)B = 0$

$A = 50(45-\omega^2)/(2025 - 54\omega^2 + \omega^4), \quad B = 300\omega/(2025 - 54\omega^2 + \omega^4)$

$C(\omega) = 50/\sqrt{2025 - 54\omega^2 + \omega^4}$ so, to find its maximum value, we calculate the derivative

$$C'(\omega) = \frac{-100\,\omega(-27+\omega^2)}{(2025 - 54\omega^2 + \omega^4)^{3/2}}.$$

Hence the practical resonance frequency (where the derivative vanishes) is $\omega = \sqrt{27} = 3\sqrt{3}$.

19. $m = 100/32$ slugs and $k = 1200$ lb/ft, so the critical frequency is $\omega_0 = \sqrt{k/m} = \sqrt{384}$ rad/sec $= \sqrt{384}/2\pi \approx 3.12$ Hz.

21. If θ is the angular displacement from the vertical, then the (essentially horizontal) displacement of the mass is $x = L\theta$, so twice its total energy (KE + PE) is

$$m(x')^2 + kx^2 + 2mgh = mL^2(\theta')^2 + kL^2\theta^2 + 2mgL(1 - \cos\theta) = C.$$

Differentiation, substitution of $\theta \approx \sin\theta$, and simplification yields

$$\theta'' + (k/m + g/L)\theta = 0$$

so
$$\omega_0 = \sqrt{k/m + g/L}.$$

23. **(a)** In ft-lb-sec units we have $m = 1000$ and $k = 10000$, so $\omega_0 = \sqrt{10}$ rad/sec ≈ 0.50 Hz.

(b) We are given that $\omega = 2\pi/2.25 \approx 2.79$ rad/sec, and the equation $mx'' + kx = F(t)$ simplifies to

$$x'' + 10x = (1/4)\omega^2 \sin \omega t.$$

When we substitute $x(t) = A \sin \omega t$ we find that the amplitude is

$$A = \omega^2 / 4(10 - \omega^2) \approx 0.8854 \text{ ft} \approx 10.63 \text{ in.}$$

25. Substitution of the trial solution $x = A\cos\omega t + B\sin\omega t$ in the differential equation, and then collection of coefficients as usual yields the equations

$$(k - m\omega^2)A + (c\omega)B = 0, \quad -(c\omega)A + (k - m\omega^2)B = F_0$$

with coefficient determinant $\Delta = (k - m\omega^2)^2 + (c\omega)^2$ and solution $A = -(c\omega)F_0/\Delta$, $B = (k - m\omega^2)F_0/\Delta$. Hence

$$x(t) = \frac{F_0}{\sqrt{\Delta}}\left[\frac{k - m\omega^2}{\sqrt{\Delta}}\sin\omega t - \frac{c\omega}{\sqrt{\Delta}}\cos\omega t\right] = C\sin(\omega t - \alpha),$$

where $C = F_0/\sqrt{\Delta}$ and $\sin\alpha = c\omega/\sqrt{\Delta}$, $\cos\alpha = (k - m\omega^2)/\sqrt{\Delta}$.

27. The derivative of $C(\omega) = F_0/\sqrt{(k - m\omega^2)^2 + (c\omega)^2}$ is given by

$$C'(\omega) = -\frac{\omega F_0}{2} \cdot \frac{(c^2 - 2km) + 2(m\omega)^2}{\left[(k - m\omega^2)^2 + (c\omega)^2\right]^{3/2}}.$$

(a) Therefore, if $c^2 \geq 2km$, it is clear from the numerator that $C'(\omega) < 0$ for all ω, so $C(\omega)$ steadily decreases as ω increases.

(b) But if $c^2 < 2km$, then the numerator (and hence $C'(\omega)$) vanishes when $\omega = \omega_m = \sqrt{k/m - c^2/2m^2} < \sqrt{k/m} = \omega_0$. Calculation then shows that

$$C''(\omega_m) = \frac{16F_0 m^3 (c^2 - 2km)}{c^3 (4km - c^2)^{3/2}} < 0,$$

so it follows from the second-derivative test that $C(\omega_m)$ is a local maximum value.

29. We need only substitute $E_0 = ac\omega$ and $F_0 = ak$ in the result of Problem 26.

CHAPTER 6

EIGENVALUES AND EIGENVECTORS

SECTION 6.1

INTRODUCTION TO EIGENVALUES

In each of Problems 1–32 we first list the characteristic polynomial $p(\lambda) = |\mathbf{A} - \lambda\mathbf{I}|$ of the given matrix $\mathbf{A}$, and then the roots of $p(\lambda)$ — which are the eigenvalues of $\mathbf{A}$. All of the eigenvalues that appear in Problems 1–26 are integers, so each characteristic polynomial factors readily. For each eigenvalue λ_j of the matrix $\mathbf{A}$, we determine the associate eigenvector(s) by finding a basis for the solution space of the linear system $(\mathbf{A} - \lambda_j\mathbf{I})\mathbf{v} = \mathbf{0}$. We write this linear system in scalar form in terms of the components of $\mathbf{v} = \begin{bmatrix} a & b & \cdots \end{bmatrix}^T$. In most cases an associated eigenvector is then apparent. If $\mathbf{A}$ is a 2×2 matrix, for instance, then our two scalar equations will be multiples one of the other, so we can substitute a convenient numerical value for the first component a of $\mathbf{v}$ and then solve either equation for the second component b (or vice versa).

1. Characteristic polynomial: $p(\lambda) = \lambda^2 - 5\lambda + 6 = (\lambda - 2)(\lambda - 3)$

   Eigenvalues: $\lambda_1 = 2, \quad \lambda_2 = 3$

   With $\lambda_1 = 2$:
   $\begin{aligned} 2a - 2b &= 0 \\ a - b &= 0 \end{aligned}$
   $\quad \mathbf{v}_1 = \begin{bmatrix} 1 \\ 1 \end{bmatrix}$

   With $\lambda_2 = 3$:
   $\begin{aligned} a - 2b &= 0 \\ a - 2b &= 0 \end{aligned}$
   $\quad \mathbf{v}_2 = \begin{bmatrix} 2 \\ 1 \end{bmatrix}$

3. Characteristic polynomial: $p(\lambda) = \lambda^2 - 7\lambda + 10 = (\lambda - 2)(\lambda - 5)$

   Eigenvalues: $\lambda_1 = 2, \quad \lambda_2 = 5$

   With $\lambda_1 = 2$:
   $\begin{aligned} 6a - 6b &= 0 \\ 3a - 3b &= 0 \end{aligned}$
   $\quad \mathbf{v}_1 = \begin{bmatrix} 1 \\ 1 \end{bmatrix}$

   With $\lambda_2 = 5$:
   $\begin{aligned} 3a - 6b &= 0 \\ 3a - 6b &= 0 \end{aligned}$
   $\quad \mathbf{v}_2 = \begin{bmatrix} 2 \\ 1 \end{bmatrix}$

5. Characteristic polynomial: $p(\lambda) = \lambda^2 - 5\lambda + 4 = (\lambda - 1)(\lambda - 4)$

Eigenvalues: $\lambda_1 = 1$, $\lambda_2 = 4$

With $\lambda_1 = 1$: $\left.\begin{matrix} 9a - 9b = 0 \\ 6a - 6b = 0 \end{matrix}\right\}$ $\quad \mathbf{v}_1 = \begin{bmatrix} 1 \\ 1 \end{bmatrix}$

With $\lambda_2 = 4$: $\left.\begin{matrix} 6a - 9b = 0 \\ 6a - 9b = 0 \end{matrix}\right\}$ $\quad \mathbf{v}_2 = \begin{bmatrix} 3 \\ 2 \end{bmatrix}$

7. Characteristic polynomial: $p(\lambda) = \lambda^2 - 6\lambda + 8 = (\lambda - 2)(\lambda - 4)$

Eigenvalues: $\lambda_1 = 2$, $\lambda_2 = 4$

With $\lambda_1 = 2$: $\left.\begin{matrix} 8a - 8b = 0 \\ 6a - 6b = 0 \end{matrix}\right\}$ $\quad \mathbf{v}_1 = \begin{bmatrix} 1 \\ 1 \end{bmatrix}$

With $\lambda_2 = 4$: $\left.\begin{matrix} 6a - 8b = 0 \\ 6a - 8b = 0 \end{matrix}\right\}$ $\quad \mathbf{v}_2 = \begin{bmatrix} 4 \\ 3 \end{bmatrix}$

9. Characteristic polynomial: $p(\lambda) = \lambda^2 - 7\lambda + 12 = (\lambda - 3)(\lambda - 4)$

Eigenvalues: $\lambda_1 = 3$, $\lambda_2 = 4$

With $\lambda_1 = 3$: $\left.\begin{matrix} 5a - 10b = 0 \\ 2a - 4b = 0 \end{matrix}\right\}$ $\quad \mathbf{v}_1 = \begin{bmatrix} 2 \\ 1 \end{bmatrix}$

With $\lambda_2 = 4$: $\left.\begin{matrix} 4a - 10b = 0 \\ 2a - 5b = 0 \end{matrix}\right\}$ $\quad \mathbf{v}_2 = \begin{bmatrix} 5 \\ 2 \end{bmatrix}$

11. Characteristic polynomial: $p(\lambda) = \lambda^2 - 9\lambda + 20 = (\lambda - 4)(\lambda - 5)$

Eigenvalues: $\lambda_1 = 4$, $\lambda_2 = 5$

With $\lambda_1 = 4$: $\left.\begin{matrix} 15a - 10b = 0 \\ 21a - 14b = 0 \end{matrix}\right\}$ $\quad \mathbf{v}_1 = \begin{bmatrix} 2 \\ 3 \end{bmatrix}$

With $\lambda_2 = 5$: $\left.\begin{matrix} 14a - 10b = 0 \\ 21a - 15b = 0 \end{matrix}\right\}$ $\quad \mathbf{v}_2 = \begin{bmatrix} 5 \\ 7 \end{bmatrix}$

13. Characteristic polynomial: $p(\lambda) = -\lambda^3 + 3\lambda^2 - 2\lambda = -\lambda(\lambda - 1)(\lambda - 2)$

Eigenvalues: $\lambda_1 = 0$, $\lambda_2 = 1$, $\lambda_3 = 2$

With $\lambda_1 = 0$: $\begin{aligned} 2a &= 0 \\ 2a - 2b - c &= 0 \\ -2a + 6b + 3c &= 0 \end{aligned}$ $\mathbf{v}_1 = \begin{bmatrix} 0 \\ -1 \\ 2 \end{bmatrix}$

With $\lambda_2 = 1$: $\begin{aligned} a &= 0 \\ 2a - 3b - c &= 0 \\ -2a + 6b + 2c &= 0 \end{aligned}$ $\mathbf{v}_2 = \begin{bmatrix} 0 \\ -1 \\ 3 \end{bmatrix}$

With $\lambda_3 = 2$: $\begin{aligned} 0 &= 0 \\ 2a - 4b - c &= 0 \\ -2a + 6b + c &= 0 \end{aligned}$ $\mathbf{v}_3 = \begin{bmatrix} 1 \\ 0 \\ 2 \end{bmatrix}$

15. Characteristic polynomial: $p(\lambda) = -\lambda^3 + 3\lambda^2 - 2\lambda = -\lambda(\lambda-1)(\lambda-2)$

Eigenvalues: $\lambda_1 = 0$, $\lambda_2 = 1$, $\lambda_3 = 2$

With $\lambda_1 = 0$: $\begin{aligned} 2a - 2b &= 0 \\ 2a - 2b - c &= 0 \\ -2a + 2b + 3c &= 0 \end{aligned}$ $\mathbf{v}_1 = \begin{bmatrix} 1 \\ 1 \\ 0 \end{bmatrix}$

With $\lambda_2 = 1$: $\begin{aligned} a - 2b &= 0 \\ 2a - 3b - c &= 0 \\ -2a + 2b + 2c &= 0 \end{aligned}$ $\mathbf{v}_2 = \begin{bmatrix} 2 \\ 1 \\ 1 \end{bmatrix}$

With $\lambda_3 = 2$: $\begin{aligned} -2b &= 0 \\ 2a - 4b - c &= 0 \\ -2a + 2b + c &= 0 \end{aligned}$ $\mathbf{v}_3 = \begin{bmatrix} 1 \\ 0 \\ 2 \end{bmatrix}$

17. Characteristic polynomial: $p(\lambda) = -\lambda^3 + 6\lambda^2 - 11\lambda + 6 = -(\lambda-1)(\lambda-2)(\lambda-3)$

Eigenvalues: $\lambda_1 = 1$, $\lambda_2 = 2$, $\lambda_3 = 3$

With $\lambda_1 = 1$: $\begin{aligned} 2a + 5b - 2c &= 0 \\ b &= 0 \\ 2b &= 0 \end{aligned}$ $\mathbf{v}_1 = \begin{bmatrix} 1 \\ 0 \\ 1 \end{bmatrix}$

With $\lambda_2 = 2$: $\begin{aligned} a + 5b - 2c &= 0 \\ 0 &= 0 \\ 2b - c &= 0 \end{aligned}$ $\mathbf{v}_2 = \begin{bmatrix} -1 \\ 1 \\ 2 \end{bmatrix}$

With $\lambda_3 = 3$: $\left.\begin{array}{r} 5b - 2c = 0 \\ -b = 0 \\ 2b - 2c = 0 \end{array}\right\}$ $\mathbf{v}_3 = \begin{bmatrix} 1 \\ 0 \\ 0 \end{bmatrix}$

19. Characteristic polynomial: $p(\lambda) = -\lambda^3 + 5\lambda^2 - 7\lambda + 3 = -(\lambda-1)^2(\lambda-3)$

Eigenvalues: $\lambda_1 = \lambda_2 = 1, \quad \lambda_3 = 3$

With $\lambda_1 = 1$: $\left.\begin{array}{r} 2a + 6b - 2c = 0 \\ 0 = 0 \\ 0 = 0 \end{array}\right\}$ $\mathbf{v}_1 = \begin{bmatrix} 1 \\ 0 \\ 1 \end{bmatrix}, \quad \mathbf{v}_2 = \begin{bmatrix} -3 \\ 1 \\ 0 \end{bmatrix}$

The eigenspace of $\lambda_1 = 1$ is 2-dimensional. We get the eigenvector $\mathbf{v}_1$ with $b = 0$, $c = 1$, and the eigenvector $\mathbf{v}_2$ with $b = 1$, $c = 0$.

With $\lambda_3 = 3$: $\left.\begin{array}{r} 6b - 2c = 0 \\ -2b = 0 \\ -2c = 0 \end{array}\right\}$ $\mathbf{v}_3 = \begin{bmatrix} 1 \\ 0 \\ 0 \end{bmatrix}$

21. Characteristic polynomial: $p(\lambda) = -\lambda^3 + 5\lambda^2 - 8\lambda + 4 = -(\lambda-1)(\lambda-2)^2$

Eigenvalues: $\lambda_1 = 1, \quad \lambda_2 = \lambda_3 = 2$

With $\lambda_1 = 1$: $\left.\begin{array}{r} 3a - 3b + c = 0 \\ 2a - 2b + c = 0 \\ c = 0 \end{array}\right\}$ $\mathbf{v}_1 = \begin{bmatrix} 1 \\ 1 \\ 0 \end{bmatrix}$

With $\lambda_2 = 2$: $\left.\begin{array}{r} 2a - 3b + c = 0 \\ 2a - 3b + c = 0 \\ 0 = 0 \end{array}\right\}$ $\mathbf{v}_2 = \begin{bmatrix} 3 \\ 2 \\ 0 \end{bmatrix} \quad \mathbf{v}_3 = \begin{bmatrix} -1 \\ 0 \\ 2 \end{bmatrix}$

The eigenspace of $\lambda_2 = 2$ is 2-dimensional. We get the eigenvector $\mathbf{v}_2$ with $b = 2$, $c = 0$, and the eigenvector $\mathbf{v}_3$ with $b = 0$, $c = 2$.

23. Characteristic polynomial: $p(\lambda) = (\lambda-1)(\lambda-2)(\lambda-3)(\lambda-4)$

Eigenvalues: $\lambda_1 = 1, \quad \lambda_2 = 2, \quad \lambda_3 = 3, \quad \lambda_4 = 4$

With $\lambda_1 = 1$: $\left.\begin{array}{r} 2b + 2c + 2d = 0 \\ b + 2c + 2d = 0 \\ 2c + 2d = 0 \\ 3d = 0 \end{array}\right\}$ $\mathbf{v}_1 = \begin{bmatrix} 1 \\ 0 \\ 0 \\ 0 \end{bmatrix}$

With $\lambda_2 = 2$:
$$\left.\begin{array}{r}-a+2b+2c+2d = 0\\ 2c+2d = 0\\ c+2d = 0\\ 2d = 0\end{array}\right\} \quad \mathbf{v}_2 = \begin{bmatrix}2\\1\\0\\0\end{bmatrix}$$

With $\lambda_3 = 3$:
$$\left.\begin{array}{r}-2a+2b+2c+2d = 0\\ b+2c+2d = 0\\ 2d = 0\\ d = 0\end{array}\right\} \quad \mathbf{v}_3 = \begin{bmatrix}3\\2\\1\\0\end{bmatrix}$$

With $\lambda_4 = 4$:
$$\left.\begin{array}{r}-3a+2b+2c+2d = 0\\ -2b+2c+2d = 0\\ -c+2d = 0\\ 0 = 0\end{array}\right\} \quad \mathbf{v}_4 = \begin{bmatrix}4\\3\\2\\1\end{bmatrix}$$

25. Characteristic polynomial: $p(\lambda) = (\lambda-1)^2(\lambda-2)^2$

Eigenvalues: $\lambda_1 = \lambda_2 = 1, \quad \lambda_3 = \lambda_4 = 2$

With $\lambda_1 = 1$:
$$\left.\begin{array}{r}c = 0\\ c = 0\\ c = 0\\ d = 0\end{array}\right\} \quad \mathbf{v}_1 = \begin{bmatrix}1\\0\\0\\0\end{bmatrix}, \quad \mathbf{v}_2 = \begin{bmatrix}0\\1\\0\\0\end{bmatrix}$$

The eigenspace of $\lambda_1 = 1$ is 2-dimensional. We note that $c = d = 0$, but a and b are arbitrary.

With $\lambda_3 = 2$:
$$\left.\begin{array}{r}-a+c = 0\\ -b+c = 0\\ 0 = 0\\ 0 = 0\end{array}\right\} \quad \mathbf{v}_3 = \begin{bmatrix}0\\0\\0\\1\end{bmatrix}, \quad \mathbf{v}_4 = \begin{bmatrix}1\\1\\1\\0\end{bmatrix}$$

The eigenspace of $\lambda_3 = 2$ is 2-dimensional. We get the eigenvector $\mathbf{v}_3$ with $b = 0$, $c = 1$, and the eigenvector $\mathbf{v}_4$ with $b = 1$, $c = 0$.

27. Characteristic polynomial: $p(\lambda) = \lambda^2 + 1$

Eigenvalues: $\lambda_1 = -i, \quad \lambda_2 = +i$

With $\lambda_1 = -i$:
$$\left.\begin{array}{r}ia+b = 0\\ -a+ib = 0\end{array}\right\} \quad \mathbf{v}_1 = \begin{bmatrix}i\\1\end{bmatrix}$$

With $\lambda_2 = +i$: $\begin{matrix} -ia+b = 0 \\ -a-ib = 0 \end{matrix}$ $\mathbf{v}_2 = \begin{bmatrix} -i \\ 1 \end{bmatrix}$

29. Characteristic polynomial: $p(\lambda) = \lambda^2 + 36$

   Eigenvalues: $\lambda_1 = -6i, \quad \lambda_2 = +6i$

   With $\lambda_1 = -6i$: $\begin{matrix} 6ia - 3b = 0 \\ 12a + 6ib = 0 \end{matrix}$ $\mathbf{v}_1 = \begin{bmatrix} -i \\ 2 \end{bmatrix}$

   With $\lambda_2 = +6i$: $\begin{matrix} -6ia - 3b = 0 \\ 12a - 6ib = 0 \end{matrix}$ $\mathbf{v}_2 = \begin{bmatrix} i \\ 2 \end{bmatrix}$

31. Characteristic polynomial: $p(\lambda) = \lambda^2 + 144$

    Eigenvalues: $\lambda_1 = -12i, \quad \lambda_2 = +12i$

    With $\lambda_1 = -12i$: $\begin{matrix} 12ia + 24b = 0 \\ -6a + 12ib = 0 \end{matrix}$ $\mathbf{v}_1 = \begin{bmatrix} 2i \\ 1 \end{bmatrix}$

    With $\lambda_2 = +12i$: $\begin{matrix} -12ia + 24b = 0 \\ -6a - 12ib = 0 \end{matrix}$ $\mathbf{v}_2 = \begin{bmatrix} -2i \\ 1 \end{bmatrix}$

33. If $\mathbf{Av} = \lambda\mathbf{v}$ and we assume that $\mathbf{A}^{n-1}\mathbf{v} = \lambda^{n-1}\mathbf{v}$ — meaning that λ^{n-1} is an eigenvalue of $\mathbf{A}^{n-1}$ with associated eigenvector $\mathbf{v}$, then multiplication by $\mathbf{A}$ yields

$$\mathbf{A}^n\mathbf{v} = \mathbf{A} \cdot \mathbf{A}^{n-1}\mathbf{v} = \mathbf{A} \cdot \lambda^{n-1}\mathbf{v} = \lambda^{n-1} \cdot \mathbf{Av} = \lambda^{n-1} \cdot \lambda\mathbf{v} = \lambda^n\mathbf{v}.$$

Thus λ^n is an eigenvalue of the matrix $\mathbf{A}^n$ with associated eigenvector $\mathbf{v}$.

35. **(a)** Note first that $(\mathbf{A} - \lambda\mathbf{I})^T = (\mathbf{A}^T - \lambda\mathbf{I})$ because $\mathbf{I}^T = \mathbf{I}$. Since the determinant of a square matrix equals the determinant of its transpose, it follows that

$$|\mathbf{A} - \lambda\mathbf{I}| = |\mathbf{A}^T - \lambda\mathbf{I}|.$$

Thus the matrices $\mathbf{A}$ and $\mathbf{A}^T$ have the same characteristic polynomial, and therefore have the same eigenvalues.

(b) Consider the matrix $\mathbf{A} = \begin{bmatrix} 1 & 0 \\ 1 & 1 \end{bmatrix}$ with characteristic equation $(\lambda - 1)^2 = 0$ and the single eigenvalue $\lambda = 1$. Then $\mathbf{A} - \mathbf{I} = \begin{bmatrix} 0 & 0 \\ 1 & 0 \end{bmatrix}$ and it follows that the only associated eigenvector is a multiple of $[0\ 1]^T$. The transpose $\mathbf{A}^T = \begin{bmatrix} 1 & 1 \\ 0 & 1 \end{bmatrix}$ has the same characteristic equation and eigenvalue, but we see similarly that its only eigenvector is a multiple of $[1\ 0]^T$. Thus $\mathbf{A}$ and $\mathbf{A}^T$ have the same eigenvalue but different eigenvectors.

37. If $|\mathbf{A} - \lambda\mathbf{I}| = (-1)^n \lambda^n + c_{n-1}\lambda^{n-1} + \cdots + c_1\lambda + c_0$, then substitution of $\lambda = 0$ yields $c_0 = |\mathbf{A} - 0\mathbf{I}| = |\mathbf{A}|$ for the constant term in the characteristic polynomial.

39. If the characteristic equation of the $n \times n$ matrix $\mathbf{A}$ with eigenvalues $\lambda_1, \lambda_2, \cdots, \lambda_n$ (not necessarily distinct) is written in the factored form

$$(\lambda - \lambda_1)(\lambda - \lambda_2) \cdots (\lambda - \lambda_n) = 0,$$

then it should be clear that upon multiplying out the factors the coefficient of λ^{n-1} will be $-(\lambda_1 + \lambda_2 + \cdots + \lambda_n)$. But according to Problem 38, this coefficient also equals $-(\text{trace } \mathbf{A})$. Therefore

$$\lambda_1 + \lambda_2 + \cdots + \lambda_n = \text{trace } \mathbf{A} = a_{11} + a_{22} + \cdots a_{nn}.$$

41. We find that trace $\mathbf{A} = 8$ and det $\mathbf{A} = -60$, so the characteristic polynomial of the given matrix $\mathbf{A}$ is

$$p(\lambda) = \lambda^4 - 8\lambda^3 + c_2\lambda^2 + c_1\lambda - 60.$$

Substitution of

$$\lambda = 1,\ p(1) = \det(\mathbf{A} - \mathbf{I}) = -24 \text{ and } \lambda = -1,\ p(-1) = \det(\mathbf{A} + \mathbf{I}) = -72$$

yields the equations

$$c_2 + c_1 = 43,\quad c_2 - c_1 = -21$$

that we solve readily for $c_1 = 32,\ c_2 = 11$. Hence the characteristic equation of $\mathbf{A}$ is

$$\lambda^4 - 8\lambda^3 + 11\lambda^2 + 32\lambda - 60 = 0.$$

Trying $\lambda = \pm 1, \pm 2$ in turn, we discover the eigenvalues $\lambda_1 = -2$ and $\lambda_2 = 2$. Then division of the quartic by $(\lambda^2 - 4)$ yields

$$\lambda^2 - 8\lambda + 15 = (\lambda - 3)(\lambda - 5),$$

so the other two eigenvalues are $\lambda_3 = 3$ and $\lambda_4 = 5$. We proceed to find the eigenvectors associated with these four eigenvalues.

With $\lambda_1 = -2$:
$$\left.\begin{array}{r}24a-9b-8c-8d=0\\10a-5b-14c+2d=0\\10a+10c-10d=0\\29a-9b-3c-13d=0\end{array}\right\} \rightarrow \left.\begin{array}{r}a-(1/2)d=0\\b=0\\c-(1/2)d=0\\0=0\end{array}\right\} \quad \mathbf{v}_1 = \begin{bmatrix}1\\0\\1\\2\end{bmatrix}$$

With $\lambda_2 = 2$:
$$\left.\begin{array}{r}20a-9b-8c-8d=0\\10a-9b-14c+2d=0\\10a+6c-10d=0\\29a-9b-3c-17d=0\end{array}\right\} \rightarrow \left.\begin{array}{r}a-d=0\\b-(4/3)d=0\\c=0\\0=0\end{array}\right\} \quad \mathbf{v}_2 = \begin{bmatrix}3\\4\\0\\3\end{bmatrix}$$

With $\lambda_3 = 3$:
$$\left.\begin{array}{r}19a-9b-8c-8d=0\\10a-10b-14c+2d=0\\10a+5c-10d=0\\29a-9b-3c-18d=0\end{array}\right\} \rightarrow \left.\begin{array}{r}a-(3/4)d=0\\b-(1/4)d=0\\c-(1/2)d=0\\0=0\end{array}\right\} \quad \mathbf{v}_3 = \begin{bmatrix}3\\1\\2\\4\end{bmatrix}$$

With $\lambda_4 = 5$:
$$\left.\begin{array}{r}17a-9b-8c-8d=0\\10a-12b-14c+2d=0\\10a+3c-10d=0\\29a-9b-3c-20d=0\end{array}\right\} \rightarrow \left.\begin{array}{r}a-d=0\\b-d=0\\c=0\\0=0\end{array}\right\} \quad \mathbf{v}_4 = \begin{bmatrix}1\\1\\0\\1\end{bmatrix}$$

SECTION 6.2

DIAGONALIZATION OF MATRICES

In Problems 1–28 we first find the eigenvalues and associated eigenvectors of the given $n \times n$ matrix $\mathbf{A}$. If $\mathbf{A}$ has n linearly independent eigenvectors, then we can proceed to set up the desired diagonalizing matrix $\mathbf{P} = [\mathbf{v}_1 \; \mathbf{v}_2 \; \cdots \; \mathbf{v}_n]$ and diagonal matrix $\mathbf{D}$ such that $\mathbf{P}^{-1}\mathbf{AP} = \mathbf{D}$. If you write the eigenvalues in a different order on the diagonal of $\mathbf{D}$, then naturally the eigenvector columns of $\mathbf{P}$ must be rearranged in the same order.

1. Characteristic polynomial: $p(\lambda) = \lambda^2 - 4\lambda + 3 = (\lambda-1)(\lambda-3)$

   Eigenvalues: $\lambda_1 = 1, \; \lambda_2 = 3$

With $\lambda_1 = 1$: $\begin{matrix} 4a-4b = 0 \\ 2a-2b = 0 \end{matrix}\Big\}$ $\mathbf{v}_1 = \begin{bmatrix} 1 \\ 1 \end{bmatrix}$

With $\lambda_2 = 3$: $\begin{matrix} 2a-4b = 0 \\ 2a-4b = 0 \end{matrix}\Big\}$ $\mathbf{v}_2 = \begin{bmatrix} 2 \\ 1 \end{bmatrix}$

$\mathbf{P} = \begin{bmatrix} 1 & 2 \\ 1 & 1 \end{bmatrix}, \quad \mathbf{D} = \begin{bmatrix} 1 & 0 \\ 0 & 3 \end{bmatrix}$

3.   Characteristic polynomial: $p(\lambda) = \lambda^2 - 5\lambda + 6 = (\lambda-2)(\lambda-3)$

Eigenvalues: $\lambda_1 = 2, \quad \lambda_2 = 3$

With $\lambda_1 = 2$: $\begin{matrix} 3a-3b = 0 \\ 2a-2b = 0 \end{matrix}\Big\}$ $\mathbf{v}_1 = \begin{bmatrix} 1 \\ 1 \end{bmatrix}$

With $\lambda_2 = 3$: $\begin{matrix} 2a-3b = 0 \\ 2a-3b = 0 \end{matrix}\Big\}$ $\mathbf{v}_2 = \begin{bmatrix} 3 \\ 2 \end{bmatrix}$

$\mathbf{P} = \begin{bmatrix} 1 & 3 \\ 1 & 2 \end{bmatrix}, \quad \mathbf{D} = \begin{bmatrix} 2 & 0 \\ 0 & 3 \end{bmatrix}$

5.   Characteristic polynomial: $p(\lambda) = \lambda^2 - 4\lambda + 3 = (\lambda-1)(\lambda-3)$

Eigenvalues: $\lambda_1 = 1, \quad \lambda_2 = 3$

With $\lambda_1 = 1$: $\begin{matrix} 8a-8b = 0 \\ 6a-6b = 0 \end{matrix}\Big\}$ $\mathbf{v}_1 = \begin{bmatrix} 1 \\ 1 \end{bmatrix}$

With $\lambda_2 = 3$: $\begin{matrix} 6a-8b = 0 \\ 6a-8b = 0 \end{matrix}\Big\}$ $\mathbf{v}_2 = \begin{bmatrix} 4 \\ 3 \end{bmatrix}$

$\mathbf{P} = \begin{bmatrix} 1 & 4 \\ 1 & 3 \end{bmatrix}, \quad \mathbf{D} = \begin{bmatrix} 1 & 0 \\ 0 & 3 \end{bmatrix}$

7.   Characteristic polynomial: $p(\lambda) = \lambda^2 - 3\lambda + 2 = (\lambda-1)(\lambda-2)$

Eigenvalues: $\lambda_1 = 1, \quad \lambda_2 = 2$

With $\lambda_1 = 1$: $\begin{matrix} 5a-10b = 0 \\ 2a-4b = 0 \end{matrix}\Big\}$ $\mathbf{v}_1 = \begin{bmatrix} 2 \\ 1 \end{bmatrix}$

With $\lambda_2 = 2$: $\begin{matrix} 4a-10b = 0 \\ 2a-5b = 0 \end{matrix}\Big\}$ $\mathbf{v}_2 = \begin{bmatrix} 5 \\ 2 \end{bmatrix}$

$$\mathbf{P} = \begin{bmatrix} 2 & 5 \\ 1 & 2 \end{bmatrix}, \qquad \mathbf{D} = \begin{bmatrix} 1 & 0 \\ 0 & 2 \end{bmatrix}$$

9. Characteristic polynomial: $p(\lambda) = \lambda^2 - 2\lambda + 1 = (\lambda - 1)^2$

Eigenvalues: $\lambda_1 = 1, \quad \lambda_2 = 1$

With $\lambda_1 = 1$: $\quad \begin{aligned} 4a - 2b &= 0 \\ 2a - b &= 0 \end{aligned} \bigg\} \qquad \mathbf{v}_1 = \begin{bmatrix} 2 \\ 1 \end{bmatrix}$

Because the given matrix $\mathbf{A}$ has only the single eigenvector $\mathbf{v}_1$, it is not diagonalizable.

11. Characteristic polynomial: $p(\lambda) = \lambda^2 - 4\lambda + 4 = (\lambda - 2)^2$

Eigenvalues: $\lambda_1 = 2, \quad \lambda_2 = 2$

With $\lambda_1 = 2$: $\quad \begin{aligned} 3a + b &= 0 \\ -9a - 3b &= 0 \end{aligned} \bigg\} \qquad \mathbf{v}_1 = \begin{bmatrix} -1 \\ 3 \end{bmatrix}$

Because the given matrix $\mathbf{A}$ has only the single eigenvector $\mathbf{v}_1$, it is not diagonalizable.

13. Characteristic polynomial: $p(\lambda) = -\lambda^3 + 5\lambda^2 - 8\lambda + 4 = -(\lambda - 1)(\lambda - 2)^2$

Eigenvalues: $\lambda_1 = 1, \quad \lambda_2 = 2, \quad \lambda_3 = 2$

With $\lambda_1 = 1$: $\quad \begin{aligned} 3b &= 0 \\ b &= 0 \\ c &= 0 \end{aligned} \bigg\} \qquad \mathbf{v}_1 = \begin{bmatrix} 1 \\ 0 \\ 0 \end{bmatrix}$

With $\lambda_2 = 2$: $\quad \begin{aligned} 3b - a &= 0 \\ 0 &= 0 \\ 0 &= 0 \end{aligned} \bigg\} \qquad \mathbf{v}_2 = \begin{bmatrix} 0 \\ 0 \\ 1 \end{bmatrix}, \quad \mathbf{v}_3 = \begin{bmatrix} 3 \\ 1 \\ 0 \end{bmatrix}$

The eigenspace of $\lambda_2 = 2$ is 2-dimensional. We get the eigenvector $\mathbf{v}_2$ with $b = 0, c = 1$, and the eigenvector $\mathbf{v}_3$ with $b = 1, c = 0$.

$$\mathbf{P} = \begin{bmatrix} 1 & 0 & 3 \\ 0 & 0 & 1 \\ 0 & 1 & 0 \end{bmatrix}, \qquad \mathbf{D} = \begin{bmatrix} 1 & 0 & 0 \\ 0 & 2 & 0 \\ 0 & 0 & 2 \end{bmatrix}$$

15. Characteristic polynomial: $p(\lambda) = -\lambda^3 + 2\lambda^2 - \lambda = -\lambda(\lambda - 1)^2$

Eigenvalues: $\lambda_1 = 0, \quad \lambda_2 = 1, \quad \lambda_3 = 1$

With $\lambda_1 = 0$: $\begin{aligned} 3a-3b+c &= 0 \\ 2a-2b+c &= 0 \\ c &= 0 \end{aligned}\Bigg\}$ $\mathbf{v}_1 = \begin{bmatrix} 1 \\ 1 \\ 0 \end{bmatrix}$

With $\lambda_2 = 1$: $\begin{aligned} 2a-3b+c &= 0 \\ 2a-3b+c &= 0 \\ 0 &= 0 \end{aligned}\Bigg\}$ $\mathbf{v}_2 = \begin{bmatrix} -1 \\ 0 \\ 2 \end{bmatrix}$, $\mathbf{v}_3 = \begin{bmatrix} 3 \\ 2 \\ 0 \end{bmatrix}$

The eigenspace of $\lambda_2 = 2$ is 2-dimensional. We get the eigenvector $\mathbf{v}_2$ with $b=0$, $c=2$, and the eigenvector $\mathbf{v}_3$ with $b=2$, $c=0$.

$$\mathbf{P} = \begin{bmatrix} 1 & -1 & 3 \\ 1 & 0 & 2 \\ 0 & 2 & 0 \end{bmatrix}, \quad \mathbf{D} = \begin{bmatrix} 0 & 0 & 0 \\ 0 & 1 & 0 \\ 0 & 0 & 1 \end{bmatrix}$$

17. Characteristic polynomial: $p(\lambda) = -\lambda^3 + 2\lambda^2 + \lambda - 2 = -(\lambda+1)(\lambda-1)(\lambda-2)$

Eigenvalues: $\lambda_1 = -1$, $\lambda_2 = 1$, $\lambda_3 = 2$

With $\lambda_1 = -1$: $\begin{aligned} 8a-8b+3c &= 0 \\ 6a-6b+3c &= 0 \\ 2a-2b+3c &= 0 \end{aligned}\Bigg\}$ $\mathbf{v}_1 = \begin{bmatrix} 1 \\ 1 \\ 0 \end{bmatrix}$

With $\lambda_2 = 1$: $\begin{aligned} 6a-8b+3c &= 0 \\ 6a-8b+3c &= 0 \\ 2a-2b+c &= 0 \end{aligned}\Bigg\}$ $\mathbf{v}_2 = \begin{bmatrix} -1 \\ 0 \\ 2 \end{bmatrix}$

With $\lambda_3 = 2$: $\begin{aligned} 5a-8b+3c &= 0 \\ 6a-9b+3c &= 0 \\ 2a-2b &= 0 \end{aligned}\Bigg\}$ $\mathbf{v}_3 = \begin{bmatrix} 1 \\ 1 \\ 1 \end{bmatrix}$

$$\mathbf{P} = \begin{bmatrix} 1 & -1 & 1 \\ 1 & 0 & 1 \\ 0 & 2 & 1 \end{bmatrix}, \quad \mathbf{D} = \begin{bmatrix} -1 & 0 & 0 \\ 0 & 1 & 0 \\ 0 & 0 & 2 \end{bmatrix}$$

19. Characteristic polynomial: $p(\lambda) = -\lambda^3 + 6\lambda^2 - 11\lambda + 6 = -(\lambda-1)(\lambda-2)(\lambda-3)$

Eigenvalues: $\lambda_1 = 1$, $\lambda_2 = 2$, $\lambda_3 = 3$

With $\lambda_1 = 1$:
$$\begin{aligned} b-c &= 0 \\ -2a+3b-c &= 0 \\ -4a+4b &= 0 \end{aligned}\Bigg\} \qquad \mathbf{v}_1 = \begin{bmatrix} 1 \\ 1 \\ 1 \end{bmatrix}$$

With $\lambda_2 = 2$:
$$\begin{aligned} -a+b-c &= 0 \\ -2a+2b-c &= 0 \\ -4a+4b-c &= 0 \end{aligned}\Bigg\} \qquad \mathbf{v}_2 = \begin{bmatrix} 1 \\ 1 \\ 0 \end{bmatrix}$$

With $\lambda_3 = 3$:
$$\begin{aligned} -2a+b-c &= 0 \\ -2a+b-c &= 0 \\ -4a+4b-2c &= 0 \end{aligned}\Bigg\} \qquad \mathbf{v}_3 = \begin{bmatrix} -1 \\ 0 \\ 2 \end{bmatrix}$$

$$\mathbf{P} = \begin{bmatrix} 1 & 1 & -1 \\ 1 & 1 & 0 \\ 1 & 0 & 2 \end{bmatrix}, \qquad \mathbf{D} = \begin{bmatrix} 1 & 0 & 0 \\ 0 & 2 & 0 \\ 0 & 0 & 3 \end{bmatrix}$$

21. Characteristic polynomial: $p(\lambda) = -\lambda^3 + 3\lambda^2 - 3\lambda + 1 = -(\lambda-1)^3$

Eigenvalues: $\lambda_1 = 1, \quad \lambda_2 = 1, \quad \lambda_3 = 1$

With $\lambda_1 = 1$:
$$\begin{aligned} b-a &= 0 \\ b-a &= 0 \\ b-a &= 0 \end{aligned}\Bigg\} \qquad \mathbf{v}_1 = \begin{bmatrix} 0 \\ 0 \\ 1 \end{bmatrix}, \quad \mathbf{v}_2 = \begin{bmatrix} 1 \\ 1 \\ 0 \end{bmatrix}$$

The eigenspace of $\lambda_1 = 1$ is 2-dimensional. We get the eigenvector $\mathbf{v}_1$ with $b = 0$, $c = 1$, and the eigenvector $\mathbf{v}_2$ with $b = 1$, $c = 0$. Because the given matrix **A** has only two linearly independent eigenvectors, it is not diagonalizable.

23. Characteristic polynomial: $p(\lambda) = -\lambda^3 + 4\lambda^2 - 5\lambda + 2 = -(\lambda-1)^2(\lambda-2)$

Eigenvalues: $\lambda_1 = 1, \quad \lambda_2 = 1, \quad \lambda_3 = 2$

With $\lambda_1 = 1$:
$$\begin{aligned} -3a+4b-c &= 0 \\ -3a+4b-c &= 0 \\ -a+b &= 0 \end{aligned}\Bigg\} \qquad \mathbf{v}_1 = \begin{bmatrix} 1 \\ 1 \\ 1 \end{bmatrix}$$

With $\lambda_3 = 2$:
$$\begin{aligned} -4a+4b-c &= 0 \\ -3a+3b-c &= 0 \\ -a+b-c &= 0 \end{aligned}\Bigg\} \qquad \mathbf{v}_2 = \begin{bmatrix} 1 \\ 1 \\ 0 \end{bmatrix}$$

The given matrix **A** has only the two linearly independent eigenvectors $\mathbf{v}_1$ and $\mathbf{v}_2$, and therefore is not diagonalizable.

25. Characteristic polynomial: $p(\lambda) = (\lambda+1)^2(\lambda-1)^2$

Eigenvalues: $\lambda_1 = -1, \quad \lambda_2 = -1, \quad \lambda_3 = 1, \quad \lambda_4 = 1$

With $\lambda_1 = -1$:
$$\left.\begin{array}{r} 2a - 2c = 0 \\ 2b - 2c = 0 \\ 0 = 0 \\ 0 = 0 \end{array}\right\} \quad \mathbf{v}_1 = \begin{bmatrix} 0 \\ 0 \\ 0 \\ 1 \end{bmatrix}, \quad \mathbf{v}_2 = \begin{bmatrix} 1 \\ 1 \\ 1 \\ 0 \end{bmatrix}$$

The eigenspace of $\lambda_1 = -1$ is 2-dimensional. We get the eigenvector $\mathbf{v}_1$ with $c = 0$, $d = 1$, and the eigenvector $\mathbf{v}_2$ with $c = 1$, $d = 0$.

With $\lambda_3 = 1$:
$$\left.\begin{array}{r} -2c = 0 \\ -2c = 0 \\ -2c = 0 \\ -2d = 0 \end{array}\right\} \quad \mathbf{v}_3 = \begin{bmatrix} 0 \\ 1 \\ 0 \\ 0 \end{bmatrix}, \quad \mathbf{v}_4 = \begin{bmatrix} 1 \\ 0 \\ 0 \\ 0 \end{bmatrix}$$

The eigenspace of $\lambda_3 = 1$ is also 2-dimensional. We get the eigenvector $\mathbf{v}_3$ with $a = 0$, $b = 1$, and the eigenvector $\mathbf{v}_4$ with $a = 1$, $b = 0$.

$$\mathbf{P} = \begin{bmatrix} 0 & 1 & 0 & 1 \\ 0 & 1 & 1 & 0 \\ 0 & 1 & 0 & 0 \\ 1 & 0 & 0 & 0 \end{bmatrix}, \quad \mathbf{D} = \begin{bmatrix} -1 & 0 & 0 & 0 \\ 0 & -1 & 0 & 0 \\ 0 & 0 & 1 & 0 \\ 0 & 0 & 0 & 1 \end{bmatrix}$$

27. Characteristic polynomial: $p(\lambda) = (\lambda-1)^3(\lambda-2)$

Eigenvalues: $\lambda_1 = 1, \quad \lambda_2 = 1, \quad \lambda_3 = 1, \quad \lambda_4 = 2$

With $\lambda_1 = 1$:
$$\left.\begin{array}{r} b = 0 \\ c = 0 \\ d = 0 \\ d = 0 \end{array}\right\} \quad \mathbf{v}_1 = \begin{bmatrix} 1 \\ 0 \\ 0 \\ 0 \end{bmatrix}$$

The eigenspace of $\lambda_1 = 1$ is 1-dimensional, with only a single associated eigenvector.

With $\lambda_4 = 2$:
$$\left.\begin{array}{r} b - a = 0 \\ c - b = 0 \\ d - c = 0 \\ 0 = 0 \end{array}\right\} \quad \mathbf{v}_4 = \begin{bmatrix} 1 \\ 1 \\ 1 \\ 1 \end{bmatrix}$$

The given matrix $\mathbf{A}$ has only the two linearly independent eigenvectors $\mathbf{v}_1$ and $\mathbf{v}_4$, and therefore is not diagonalizable.

29. If $\mathbf{A}$ is similar to $\mathbf{B}$ and $\mathbf{B}$ is similar to $\mathbf{C}$, so $\mathbf{A} = \mathbf{P}^{-1}\mathbf{B}\mathbf{P}$ and $\mathbf{B} = \mathbf{Q}^{-1}\mathbf{C}\mathbf{Q}$, then

$$\mathbf{A} = \mathbf{P}^{-1}(\mathbf{Q}^{-1}\mathbf{C}\mathbf{Q})\mathbf{P} = (\mathbf{P}^{-1}\mathbf{Q}^{-1})\mathbf{C}(\mathbf{Q}\mathbf{P}) = (\mathbf{Q}\mathbf{P})^{-1}\mathbf{C}(\mathbf{Q}\mathbf{P}) = \mathbf{R}^{-1}\mathbf{C}\mathbf{R}$$

with $\mathbf{R} = \mathbf{Q}\mathbf{P}$, so $\mathbf{A}$ is similar to $\mathbf{C}$.

31. If $\mathbf{A}$ is similar to $\mathbf{B}$ so $\mathbf{A} = \mathbf{P}^{-1}\mathbf{B}\mathbf{P}$ then $\mathbf{A}^{-1} = (\mathbf{P}^{-1}\mathbf{B}\mathbf{P})^{-1} = \mathbf{P}^{-1}\mathbf{B}^{-1}\mathbf{P}$, so $\mathbf{A}^{-1}$ is similar to $\mathbf{B}^{-1}$.

33. If $\mathbf{A}$ and $\mathbf{B}$ are similar with $\mathbf{A} = \mathbf{P}^{-1}\mathbf{B}\mathbf{P}$, then

$$|\mathbf{A}| = |\mathbf{P}^{-1}\mathbf{B}\mathbf{P}| = |\mathbf{P}^{-1}||\mathbf{B}||\mathbf{P}| = |\mathbf{P}|^{-1}|\mathbf{B}||\mathbf{P}| = |\mathbf{B}|.$$

Moreover, by Problem 32 the two matrices have the same eigenvalues, and by Problem 39 in Section 6.1, the trace of a square matrix with real eigenvalues is equal to the sum of those eigenvalues. Therefore trace $\mathbf{A}$ = (eigenvalue sum) = trace $\mathbf{B}$.

35. Three eigenvectors associated with three distinct eigenvalues can be arranged in six different orders as the column vectors of the diagonalizing matrix $\mathbf{P} = \begin{bmatrix} \mathbf{v}_1 & \mathbf{v}_2 & \mathbf{v}_3 \end{bmatrix}^T$.

37. If $\mathbf{A} = \mathbf{P}\mathbf{D}\mathbf{P}^{-1}$ with $\mathbf{P}$ the eigenvector matrix of $\mathbf{A}$ and $\mathbf{D}$ its diagonal matrix of eigenvalues, then $\mathbf{A}^2 = (\mathbf{P}\mathbf{D}\mathbf{P}^{-1})(\mathbf{P}\mathbf{D}\mathbf{P}^{-1}) = \mathbf{P}\mathbf{D}(\mathbf{P}^{-1}\mathbf{P})\mathbf{D}\mathbf{P}^{-1} = \mathbf{P}\mathbf{D}^2\mathbf{P}^{-1}$. Thus the same (eigenvector) matrix $\mathbf{P}$ diagonalizes $\mathbf{A}^2$, but the resulting diagonal (eigenvalue) matrix $\mathbf{D}^2$ is the square of the one for $\mathbf{A}$. The diagonal elements of $\mathbf{D}^2$ are the eigenvalues of $\mathbf{A}^2$ and the diagonal elements of $\mathbf{D}$ are the eigenvalues of $\mathbf{A}$, so the former are the squares of the latter.

39. Let the $n \times n$ matrix $\mathbf{A}$ have $k \leq n$ distinct eigenvalues $\lambda_1, \lambda_2, \cdots, \lambda_k$. Then the definition of algebraic multiplicity and the fact that all solutions of the nth degree polynomial equation $|\mathbf{A} - \lambda\mathbf{I}| = 0$ are real imply that the sum of the multiplicities of the eigenvalues equals n,

$$p_1 + p_2 + \cdots + p_k = n.$$

Now Theorem 4 in this section implies that $\mathbf{A}$ is diagonalizable if and only if

$$q_1 + q_2 + \cdots + q_k = n$$

where q_i denotes the geometric multiplicity of λ_i ($i = 1, 2, \cdots, k$). But, because $p_i \geq q_i$ for each $i = 1, 2, \cdots, k$, the two equations displayed above can both be satisfied if and only if $p_i = q_i$ for each i.

SECTION 6.3

APPLICATIONS INVOLVING POWERS OF MATRICES

In Problems 1–10 we first find the eigensystem of the given matrix $\mathbf{A}$ so as to determine its eigenvector matrix $\mathbf{P}$ and its diagonal eigenvalue matrix $\mathbf{D}$. Then we calculate the matrix power $\mathbf{A}^5 = \mathbf{PD}^5\mathbf{P}^{-1}$.

1. Characteristic polynomial: $p(\lambda) = \lambda^2 - 3\lambda + 2 = (\lambda - 1)(\lambda - 2)$

Eigenvalues: $\lambda_1 = 1, \quad \lambda_2 = 2$

With $\lambda_1 = 1$: $\quad \begin{aligned} 2a - 2b &= 0 \\ a - b &= 0 \end{aligned} \bigg\} \quad \mathbf{v}_1 = \begin{bmatrix} 1 \\ 1 \end{bmatrix}$

With $\lambda_2 = 2$: $\quad \begin{aligned} a - 2b &= 0 \\ a - 2b &= 0 \end{aligned} \bigg\} \quad \mathbf{v}_2 = \begin{bmatrix} 2 \\ 1 \end{bmatrix}$

$$\mathbf{P} = \begin{bmatrix} 1 & 2 \\ 1 & 1 \end{bmatrix}, \quad \mathbf{D} = \begin{bmatrix} 1 & 0 \\ 0 & 2 \end{bmatrix}, \quad \mathbf{P}^{-1} = \begin{bmatrix} -1 & 2 \\ 1 & -1 \end{bmatrix}$$

$$\mathbf{A}^5 = \begin{bmatrix} 1 & 2 \\ 1 & 1 \end{bmatrix}\begin{bmatrix} 1 & 0 \\ 0 & 32 \end{bmatrix}\begin{bmatrix} -1 & 2 \\ 1 & -1 \end{bmatrix} = \begin{bmatrix} 63 & -62 \\ 31 & -30 \end{bmatrix}$$

3. Characteristic polynomial: $p(\lambda) = \lambda^2 - 2\lambda = \lambda(\lambda - 2)$

Eigenvalues: $\lambda_1 = 0, \quad \lambda_2 = 2$

With $\lambda_1 = 0$: $\quad \begin{aligned} 6a - 6b &= 0 \\ 4a - 4b &= 0 \end{aligned} \bigg\} \quad \mathbf{v}_1 = \begin{bmatrix} 1 \\ 1 \end{bmatrix}$

With $\lambda_2 = 2$: $\quad \begin{aligned} 4a - 6b &= 0 \\ 4a - 6b &= 0 \end{aligned} \bigg\} \quad \mathbf{v}_2 = \begin{bmatrix} 3 \\ 2 \end{bmatrix}$

$$\mathbf{P} = \begin{bmatrix} 1 & 3 \\ 1 & 2 \end{bmatrix}, \quad \mathbf{D} = \begin{bmatrix} 0 & 0 \\ 0 & 2 \end{bmatrix}, \quad \mathbf{P}^{-1} = \begin{bmatrix} -2 & 3 \\ 1 & -1 \end{bmatrix}$$

$$\mathbf{A}^5 = \begin{bmatrix} 1 & 3 \\ 1 & 2 \end{bmatrix}\begin{bmatrix} 0 & 0 \\ 0 & 32 \end{bmatrix}\begin{bmatrix} -2 & 3 \\ 1 & -1 \end{bmatrix} = \begin{bmatrix} 96 & -96 \\ 64 & -64 \end{bmatrix}$$

5. Characteristic polynomial: $p(\lambda) = \lambda^2 - 3\lambda + 2 = (\lambda - 1)(\lambda - 2)$

Eigenvalues: $\lambda_1 = 1, \quad \lambda_2 = 2$

With $\lambda_1 = 1$: $\begin{aligned} 4a - 4b &= 0 \\ 3a - 3b &= 0 \end{aligned}\Big\}$ $\mathbf{v}_1 = \begin{bmatrix} 1 \\ 1 \end{bmatrix}$

With $\lambda_2 = 2$: $\begin{aligned} 3a - 4b &= 0 \\ 3a - 4b &= 0 \end{aligned}\Big\}$ $\mathbf{v}_2 = \begin{bmatrix} 4 \\ 3 \end{bmatrix}$

$$\mathbf{P} = \begin{bmatrix} 1 & 4 \\ 1 & 3 \end{bmatrix}, \quad \mathbf{D} = \begin{bmatrix} 1 & 0 \\ 0 & 2 \end{bmatrix}, \quad \mathbf{P}^{-1} = \begin{bmatrix} -3 & 4 \\ 1 & -1 \end{bmatrix}$$

$$\mathbf{A}^5 = \begin{bmatrix} 1 & 4 \\ 1 & 3 \end{bmatrix} \begin{bmatrix} 1 & 0 \\ 0 & 32 \end{bmatrix} \begin{bmatrix} -3 & 4 \\ 1 & -1 \end{bmatrix} = \begin{bmatrix} 125 & -124 \\ 93 & -92 \end{bmatrix}$$

7. Characteristic polynomial: $p(\lambda) = -(\lambda - 1)(\lambda - 2)^2$

   Eigenvalues: $\lambda_1 = 1, \quad \lambda_2 = 2, \quad \lambda_3 = 2$

   With $\lambda_1 = 1$: $\begin{aligned} 3b &= 0 \\ b &= 0 \\ c &= 0 \end{aligned}\Bigg\}$ $\mathbf{v}_1 = \begin{bmatrix} 1 \\ 0 \\ 0 \end{bmatrix}$

   With $\lambda_2 = 2$: $\begin{aligned} 3b - a &= 0 \\ 0 &= 0 \\ 0 &= 0 \end{aligned}\Bigg\}$ $\mathbf{v}_2 = \begin{bmatrix} 0 \\ 0 \\ 1 \end{bmatrix}, \quad \mathbf{v}_3 = \begin{bmatrix} 3 \\ 1 \\ 0 \end{bmatrix}$

$$\mathbf{P} = \begin{bmatrix} 1 & 0 & 3 \\ 0 & 0 & 1 \\ 0 & 1 & 0 \end{bmatrix}, \quad \mathbf{D} = \begin{bmatrix} 1 & 0 & 0 \\ 0 & 2 & 0 \\ 0 & 0 & 2 \end{bmatrix}, \quad \mathbf{P}^{-1} = \begin{bmatrix} 1 & -3 & 0 \\ 0 & 0 & 1 \\ 0 & 1 & 0 \end{bmatrix}$$

$$\mathbf{A}^5 = \begin{bmatrix} 1 & 0 & 3 \\ 0 & 0 & 1 \\ 0 & 1 & 0 \end{bmatrix} \begin{bmatrix} 1 & 0 & 0 \\ 0 & 32 & 0 \\ 0 & 0 & 32 \end{bmatrix} \begin{bmatrix} 1 & -3 & 0 \\ 0 & 0 & 1 \\ 0 & 1 & 0 \end{bmatrix} = \begin{bmatrix} 1 & 93 & 0 \\ 0 & 32 & 0 \\ 0 & 0 & 32 \end{bmatrix}$$

9. Characteristic polynomial: $p(\lambda) = -(\lambda - 1)(\lambda - 2)^2$

   Eigenvalues: $\lambda_1 = 1, \quad \lambda_2 = 2, \quad \lambda_3 = 2$

   With $\lambda_1 = 1$: $\begin{aligned} c - 3b &= 0 \\ b &= 0 \\ c &= 0 \end{aligned}\Bigg\}$ $\mathbf{v}_1 = \begin{bmatrix} 1 \\ 0 \\ 0 \end{bmatrix}$

With $\lambda_2 = 2$: $\begin{matrix} -a - 3b + c = 0 \\ 0 = 0 \\ 0 = 0 \end{matrix}$ $\mathbf{v}_2 = \begin{bmatrix} 1 \\ 0 \\ 1 \end{bmatrix}$, $\mathbf{v}_3 = \begin{bmatrix} -3 \\ 1 \\ 0 \end{bmatrix}$

$$\mathbf{P} = \begin{bmatrix} 1 & 1 & -3 \\ 0 & 0 & 1 \\ 0 & 1 & 0 \end{bmatrix}, \quad \mathbf{D} = \begin{bmatrix} 1 & 0 & 0 \\ 0 & 2 & 0 \\ 0 & 0 & 2 \end{bmatrix}, \quad \mathbf{P}^{-1} = \begin{bmatrix} 1 & 3 & -1 \\ 0 & 0 & 1 \\ 0 & 1 & 0 \end{bmatrix}$$

$$\mathbf{A}^5 = \begin{bmatrix} 1 & 1 & -3 \\ 0 & 0 & 1 \\ 0 & 1 & 0 \end{bmatrix} \begin{bmatrix} 1 & 0 & 0 \\ 0 & 32 & 0 \\ 0 & 0 & 32 \end{bmatrix} \begin{bmatrix} 1 & 3 & -1 \\ 0 & 0 & 1 \\ 0 & 1 & 0 \end{bmatrix} = \begin{bmatrix} 1 & -93 & 31 \\ 0 & 32 & 0 \\ 0 & 0 & 32 \end{bmatrix}$$

11. Characteristic polynomial: $p(\lambda) = -\lambda^3 + \lambda = -\lambda(\lambda+1)(\lambda-1)$

Eigenvalues: $\lambda_1 = -1$, $\lambda_2 = 0$, $\lambda_3 = 1$

With $\lambda_1 = -1$: $\begin{matrix} 2a = 0 \\ 6a + 6b + 2c = 0 \\ 21a - 15b - 5c = 0 \end{matrix}$ $\mathbf{v}_1 = \begin{bmatrix} 0 \\ -1 \\ 3 \end{bmatrix}$

With $\lambda_2 = 0$: $\begin{matrix} a = 0 \\ 6a + 5b + 2c = 0 \\ 21a - 15b - 6c = 0 \end{matrix}$ $\mathbf{v}_2 = \begin{bmatrix} 0 \\ -2 \\ 5 \end{bmatrix}$

With $\lambda_3 = 1$: $\begin{matrix} 0 = 0 \\ 6a + 4b + 2c = 0 \\ 21a - 15b - 7c = 0 \end{matrix}$ $\mathbf{v}_3 = \begin{bmatrix} -1 \\ -42 \\ 87 \end{bmatrix}$

$$\mathbf{P} = \begin{bmatrix} 0 & 0 & -1 \\ -1 & -2 & -42 \\ 3 & 5 & 87 \end{bmatrix}, \quad \mathbf{D} = \begin{bmatrix} -1 & 0 & 0 \\ 0 & 0 & 0 \\ 0 & 0 & 1 \end{bmatrix}, \quad \mathbf{P}^{-1} = \begin{bmatrix} -36 & 5 & 2 \\ 39 & -3 & -1 \\ -1 & 0 & 0 \end{bmatrix}$$

$$\mathbf{A}^{10} = \begin{bmatrix} 0 & 0 & -1 \\ -1 & -2 & -42 \\ 3 & 5 & 87 \end{bmatrix} \begin{bmatrix} 1 & 0 & 0 \\ 0 & 0 & 0 \\ 0 & 0 & 1 \end{bmatrix} \begin{bmatrix} -36 & 5 & 2 \\ 39 & -3 & -1 \\ -1 & 0 & 0 \end{bmatrix} = \begin{bmatrix} 1 & 0 & 0 \\ 78 & -5 & -2 \\ -195 & 15 & 6 \end{bmatrix}$$

13. Characteristic polynomial: $p(\lambda) = -\lambda^3 + \lambda = -\lambda(\lambda+1)(\lambda-1)$

Eigenvalues: $\lambda_1 = -1$, $\lambda_2 = 0$, $\lambda_3 = 1$

With $\lambda_1 = -1$:
$$\left.\begin{array}{r}2a-b+c = 0\\2a-b+c = 0\\4a-4b+2c = 0\end{array}\right\} \qquad \mathbf{v}_1 = \begin{bmatrix}-1\\0\\2\end{bmatrix}$$

With $\lambda_2 = 0$:
$$\left.\begin{array}{r}a-b+c = 0\\2a-2b+c = 0\\4a-4b+c = 0\end{array}\right\} \qquad \mathbf{v}_2 = \begin{bmatrix}1\\1\\0\end{bmatrix}$$

With $\lambda_3 = 1$:
$$\left.\begin{array}{r}-b+c = 0\\2a-3b+c = 0\\4a-4b = 0\end{array}\right\} \qquad \mathbf{v}_3 = \begin{bmatrix}1\\1\\1\end{bmatrix}$$

$$\mathbf{P} = \begin{bmatrix}-1 & 1 & 1\\0 & 1 & 1\\2 & 0 & 1\end{bmatrix},\quad \mathbf{D} = \begin{bmatrix}-1 & 0 & 0\\0 & 0 & 0\\0 & 0 & 1\end{bmatrix},\quad \mathbf{P}^{-1} = \begin{bmatrix}-1 & 1 & 0\\-2 & 3 & -1\\2 & -2 & 1\end{bmatrix}$$

$$\mathbf{A}^{10} = \begin{bmatrix}-1 & 1 & 1\\0 & 1 & 1\\2 & 0 & 1\end{bmatrix}\begin{bmatrix}1 & 0 & 0\\0 & 0 & 0\\0 & 0 & 1\end{bmatrix}\begin{bmatrix}-1 & 1 & 0\\-2 & 3 & -1\\2 & -2 & 1\end{bmatrix} = \begin{bmatrix}3 & -3 & 1\\2 & -2 & 1\\0 & 0 & 1\end{bmatrix}$$

15. $p(\lambda) = \lambda^2 - 3\lambda + 2$ so $\mathbf{A}^2 - 3\mathbf{A} + 2\mathbf{I} = 0$

$$\mathbf{A}^2 = 3\mathbf{A} - 2\mathbf{I} = 3\begin{bmatrix}5 & -4\\3 & -2\end{bmatrix} - 2\begin{bmatrix}1 & 0\\0 & 1\end{bmatrix} = \begin{bmatrix}13 & -12\\9 & -8\end{bmatrix}$$

$$\mathbf{A}^3 = 3\mathbf{A}^2 - 2\mathbf{A} = 3\begin{bmatrix}13 & -12\\9 & -8\end{bmatrix} - 2\begin{bmatrix}5 & -4\\3 & -2\end{bmatrix} = \begin{bmatrix}29 & -28\\21 & -20\end{bmatrix}$$

$$\mathbf{A}^4 = 3\mathbf{A}^3 - 2\mathbf{A}^2 = 3\begin{bmatrix}29 & -28\\21 & -20\end{bmatrix} - 2\begin{bmatrix}13 & -12\\9 & -8\end{bmatrix} = \begin{bmatrix}61 & -60\\45 & -44\end{bmatrix}$$

$$\mathbf{A}^{-1} = \frac{1}{2}(-\mathbf{A} + 3\mathbf{I}) = \frac{1}{2}\left(-\begin{bmatrix}5 & -4\\3 & -2\end{bmatrix} + 3\begin{bmatrix}1 & 0\\0 & 1\end{bmatrix}\right) = \frac{1}{2}\begin{bmatrix}-2 & 4\\-3 & 5\end{bmatrix}$$

17. $p(\lambda) = -\lambda^3 + 5\lambda^2 - 8\lambda + 4$ so $-\mathbf{A}^3 + 5\mathbf{A}^2 - 8\mathbf{A} + 4\mathbf{I} = 0$

$$\mathbf{A}^2 = \begin{bmatrix}1 & 9 & 0\\0 & 4 & 0\\0 & 0 & 4\end{bmatrix},\qquad \mathbf{A}^3 = 5\mathbf{A}^2 - 8\mathbf{A} + 4\mathbf{I} = \begin{bmatrix}1 & 21 & 0\\0 & 8 & 0\\0 & 0 & 8\end{bmatrix}$$

$$\mathbf{A}^4 = 5\mathbf{A}^3 - 8\mathbf{A}^2 + 4\mathbf{A} = \begin{bmatrix} 1 & 45 & 0 \\ 0 & 16 & 0 \\ 0 & 0 & 16 \end{bmatrix}$$

$$\mathbf{A}^{-1} = \frac{1}{4}(\mathbf{A}^2 - 5\mathbf{A} + 8\mathbf{I}) = \frac{1}{2}\begin{bmatrix} 2 & -3 & 0 \\ 0 & 1 & 0 \\ 0 & 0 & 1 \end{bmatrix}$$

19. $p(\lambda) = -\lambda^3 + 5\lambda^2 - 8\lambda + 4$ so $-\mathbf{A}^3 + 5\mathbf{A}^2 - 8\mathbf{A} + 4\mathbf{I} = \mathbf{0}$

$$\mathbf{A}^2 = \begin{bmatrix} 1 & -9 & 3 \\ 0 & 4 & 0 \\ 0 & 0 & 4 \end{bmatrix}, \qquad \mathbf{A}^3 = 5\mathbf{A}^2 - 8\mathbf{A} + 4\mathbf{I} = \begin{bmatrix} 1 & -21 & 7 \\ 0 & 8 & 0 \\ 0 & 0 & 8 \end{bmatrix}$$

$$\mathbf{A}^4 = 5\mathbf{A}^3 - 8\mathbf{A}^2 + 4\mathbf{A} = \begin{bmatrix} 1 & -45 & 15 \\ 0 & 16 & 0 \\ 0 & 0 & 16 \end{bmatrix}$$

$$\mathbf{A}^{-1} = \frac{1}{4}(\mathbf{A}^2 - 5\mathbf{A} + 8\mathbf{I}) = \frac{1}{2}\begin{bmatrix} 2 & 3 & -1 \\ 0 & 1 & 0 \\ 0 & 0 & 1 \end{bmatrix}$$

21. $p(\lambda) = -\lambda^3 + \lambda$ so $-\mathbf{A}^3 + \mathbf{A} = \mathbf{0}$

$$\mathbf{A}^2 = \begin{bmatrix} 1 & 0 & 0 \\ 78 & -5 & -2 \\ -195 & 15 & 6 \end{bmatrix}, \qquad \mathbf{A}^3 = \mathbf{A} = \begin{bmatrix} 1 & 0 & 0 \\ 6 & 5 & 2 \\ 21 & -15 & -6 \end{bmatrix}$$

$$\mathbf{A}^4 = \mathbf{A}^2 = \begin{bmatrix} 1 & 0 & 0 \\ 78 & -5 & -2 \\ -195 & 15 & 6 \end{bmatrix}$$

Because $\lambda = 0$ is an eigenvalue, $\mathbf{A}$ is singular and $\mathbf{A}^{-1}$ does not exist.

23. $p(\lambda) = -\lambda^3 + \lambda$ so $-\mathbf{A}^3 + \mathbf{A} = \mathbf{0}$

$$\mathbf{A}^2 = \begin{bmatrix} 3 & -3 & 1 \\ 2 & -2 & 1 \\ 0 & 0 & 1 \end{bmatrix}, \qquad \mathbf{A}^3 = \mathbf{A} = \begin{bmatrix} 1 & -1 & 1 \\ 2 & -2 & 1 \\ 4 & -4 & 1 \end{bmatrix}$$

$$\mathbf{A}^4 = \mathbf{A}^2 = \begin{bmatrix} 3 & -3 & 1 \\ 2 & -2 & 1 \\ 0 & 0 & 1 \end{bmatrix}$$

Because $\lambda = 0$ is an eigenvalue, $\mathbf{A}$ is singular and $\mathbf{A}^{-1}$ does not exist.

In Problems 25–30 we first find the eigensystem of the given transition matrix $\mathbf{A}$ so as to determine its eigenvector matrix $\mathbf{P}$ and its diagonal eigenvalue matrix $\mathbf{D}$. Then we determine how the matrix power $\mathbf{A}^k = \mathbf{P}\mathbf{D}^k\mathbf{P}^{-1}$ behaves as $k \to \infty$. For simpler calculations of eigenvalues and eigenvectors, we write the entries of $\mathbf{A}$ in fractional rather than decimal form.

25. Characteristic polynomial: $p(\lambda) = \lambda^2 - \dfrac{9}{5}\lambda + \dfrac{4}{5} = \dfrac{1}{5}(\lambda - 1)(5\lambda - 4)$

Eigenvalues: $\lambda_1 = 1, \quad \lambda_2 = \dfrac{4}{5}$

With $\lambda_1 = 1$:
$$\left. \begin{array}{r} -\dfrac{1}{10}a + \dfrac{1}{10}b = 0 \\ \dfrac{1}{10}a - \dfrac{1}{10}b = 0 \end{array} \right\} \qquad \mathbf{v}_1 = \begin{bmatrix} 1 \\ 1 \end{bmatrix}$$

With $\lambda_2 = \dfrac{4}{5}$:
$$\left. \begin{array}{r} \dfrac{1}{10}a + \dfrac{1}{10}b = 0 \\ \dfrac{1}{10}a + \dfrac{1}{10}b = 0 \end{array} \right\} \qquad \mathbf{v}_2 = \begin{bmatrix} -1 \\ 1 \end{bmatrix}$$

$$\mathbf{P} = \begin{bmatrix} 1 & -1 \\ 1 & 1 \end{bmatrix}, \quad \mathbf{D} = \begin{bmatrix} 1 & 0 \\ 0 & 4/5 \end{bmatrix}, \quad \mathbf{P}^{-1} = \dfrac{1}{2}\begin{bmatrix} 1 & 1 \\ -1 & 1 \end{bmatrix}$$

$$\mathbf{x}_k = \mathbf{A}^k \mathbf{x}_0 = \begin{bmatrix} 1 & -1 \\ 1 & 1 \end{bmatrix}\begin{bmatrix} 1 & 0 \\ 0 & 4/5 \end{bmatrix}^k \dfrac{1}{2}\begin{bmatrix} 1 & 1 \\ -1 & 1 \end{bmatrix} \mathbf{x}_0$$

$$\to \begin{bmatrix} 1 & -1 \\ 1 & 1 \end{bmatrix}\begin{bmatrix} 1 & 0 \\ 0 & 0 \end{bmatrix}\dfrac{1}{2}\begin{bmatrix} 1 & 1 \\ -1 & 1 \end{bmatrix}\mathbf{x}_0 = \dfrac{1}{2}\begin{bmatrix} 1 & 1 \\ 1 & 1 \end{bmatrix}\begin{bmatrix} C_0 \\ S_0 \end{bmatrix} = (C_0 + S_0)\begin{bmatrix} 1/2 \\ 1/2 \end{bmatrix}$$

as $k \to \infty$. Thus the long-term distribution of population is 50% city, 50% suburban.

27. Characteristic polynomial: $p(\lambda) = \lambda^2 - \dfrac{8}{5}\lambda + \dfrac{3}{5} = \dfrac{1}{5}(\lambda - 1)(5\lambda - 3)$

Eigenvalues: $\lambda_1 = 1, \quad \lambda_2 = \dfrac{3}{5}$

With $\lambda_1 = 1$:
$$\left. \begin{array}{r} -\dfrac{1}{4}a + \dfrac{3}{20}b = 0 \\ \dfrac{1}{4}a - \dfrac{3}{20}b = 0 \end{array} \right\} \qquad \mathbf{v}_1 = \begin{bmatrix} 3 \\ 5 \end{bmatrix}$$

Section 6.3

With $\lambda_2 = \dfrac{3}{5}$:
$$\begin{aligned}\dfrac{3}{20}a + \dfrac{3}{20}b &= 0 \\ \dfrac{1}{4}a + \dfrac{1}{4}b &= 0\end{aligned}\Bigg\} \qquad \mathbf{v}_2 = \begin{bmatrix}-1\\1\end{bmatrix}$$

$$\mathbf{P} = \begin{bmatrix}3 & -1\\5 & 1\end{bmatrix}, \quad \mathbf{D} = \begin{bmatrix}1 & 0\\0 & 3/5\end{bmatrix}, \quad \mathbf{P}^{-1} = \dfrac{1}{8}\begin{bmatrix}1 & 1\\-5 & 3\end{bmatrix}$$

$$\mathbf{x}_k = \mathbf{A}^k\mathbf{x}_0 = \begin{bmatrix}3 & -1\\5 & 1\end{bmatrix}\begin{bmatrix}1 & 0\\0 & 3/5\end{bmatrix}^k \dfrac{1}{8}\begin{bmatrix}1 & 1\\-5 & 3\end{bmatrix}\mathbf{x}_0$$

$$\to \begin{bmatrix}3 & -1\\5 & 1\end{bmatrix}\begin{bmatrix}1 & 0\\0 & 0\end{bmatrix}\dfrac{1}{8}\begin{bmatrix}1 & 1\\-5 & 3\end{bmatrix}\mathbf{x}_0 = \dfrac{1}{8}\begin{bmatrix}3 & 3\\5 & 5\end{bmatrix}\begin{bmatrix}C_0\\S_0\end{bmatrix} = (C_0+S_0)\begin{bmatrix}3/8\\5/8\end{bmatrix}$$

as $k \to \infty$. Thus the long-term distribution of population is 3/8 city, 5/8 suburban.

29. Characteristic polynomial: $p(\lambda) = \lambda^2 - \dfrac{37}{20}\lambda + \dfrac{17}{20} = \dfrac{1}{20}(\lambda-1)(20\lambda-17)$

Eigenvalues: $\lambda_1 = 1, \quad \lambda_2 = \dfrac{17}{20}$

With $\lambda_1 = 1$:
$$\begin{aligned}-\dfrac{1}{10}a + \dfrac{1}{20}b &= 0 \\ \dfrac{1}{10}a - \dfrac{1}{20}b &= 0\end{aligned}\Bigg\} \qquad \mathbf{v}_1 = \begin{bmatrix}1\\2\end{bmatrix}$$

With $\lambda_2 = \dfrac{17}{20}$:
$$\begin{aligned}\dfrac{1}{20}a + \dfrac{1}{20}b &= 0 \\ \dfrac{1}{10}a + \dfrac{1}{10}b &= 0\end{aligned}\Bigg\} \qquad \mathbf{v}_2 = \begin{bmatrix}-1\\1\end{bmatrix}$$

$$\mathbf{P} = \begin{bmatrix}1 & -1\\2 & 1\end{bmatrix}, \quad \mathbf{D} = \begin{bmatrix}1 & 0\\0 & 17/20\end{bmatrix}, \quad \mathbf{P}^{-1} = \dfrac{1}{3}\begin{bmatrix}1 & 1\\-2 & 1\end{bmatrix}$$

$$\mathbf{x}_k = \mathbf{A}^k\mathbf{x}_0 = \begin{bmatrix}1 & -1\\2 & 1\end{bmatrix}\begin{bmatrix}1 & 0\\0 & 17/20\end{bmatrix}^k \dfrac{1}{3}\begin{bmatrix}1 & 1\\-2 & 1\end{bmatrix}\mathbf{x}_0$$

$$\to \begin{bmatrix}1 & -1\\2 & 1\end{bmatrix}\begin{bmatrix}1 & 0\\0 & 0\end{bmatrix}\dfrac{1}{3}\begin{bmatrix}1 & 1\\-2 & 1\end{bmatrix}\mathbf{x}_0 = \dfrac{1}{3}\begin{bmatrix}1 & 1\\2 & 2\end{bmatrix}\begin{bmatrix}C_0\\S_0\end{bmatrix} = (C_0+S_0)\begin{bmatrix}1/3\\2/3\end{bmatrix}$$

as $k \to \infty$. Thus the long-term distribution of population is 1/3 city, 2/3 suburban.

31. Characteristic polynomial: $p(\lambda) = \lambda^2 - \frac{9}{5}\lambda + \frac{4}{5} = \frac{1}{5}(\lambda - 1)(5\lambda - 4)$

Eigenvalues: $\lambda_1 = 1$, $\lambda_2 = \frac{4}{5}$

With $\lambda_1 = 1$: $\begin{aligned} -\frac{2}{5}a + \frac{1}{2}b &= 0 \\ -\frac{4}{25}a + \frac{1}{5}b &= 0 \end{aligned}$ $\quad \mathbf{v}_1 = \begin{bmatrix} 5 \\ 4 \end{bmatrix}$

With $\lambda_2 = \frac{4}{5}$: $\begin{aligned} -\frac{1}{5}a + \frac{1}{2}b &= 0 \\ -\frac{4}{25}a + \frac{2}{5}b &= 0 \end{aligned}$ $\quad \mathbf{v}_2 = \begin{bmatrix} 5 \\ 2 \end{bmatrix}$

$\mathbf{P} = \begin{bmatrix} 5 & 5 \\ 4 & 2 \end{bmatrix}$, $\mathbf{D} = \begin{bmatrix} 1 & 0 \\ 0 & 4/5 \end{bmatrix}$, $\mathbf{P}^{-1} = \frac{1}{10}\begin{bmatrix} -2 & 5 \\ 4 & -5 \end{bmatrix}$

$\mathbf{x}_k = \mathbf{A}^k \mathbf{x}_0 = \begin{bmatrix} 5 & 5 \\ 4 & 2 \end{bmatrix}\begin{bmatrix} 1 & 0 \\ 0 & 4/5 \end{bmatrix}^k \frac{1}{10}\begin{bmatrix} -2 & 5 \\ 4 & -5 \end{bmatrix}\mathbf{x}_0$

$\to \begin{bmatrix} 5 & 5 \\ 4 & 2 \end{bmatrix}\begin{bmatrix} 1 & 0 \\ 0 & 0 \end{bmatrix}\frac{1}{10}\begin{bmatrix} -2 & 5 \\ 4 & -5 \end{bmatrix}\mathbf{x}_0 = \frac{1}{10}\begin{bmatrix} -10 & 25 \\ -8 & 20 \end{bmatrix}\begin{bmatrix} F_0 \\ R_0 \end{bmatrix} = \begin{bmatrix} 2.5R_0 - F_0 \\ 2R_0 - 0.8F_0 \end{bmatrix}$

as $k \to \infty$. Thus the fox-rabbit population approaches a stable situation with $2.5R_0 - F_0$ foxes and $2R_0 - 0.8F_0$ rabbits.

33. Characteristic polynomial: $p(\lambda) = \lambda^2 - \frac{9}{5}\lambda + \frac{323}{400} = \frac{1}{400}(20\lambda - 19)(20\lambda - 17)$

Eigenvalues: $\lambda_1 = \frac{19}{20}$, $\lambda_2 = \frac{17}{20}$

With $\lambda_1 = \frac{19}{20}$: $\begin{aligned} -\frac{7}{20}a + \frac{1}{2}b &= 0 \\ -\frac{7}{40}a + \frac{1}{4}b &= 0 \end{aligned}$ $\quad \mathbf{v}_1 = \begin{bmatrix} 10 \\ 7 \end{bmatrix}$

With $\lambda_2 = \frac{17}{20}$: $\begin{aligned} -\frac{1}{4}a + \frac{1}{2}b &= 0 \\ -\frac{7}{40}a + \frac{7}{20}b &= 0 \end{aligned}$ $\quad \mathbf{v}_2 = \begin{bmatrix} 2 \\ 1 \end{bmatrix}$

$\mathbf{P} = \begin{bmatrix} 10 & 2 \\ 7 & 1 \end{bmatrix}$, $\mathbf{D} = \begin{bmatrix} 19/20 & 0 \\ 0 & 17/20 \end{bmatrix}$, $\mathbf{P}^{-1} = \frac{1}{4}\begin{bmatrix} -1 & 2 \\ 7 & -10 \end{bmatrix}$

$$\mathbf{x}_k = \mathbf{A}^k \mathbf{x}_0 = \begin{bmatrix} 10 & 2 \\ 7 & 1 \end{bmatrix} \begin{bmatrix} 19/20 & 0 \\ 0 & 17/20 \end{bmatrix}^k \frac{1}{4} \begin{bmatrix} -1 & 2 \\ 7 & -10 \end{bmatrix} \mathbf{x}_0$$

$$\rightarrow \begin{bmatrix} 10 & 2 \\ 7 & 1 \end{bmatrix} \begin{bmatrix} 0 & 0 \\ 0 & 0 \end{bmatrix} \frac{1}{4} \begin{bmatrix} -1 & 2 \\ 7 & -10 \end{bmatrix} \mathbf{x}_0 = \begin{bmatrix} 0 & 0 \\ 0 & 0 \end{bmatrix} \begin{bmatrix} F_0 \\ R_0 \end{bmatrix} = \begin{bmatrix} 0 \\ 0 \end{bmatrix}$$

as $k \rightarrow \infty$. Thus the fox and rabbit population both die out.

35. The fact that each $|\lambda| = 1$, so $\lambda = \pm 1$, implies that $\mathbf{D}^n = \mathbf{I}$ if n is even, in which case $\mathbf{A}^n = \mathbf{P}\mathbf{D}^n\mathbf{P}^{-1} = \mathbf{P}\mathbf{I}\mathbf{P}^{-1} = \mathbf{I}$.

37. We find immediately that $\mathbf{A}^2 = -\mathbf{I}$, so $\mathbf{A}^3 = \mathbf{A}^2\mathbf{A} = -\mathbf{I}\mathbf{A} = -\mathbf{A}$, $\mathbf{A}^4 = \mathbf{A}^3\mathbf{A} = -\mathbf{A}^2 = \mathbf{I}$, and so forth.

39. The characteristic equation of $\mathbf{A}$ is

$$(p-\lambda)(q-\lambda) = (1-p)(1-q) = \lambda^2 - (p+q)\lambda + (p+q-1)$$
$$= (\lambda-1)[\lambda-(p+q-1)],$$

so the eigenvalues of A are $\lambda_1 = 1$ and $\lambda_2 = p+q-1$.

CHAPTER 7

LINEAR SYSTEMS OF DIFFERENTIAL EQUATIONS

SECTION 7.1

FIRST-ORDER SYSTEMS AND APPLICATIONS

1. Let $x_1 = x$ and $x_2 = x_1' = x'$, so $x_2' = x'' = -7x - 3x' + t^2$.

   Equivalent system:
   $$x_1' = x_2, \qquad x_2' = -7x_1 - 3x_2 + t^2$$

3. Let $x_1 = x$ and $x_2 = x_1' = x'$, so $x_2' = x'' = \left[(1-t^2)x - tx'\right]/t^2$.

   Equivalent system:
   $$x_1' = x_2, \qquad t^2 x_2' = (1-t^2)x_1 - tx_2$$

5. Let $x_1 = x$, $x_2 = x_1' = x'$, $x_3 = x_2' = x''$, so $x_3' = x''' = (x')^2 + \cos x$.

   Equivalent system:
   $$x_1' = x_2, \qquad x_2' = x_3, \qquad x_3' = x_2^2 + \cos x_1$$

7. Let $x_1 = x$, $x_2 = x_1' = x'$, $y_1 = y$, $y_2 = y_1' = y'$ so $x_2' = x'' = -kx/(x^2+y^2)^{3/2}$, $y_2' = y'' = -ky/(x^2+y^2)^{3/2}$.

   Equivalent system:
   $$x_1' = x_2, \qquad x_2' = -kx_1/\left(x_1^2 + y_1^2\right)^{3/2}$$
   $$y_1' = y_2, \qquad y_2' = -ky_1/\left(x_1^2 + y_1^2\right)^{3/2}$$

9. Let $x_1 = x$, $x_2 = x_1' = x'$, $y_1 = y$, $y_2 = y_1' = y'$, $z_1 = z$, $z_2 = z_1' = z'$, so $x_2' = x'' = 3x - y + 2z$, $y_2' = y'' = x + y - 4z$, $z_2' = z'' = 5x - y - z$.

Equivalent system:

$x_1' = x_2,$ $x_2' = 3x_1 - y_1 + 2z_1$

$y_1' = y_2,$ $y_2' = x_1 + y_1 - 4z_1$

$z_1' = z_2,$ $z_2' = 5x_1 - y_1 - z_1$

11. The computation $x'' = y' = -x$ yields the single linear second-order equation $x'' + x = 0$ with characteristic equation $r^2 + 1 = 0$ and general solution

$$x(t) = A\cos t + B\sin t.$$

Then the original first equation $y = x'$ gives

$$y(t) = B\cos t - A\sin t.$$

13. The computation $x'' = -2y' = -4x$ yields the single linear second-order equation $x'' + 4x = 0$ with characteristic equation $r^2 + 4 = 0$ and general solution

$$x(t) = A\cos 2t + B\sin 2t.$$

Then the original first equation $y = -x'/2$ gives

$$y(t) = -B\cos 2t + A\sin 2t.$$

Finally, the condition $x(0) = 1$ implies that $A = 1$, and then the condition $y(0) = 0$ gives $B = 0$. Hence the desired particular solution is given by

$$x(t) = \cos 2t, \qquad y(t) = \sin 2t.$$

15. The computation $x'' = y'/2 = -4x$ yields the single linear second-order equation $x'' + 4x = 0$ with characteristic equation $r^2 + 4 = 0$ and general solution

$$x(t) = A\cos 2t + B\sin 2t.$$

Then the original first equation $y = 2x'$ gives

$$y(t) = 4B\cos 2t - 4A\sin 2t.$$

17. The computation $x'' = y' = 6x - y = 6x - x'$ yields the single linear second-order equation $x'' + x' - 6x = 0$ with characteristic equation $r^2 + r - 6 = 0$ and characteristic roots $r = -3$ and 2, so the general solution

$$x(t) = A e^{-3t} + B e^{2t}.$$

Then the original first equation $y = x'$ gives

$$y(t) = -3A\,e^{-3t} + 2B\,e^{2t}.$$

Finally, the initial conditions

$$x(0) = A + B = 1, \quad y(0) = -3A + 2B = 2$$

imply that $A = 0$ and $B = 1$, so the desired particular solution is given by

$$x(t) = e^{2t}, \qquad y(t) = 2\,e^{2t}.$$

19. The computation $x'' = -y' = -13x - 4y = -13x + 4x'$ yields the single linear second-order equation $x'' - 4x' + 13x = 0$ with characteristic equation $r^2 - 4r + 13 = 0$ and characteristic roots $r = 2 \pm 3i$, hence the general solution is

$$x(t) = e^{2t}(A\cos 3t + B\sin 3t).$$

The initial condition $x(0) = 0$ then gives $A = 0$, so $x(t) = B\,e^{2t}\sin 3t$. Then the original first equation $y = -x'$ gives

$$y(t) = -e^{2t}(3B\cos 3t + 2B\sin 3t).$$

Finally, the initial condition $y(0) = 3$ gives $B = -1$, so the desired particular solution is given by

$$x(t) = -e^{2t}\sin 3t, \qquad y(t) = e^{2t}(3\cos 3t + 2\sin 3t).$$

21. (a) Substituting the general solution found in Problem 11 we get

$$x^2 + y^2 = (A\cos t + B\sin t)^2 + (B\cos t - A\sin t)^2$$
$$= (A^2 + B^2)(\cos^2 t + \sin^2 t) = A^2 + B^2$$
$$x^2 + y^2 = C^2,$$

the equation of a circle of radius $C = (A^2 + B^2)^{1/2}$.

(b) Substituting the general solution found in Problem 12 we get

$$x^2 - y^2 = (Ae^t + Be^{-t})^2 - (Ae^t - Be^{-t})^2 = 4AB,$$

the equation of a hyperbola.

23. When we solve Equations (20) and (21) in the text for e^{-t} and e^{2t} we get

$$2x - y = 3Ae^{-t} \quad \text{and} \quad x + y = 3Be^{2t}.$$

Section 7.1

Hence
$$(2x-y)^2(x+y) = (3Ae^{-t})^2(3Be^{2t}) = 27A^2B = C.$$

Clearly $y = 2x$ or $y = -x$ if $C = 0$, and expansion gives the equation $4x^3 - 3xy^2 + y^3 = C$.

25. Looking at Fig. 7.1.10 in the text, we see that

$$my_1'' = -T\sin\theta_1 + T\sin\theta_2 \approx -T\tan\theta_1 + T\tan\theta_2 = -Ty_1/L + T(y_2-y_1)/L,$$
$$my_2'' = -T\sin\theta_2 - T\sin\theta_3 \approx -T\tan\theta_2 - T\tan\theta_3 = -T(y_2-y_1)/L - Ty_2/L.$$

We get the desired equations when we multiply each of these equations by L/T and set $k = mL/T$.

27. If θ is the polar angular coordinate of the point (x,y) and we write $F = k/(x^2+y^2) = k/r^2$, then Newton's second law gives

$$mx'' = -F\cos\theta = -(k/r^2)(x/r) = -kx/r^3,$$
$$my'' = -F\sin\theta = -(k/r^2)(y/r) = -ky/r^3,$$

29. If $\mathbf{r} = (x,y,z)$ is the particle's position vector, then Newton's law $m\mathbf{r}'' = \mathbf{F}$ gives

$$m\mathbf{r}'' = q\mathbf{v}\times\mathbf{B} = q\begin{vmatrix} \mathbf{i} & \mathbf{j} & \mathbf{k} \\ x' & y' & z' \\ 0 & 0 & B \end{vmatrix} = +qBy'\mathbf{i} - qBx'\mathbf{j} = qB(-y', x', 0).$$

SECTION 7.2

MATRICES AND LINEAR SYSTEMS

1. $(\mathbf{AB})' = \begin{bmatrix} t - 4t^2 + 6t^3 & t + t^2 - 4t^3 + 8t^4 \\ 3t + t^3 - t^4 & 4t^2 + t^3 + t^4 \end{bmatrix}' = \begin{bmatrix} 1 - 8t + 18t^2 & 1 + 2t - 12t^2 + 32t^3 \\ 3 + 3t^2 - 4t^3 & 8t + 3t^2 + 4t^3 \end{bmatrix}$

$\mathbf{A}'\mathbf{B} + \mathbf{A}\mathbf{B}' = \begin{bmatrix} 1 & 2 \\ 3t^2 & -\dfrac{1}{t^2} \end{bmatrix}\begin{bmatrix} 1-t & 1+t \\ 3t^2 & 4t^3 \end{bmatrix} + \begin{bmatrix} t & 2t-1 \\ t^3 & \dfrac{1}{t} \end{bmatrix}\begin{bmatrix} -1 & 1 \\ 6t & 12t^2 \end{bmatrix}$

$= \begin{bmatrix} 1 - t + 6t^2 & 1 + t + 8t^3 \\ -3 + 3t^2 - 3t^3 & -4t + 3t^2 + 3t^3 \end{bmatrix} + \begin{bmatrix} -7t + 12t^2 & t - 12t^2 + 24t^3 \\ 6 - t^3 & 12t + t^3 \end{bmatrix}$

$$= \begin{bmatrix} 1-8t+18t^2 & 1+2t-12t^2+32t^3 \\ 3+3t^2-4t^3 & 8t+3t^2+4t^3 \end{bmatrix}$$

3. $\mathbf{x} = \begin{bmatrix} x \\ y \end{bmatrix}$, $\quad \mathbf{P}(t) = \begin{bmatrix} 0 & -3 \\ 3 & 0 \end{bmatrix}$, $\quad \mathbf{f}(t) = \begin{bmatrix} 0 \\ 0 \end{bmatrix}$

5. $\mathbf{x} = \begin{bmatrix} x \\ y \end{bmatrix}$, $\quad \mathbf{P}(t) = \begin{bmatrix} 2 & 4 \\ 5 & -1 \end{bmatrix}$, $\quad \mathbf{f}(t) = \begin{bmatrix} 3e^t \\ -t^2 \end{bmatrix}$

7. $\mathbf{x} = \begin{bmatrix} x \\ y \\ z \end{bmatrix}$, $\quad \mathbf{P}(t) = \begin{bmatrix} 0 & 1 & 1 \\ 1 & 0 & 1 \\ 1 & 1 & 0 \end{bmatrix}$, $\quad \mathbf{f}(t) = \begin{bmatrix} 0 \\ 0 \\ 0 \end{bmatrix}$

9. $\mathbf{x} = \begin{bmatrix} x \\ y \\ z \end{bmatrix}$, $\quad \mathbf{P}(t) = \begin{bmatrix} 3 & -4 & 1 \\ 1 & 0 & -3 \\ 0 & 6 & -7 \end{bmatrix}$, $\quad \mathbf{f}(t) = \begin{bmatrix} t \\ t^2 \\ t^3 \end{bmatrix}$

11. $\mathbf{x} = \begin{bmatrix} x_1 \\ x_2 \\ x_3 \\ x_4 \end{bmatrix}$, $\quad \mathbf{P}(t) = \begin{bmatrix} 0 & 1 & 0 & 0 \\ 0 & 0 & 2 & 0 \\ 0 & 0 & 0 & 3 \\ 4 & 0 & 0 & 0 \end{bmatrix}$, $\quad \mathbf{f}(t) = \begin{bmatrix} 0 \\ 0 \\ 0 \\ 0 \end{bmatrix}$

13. $W(t) = \begin{vmatrix} 2e^t & e^{2t} \\ -3e^t & -e^{2t} \end{vmatrix} = e^{3t} \neq 0$

$$\mathbf{x}_1' = \begin{bmatrix} 2e^t \\ -3e^t \end{bmatrix}' = \begin{bmatrix} 2e^t \\ -3e^t \end{bmatrix} = \begin{bmatrix} 4 & 2 \\ -3 & -1 \end{bmatrix} \begin{bmatrix} 2e^t \\ -3e^t \end{bmatrix} = \mathbf{Ax}_1$$

$$\mathbf{x}_2' = \begin{bmatrix} e^{2t} \\ -e^{2t} \end{bmatrix}' = \begin{bmatrix} 2e^{2t} \\ -2e^{2t} \end{bmatrix} = \begin{bmatrix} 4 & 2 \\ -3 & -1 \end{bmatrix} \begin{bmatrix} e^{2t} \\ -e^{2t} \end{bmatrix} = \mathbf{Ax}_2$$

$$\mathbf{x}(t) = c_1\mathbf{x}_1 + c_2\mathbf{x}_2 = c_1 \begin{bmatrix} 2e^t \\ -3e^t \end{bmatrix} + c_2 \begin{bmatrix} e^{2t} \\ -e^{2t} \end{bmatrix} = \begin{bmatrix} 2c_1 e^t + c_2 e^{2t} \\ -3c_1 e^t - c_2 e^{2t} \end{bmatrix}$$

In most of Problems 14–22, we omit the verifications of the given solutions. In each case, this is simply a matter of calculating both the derivative $\mathbf{x}_i'$ of the given solution vector and the product $\mathbf{Ax}_i$ (where $\mathbf{A}$ is the coefficient matrix in the given differential equation) to verify that $\mathbf{x}_i' = \mathbf{Ax}_i$ (just as in the verification of the solutions $\mathbf{x}_1$ and $\mathbf{x}_2$ in Problem 13 above).

15. $W(t) = \begin{vmatrix} e^{2t} & e^{-2t} \\ e^{2t} & 5e^{-2t} \end{vmatrix} = 4 \neq 0$

$\mathbf{x}(t) = c_1\mathbf{x}_1 + c_2\mathbf{x}_2 = c_1\begin{bmatrix} 1 \\ 1 \end{bmatrix}e^{2t} + c_2\begin{bmatrix} 1 \\ 5 \end{bmatrix}e^{-2t} = \begin{bmatrix} c_1e^{2t} + c_2e^{-2t} \\ c_1e^{2t} + 5c_2e^{-2t} \end{bmatrix}$

17. $W(t) = \begin{vmatrix} 3e^{2t} & e^{-5t} \\ 2e^{2t} & 3e^{-5t} \end{vmatrix} = 7e^{-3t} \neq 0$

$\mathbf{x}(t) = c_1\mathbf{x}_1 + c_2\mathbf{x}_2 = c_1\begin{bmatrix} 3e^{2t} \\ 2e^{2t} \end{bmatrix} + c_2\begin{bmatrix} e^{-5t} \\ 3e^{-5t} \end{bmatrix} = \begin{bmatrix} 3c_1e^{2t} + c_2e^{-5t} \\ 2c_1e^{2t} + 3c_2e^{-5t} \end{bmatrix}$

19. $W(t) = \begin{vmatrix} e^{2t} & e^{-t} & 0 \\ e^{2t} & 0 & e^{-t} \\ e^{2t} & -e^{-t} & -e^{-t} \end{vmatrix} = 3 \neq 0$

$\mathbf{x}(t) = c_1\mathbf{x}_1 + c_2\mathbf{x}_2 + c_3\mathbf{x}_3 = c_1\begin{bmatrix} 1 \\ 1 \\ 1 \end{bmatrix}e^{2t} + c_2\begin{bmatrix} 1 \\ 0 \\ -1 \end{bmatrix}e^{-t} + c_3\begin{bmatrix} 0 \\ 1 \\ -1 \end{bmatrix}e^{-t} = \begin{bmatrix} c_1e^{2t} + c_2e^{-t} \\ c_1e^{2t} + c_3e^{-t} \\ c_1e^{2t} - c_2e^{-t} - c_3e^{-t} \end{bmatrix}$

$\mathbf{x}_1' = \begin{bmatrix} 2 \\ 2 \\ 2 \end{bmatrix}e^t = \begin{bmatrix} 0 & 1 & 1 \\ 1 & 0 & 1 \\ 1 & 1 & 0 \end{bmatrix}\begin{bmatrix} 1 \\ 1 \\ 1 \end{bmatrix}e^t = \mathbf{A}\mathbf{x}_1$

$\mathbf{x}_2' = \begin{bmatrix} -1 \\ 0 \\ 1 \end{bmatrix}e^{-t} = \begin{bmatrix} 0 & 1 & 1 \\ 1 & 0 & 1 \\ 1 & 1 & 0 \end{bmatrix}\begin{bmatrix} 1 \\ 0 \\ -1 \end{bmatrix}e^{-t} = \mathbf{A}\mathbf{x}_2$

$\mathbf{x}_3' = \begin{bmatrix} 0 \\ -1 \\ 1 \end{bmatrix}e^{-t} = \begin{bmatrix} 0 & 1 & 1 \\ 1 & 0 & 1 \\ 1 & 1 & 0 \end{bmatrix}\begin{bmatrix} 0 \\ 1 \\ -1 \end{bmatrix}e^{-t} = \mathbf{A}\mathbf{x}_3$

21. $W(t) = \begin{vmatrix} 3e^{-2t} & e^t & e^{3t} \\ -2e^{-2t} & -e^t & -e^{3t} \\ 2e^{-2t} & e^t & 0 \end{vmatrix} = e^{2t} \neq 0$

$$\mathbf{x}(t) = c_1\mathbf{x}_1 + c_2\mathbf{x}_2 + c_3\mathbf{x}_3 = c_1\begin{bmatrix}3\\-2\\2\end{bmatrix}e^{-2t} + c_2\begin{bmatrix}1\\-1\\1\end{bmatrix}e^{t} + c_3\begin{bmatrix}1\\-1\\0\end{bmatrix}e^{3t} = \begin{bmatrix}3c_1e^{-2t} + c_2e^{t} + c_3e^{3t}\\-2c_1e^{-2t} - c_2e^{t} - c_3e^{3t}\\2c_1e^{-2t} + c_2e^{t}\end{bmatrix}$$

In Problems 23–26 (and similarly in Problems 27–32) we give first the scalar components $x_1(t)$ and $x_2(t)$ of a general solution, then the equations in the coefficients c_1 and c_2 that are obtained when the given initial conditions are imposed, and finally the resulting particular solution of the given system.

23. $x_1(t) = c_1e^{3t} + 2c_2e^{-2t}, \quad x_2(t) = 3c_1e^{3t} + c_2e^{-2t}$

$c_1 + 2c_2 = 0, \quad 3c_1 + c_2 = 5$

$x_1(t) = 2e^{3t} - 2e^{-2t}, \quad x_2(t) = 6e^{3t} - e^{-2t}$

25. $x_1(t) = c_1e^{3t} + c_2e^{2t}, \quad x_2(t) = -c_1e^{3t} - 2c_2e^{2t}$

$c_1 + c_2 = 11, \quad -c_1 - 2c_2 = -7$

$x_1(t) = 15e^{3t} - 4e^{2t}, \quad x_2(t) = -15e^{3t} + 8e^{2t}$

27. $x_1(t) = 2c_1e^{t} - 2c_2e^{3t} + 2c_3e^{5t}, \quad x_2(t) = 2c_1e^{t} - 2c_3e^{5t}, \quad x_3(t) = c_1e^{t} + c_2e^{3t} + c_3e^{5t}$

$2c_1 - 2c_2 + 2c_3 = 0, \quad 2c_1 - 2c_3 = 0, \quad c_1 + c_2 + c_3 = 4$

$x_1(t) = 2e^{t} - 4e^{3t} + 2e^{5t}, \quad x_2(t) = 2e^{t} - 2e^{5t}, \quad x_3(t) = e^{t} + 2e^{3t} + e^{5t}$

29. $x_1(t) = 3c_1e^{-2t} + c_2e^{t} + c_3e^{3t}, \quad x_2(t) = -2c_1e^{-2t} - c_2e^{t} - c_3e^{3t}, \quad x_3(t) = 2c_1e^{-2t} + c_2e^{t}$

$3c_1 + c_2 + c_3 = 1, \quad -2c_1 - c_2 - c_3 = 2, \quad 2c_1 + c_2 = 3$

$x_1(t) = 9e^{-2t} - 3e^{t} - 5e^{3t}, \quad x_2(t) = -6e^{-2t} + 3e^{t} + 5e^{3t}, \quad x_3(t) = 6e^{-2t} - 3e^{t}$

31. $x_1(t) = c_1e^{-t} + c_4e^{t}, \quad x_2(t) = c_3e^{t}, \quad x_3(t) = c_2e^{-t} + 3c_4e^{t}, \quad x_4(t) = c_1e^{-t} - 2c_3e^{t}$

$c_1 + c_4 = 1, \quad c_3 = 1, \quad c_2 + 3c_4 = 1, \quad c_1 - 2c_3 = 1$

$x_1(t) = 3e^{-t} - 2e^{t}, \quad x_2(t) = e^{t}, \quad x_3(t) = 7e^{-t} - 6e^{t}, \quad x_4(t) = 3e^{-t} - 2e^{t}$

33. (a) $\mathbf{x}_2 = t\mathbf{x}_1$, so neither is a constant multiple of the other.

(b) $W(\mathbf{x}_1, \mathbf{x}_2) = 0$, whereas Theorem 2 would imply that $W \neq 0$ if $\mathbf{x}_1$ and $\mathbf{x}_2$ were independent solutions of a system of the indicated form.

35. Suppose $W(a) = x_{11}(a)x_{22}(a) - x_{12}(a)x_{21}(a) = 0$. Then the coefficient determinant of the homogeneous linear system $c_1x_{11}(a) + c_2x_{12}(a) = 0, \quad c_1x_{21}(a) + c_2x_{22}(a) = 0$ vanishes. The system therefore has a non-trivial solution $\{c_1, c_2\}$ such that

$c_1\mathbf{x}_1(a)+c_2\mathbf{x}_2(a) = \mathbf{0}$. Then $\mathbf{x}(t) = c_1\mathbf{x}_1(t)+c_2\mathbf{x}_2(t)$ is a solution of $\mathbf{x}' = \mathbf{Px}$ such that $\mathbf{x}(a) = \mathbf{0}$. It therefore follows (by uniqueness of solutions) that $\mathbf{x}(t) \equiv \mathbf{0}$, that is, $c_1\mathbf{x}_1(t)+c_2\mathbf{x}_2(t) \equiv \mathbf{0}$ with c_1 and c_2 not both zero. Thus the solution vectors $\mathbf{x}_1$ and $\mathbf{x}_2$ are linearly dependent.

37. Suppose that $c_1\mathbf{x}_1(t)+c_2\mathbf{x}_2(t)+\cdots+c_n\mathbf{x}_n(t) \equiv \mathbf{0}$. Then the ith scalar component of this vector equation is $c_1 x_{i1}(t)+c_2 x_{i2}(t)+\cdots+c_n x_{in}(t) \equiv 0$. Hence the fact that the scalar functions $x_{i1}(t), x_{i2}(t), \cdots, x_{in}(t)$ are linear linearly independent implies that $c_1 = c_2 = \cdots c_n = 0$. Consequently the vector functions $\mathbf{x}_1(t), \mathbf{x}_2(t), \cdots, \mathbf{x}_n(t)$ are linearly independent.

SECTION 7.3

THE EIGENVALUE METHOD FOR LINEAR SYSTEMS

In each of Problems 1–16 we give the characteristic equation, the eigenvalues λ_1 and λ_2 of the coefficient matrix of the given system, the corresponding equations determining the associated eigenvectors $\mathbf{v}_1 = [a_1 \ b_1]^T$ and $\mathbf{v}_2 = [a_2 \ b_2]^T$, these eigenvectors, and the resulting scalar components $x_1(t)$ and $x_2(t)$ of a general solution $\mathbf{x}(t) = c_1\mathbf{v}_1 e^{\lambda_1 t}+c_2\mathbf{v}_2 e^{\lambda_2 t}$ of the system.

1. Characteristic equation $\quad \lambda^2 - 2\lambda - 3 = 0$

   Eigenvalues $\quad \lambda_1 = -1$ and $\lambda_2 = 3$

   Eigenvector equations $\quad \begin{bmatrix} 2 & 2 \\ 2 & 2 \end{bmatrix}\begin{bmatrix} a_1 \\ b_1 \end{bmatrix} = \begin{bmatrix} 0 \\ 0 \end{bmatrix}$ and $\begin{bmatrix} -2 & 2 \\ 2 & -2 \end{bmatrix}\begin{bmatrix} a_2 \\ b_2 \end{bmatrix} = \begin{bmatrix} 0 \\ 0 \end{bmatrix}$

   Eigenvectors $\quad \mathbf{v}_1 = [1 \ -1]^T$ and $\mathbf{v}_2 = [1 \ 1]^T$

   $x_1(t) = c_1 e^{-t} + c_2 e^{3t}, \quad x_2(t) = -c_1 e^{-t} + c_2 e^{3t}$

3. Characteristic equation $\quad \lambda^2 - 5\lambda - 6 = 0$

   Eigenvalues $\quad \lambda_1 = -1$ and $\lambda_2 = 6$

   Eigenvector equations $\quad \begin{bmatrix} 4 & 4 \\ 3 & 3 \end{bmatrix}\begin{bmatrix} a_1 \\ b_1 \end{bmatrix} = \begin{bmatrix} 0 \\ 0 \end{bmatrix}$ and $\begin{bmatrix} -3 & 4 \\ 3 & -4 \end{bmatrix}\begin{bmatrix} a_2 \\ b_2 \end{bmatrix} = \begin{bmatrix} 0 \\ 0 \end{bmatrix}$

   Eigenvectors $\quad \mathbf{v}_1 = [1 \ -1]^T$ and $\mathbf{v}_2 = [4 \ 3]^T$

$$x_1(t) = c_1 e^{-t} + 4c_2 e^{6t}, \quad x_2(t) = -c_1 e^{-t} + 3c_2 e^{6t}$$

The equations
$$x_1(0) = c_1 + 4c_2 = 1$$
$$x_2(0) = -c_1 + 3c_2 = 1$$

yield $c_1 = -1/7$ and $c_2 = 2/7$, so the desired particular solution is given by

$$x_1(t) = (-e^{-t} + 8e^{6t})/7, \quad x_2(t) = (e^{-t} + 6e^{6t})/7.$$

5. Characteristic equation $\lambda^2 - 4\lambda - 5 = 0$

   Eigenvalues $\lambda_1 = -1$ and $\lambda_2 = 5$

   Eigenvector equations $\begin{bmatrix} 7 & -7 \\ 1 & -1 \end{bmatrix}\begin{bmatrix} a_1 \\ b_1 \end{bmatrix} = \begin{bmatrix} 0 \\ 0 \end{bmatrix}$ and $\begin{bmatrix} 1 & -7 \\ 1 & -7 \end{bmatrix}\begin{bmatrix} a_2 \\ b_2 \end{bmatrix} = \begin{bmatrix} 0 \\ 0 \end{bmatrix}$

   Eigenvectors $\mathbf{v}_1 = [1 \quad 1]^T$ and $\mathbf{v}_2 = [7 \quad 1]^T$

   $$x_1(t) = c_1 e^{-t} + 7c_2 e^{5t}, \quad x_2(t) = c_1 e^{-t} + c_2 e^{5t}$$

7. Characteristic equation $\lambda^2 + 8\lambda - 9 = 0$

   Eigenvalues $\lambda_1 = 1$ and $\lambda_2 = -9$

   Eigenvector equations $\begin{bmatrix} -4 & 4 \\ 6 & -6 \end{bmatrix}\begin{bmatrix} a_1 \\ b_1 \end{bmatrix} = \begin{bmatrix} 0 \\ 0 \end{bmatrix}$ and $\begin{bmatrix} 6 & 4 \\ 6 & 4 \end{bmatrix}\begin{bmatrix} a_2 \\ b_2 \end{bmatrix} = \begin{bmatrix} 0 \\ 0 \end{bmatrix}$

   Eigenvectors $\mathbf{v}_1 = [1 \quad 1]^T$ and $\mathbf{v}_2 = [2 \quad -3]^T$

   $$x_1(t) = c_1 e^{t} + 2c_2 e^{-9t}, \quad x_2(t) = c_1 e^{t} - 3c_2 e^{-9t}$$

9. Characteristic equation $\lambda^2 + 16 = 0$

   Eigenvalue $\lambda = 4i$

   Eigenvector equation $\begin{bmatrix} 2-4i & -5 \\ 4 & -2-4i \end{bmatrix}\begin{bmatrix} a \\ b \end{bmatrix} = \begin{bmatrix} 0 \\ 0 \end{bmatrix}$

   Eigenvector $\mathbf{v} = [5 \quad 2-4i]^T$

   The real and imaginary parts of

   $$\mathbf{x}(t) = \mathbf{v} e^{4it} = \begin{bmatrix} 5\cos 4t + 5i\sin 4t \\ (2\cos 4t + 4\sin 4t) + i(2\sin 4t - 4\cos 4t) \end{bmatrix}$$

   yield the general solution

$$x_1(t) = 5c_1\cos 4t + 5c_2\sin 4t$$
$$x_2(t) = c_1(2\cos 4t + 4\sin 4t) + c_2(2\sin 4t - 4\cos 4t).$$

The initial conditions $x_1(0) = 2$ and $x_2(0) = 3$ give $c_1 = 2/5$ and $c_2 = -11/20$, so the desired particular solution is

$$x_1(t) = 2\cos 4t - (11/4)\sin 4t$$
$$x_2(t) = 3\cos 4t + (1/2)\sin 4t.$$

11. Characteristic equation $\lambda^2 - 2\lambda + 5 = 0$

Eigenvalue $\lambda = 1 - 2i$

Eigenvector equation $\begin{bmatrix} 2i & -2 \\ 2 & 2i \end{bmatrix} \begin{bmatrix} a \\ b \end{bmatrix} = \begin{bmatrix} 0 \\ 0 \end{bmatrix}$

Eigenvector $\mathbf{v} = [1 \quad i]^T$

The real and imaginary parts of

$$\mathbf{x}(t) = [1 \quad i]^T e^t(\cos 2t - i\sin 2t)$$
$$= e^t[\cos 2t \quad \sin 2t]^T + ie^t[-\sin 2t \quad \cos 2t]^T$$

yield the general solution

$$x_1(t) = e^t(c_1\cos 2t - c_2\sin 2t)$$
$$x_2(t) = e^t(c_1\sin 2t + c_2\cos 2t).$$

The particular solution with $x_1(0) = 0$ and $x_2(0) = 4$ is obtained with $c_1 = 0$ and $c_2 = 4$, so
$$x_1(t) = -4e^t\sin 2t, \qquad x_2(t) = 4e^t\cos 2t.$$

13. Characteristic equation $\lambda^2 - 4\lambda + 13 = 0$

Eigenvalue $\lambda = 2 - 3i$

Eigenvector equation $\begin{bmatrix} 3+3i & -9 \\ 2 & -3-3i \end{bmatrix} \begin{bmatrix} a \\ b \end{bmatrix} = \begin{bmatrix} 0 \\ 0 \end{bmatrix}$

Eigenvector $\mathbf{v} = [3 \quad 1+i]^T$

$$\mathbf{x}(t) = \mathbf{v}e^{(2-3i)t} = e^{2t}\begin{bmatrix} 3\cos 3t - 3i\sin 3t \\ (\cos 3t + \sin 3t) + i(\cos 3t - \sin 3t) \end{bmatrix}$$

$x_1(t) = 3e^{2t}(c_1\cos 3t - c_2\sin 3t)$

$x_2(t) = e^{2t}[(c_1+c_2)\cos 3t + (c_1-c_2)\sin 3t]$.

15. Characteristic equation $\lambda^2 - 10\lambda + 41 = 0$

Eigenvalue $\lambda = 5 - 4i$

Eigenvector equation $\begin{bmatrix} 2-4i & -5 \\ 4 & -2-4i \end{bmatrix}\begin{bmatrix} a \\ b \end{bmatrix} = \begin{bmatrix} 0 \\ 0 \end{bmatrix}$

Eigenvector $\mathbf{v} = [5 \quad 2+4i]^T$

$\mathbf{x}(t) = \mathbf{v}e^{(5-4i)t} = e^{5t}\begin{bmatrix} 5\cos 4t - 5i\sin 4t \\ (2\cos 4t + 4\sin 4t) + i(4\cos 4t - 2\sin 4t) \end{bmatrix}$

$x_1(t) = 5e^{5t}(c_1\cos 4t - c_2\sin 4t)$

$x_2(t) = e^{5t}[(2c_1+4c_2)\cos 4t + (4c_1-2c_2)\sin 4t]$

17. Characteristic equation $-\lambda^3 + 15\lambda^2 - 54\lambda = 0$

Eigenvalues $\lambda_1 = 9, \lambda_2 = 6, \lambda_3 = 0$

Eigenvector equations

$\begin{bmatrix} -5 & 1 & 4 \\ 1 & -2 & 1 \\ 4 & 1 & -5 \end{bmatrix}\begin{bmatrix} a_1 \\ b_1 \\ c_1 \end{bmatrix} = \begin{bmatrix} 0 \\ 0 \\ 0 \end{bmatrix}, \quad \begin{bmatrix} -2 & 1 & 4 \\ 1 & 1 & 1 \\ 4 & 1 & -2 \end{bmatrix}\begin{bmatrix} a_2 \\ b_2 \\ c_2 \end{bmatrix} = \begin{bmatrix} 0 \\ 0 \\ 0 \end{bmatrix}, \quad \begin{bmatrix} 4 & 1 & 4 \\ 1 & 7 & 1 \\ 4 & 1 & 4 \end{bmatrix}\begin{bmatrix} a_3 \\ b_3 \\ c_3 \end{bmatrix} = \begin{bmatrix} 0 \\ 0 \\ 0 \end{bmatrix}$

Eigenvectors $\mathbf{v}_1 = [1 \ 1 \ 1]^T, \quad \mathbf{v}_2 = [1 \ -2 \ 1]^T, \quad \mathbf{v}_3 = [1 \ 0 \ -1]^T$

$x_1(t) = c_1e^{9t} + c_2e^{6t} + c_3$

$x_2(t) = c_1e^{9t} - 2c_2e^{6t}$

$x_3(t) = c_1e^{9t} + c_2e^{6t} - c_3$

19. Characteristic equation $-\lambda^3 + 12\lambda^2 - 45\lambda + 54 = 0$

Eigenvalues $\lambda_1 = 6, \lambda_2 = 3, \lambda_3 = 3$

Eigenvector equations

$\begin{bmatrix} -2 & 1 & 1 \\ 1 & -2 & 1 \\ 1 & 1 & -2 \end{bmatrix}\begin{bmatrix} a_1 \\ b_1 \\ c_1 \end{bmatrix} = \begin{bmatrix} 0 \\ 0 \\ 0 \end{bmatrix}, \quad \begin{bmatrix} 1 & 1 & 1 \\ 1 & 1 & 1 \\ 1 & 1 & 1 \end{bmatrix}\begin{bmatrix} a_2 \\ b_2 \\ c_2 \end{bmatrix} = \begin{bmatrix} 0 \\ 0 \\ 0 \end{bmatrix}, \quad \begin{bmatrix} 1 & 1 & 1 \\ 1 & 1 & 1 \\ 1 & 1 & 1 \end{bmatrix}\begin{bmatrix} a_3 \\ b_3 \\ c_3 \end{bmatrix} = \begin{bmatrix} 0 \\ 0 \\ 0 \end{bmatrix}$

Eigenvectors $\mathbf{v}_1 = [1 \ 1 \ 1]^T, \quad \mathbf{v}_2 = [1 \ -2 \ 1]^T, \quad \mathbf{v}_3 = [1 \ 0 \ -1]^T$

$$x_1(t) = c_1 e^{6t} + c_2 e^{3t} + c_3 e^{3t}$$
$$x_2(t) = c_1 e^{6t} - 2c_2 e^{3t}$$
$$x_3(t) = c_1 e^{6t} + c_2 e^{3t} - c_3 e^{3t}$$

21. Characteristic equation $-\lambda^3 + \lambda = 0$

Eigenvalues $\lambda_1 = 0, \ \lambda_2 = 1, \ \lambda_3 = -1$

Eigenvector equations

$$\begin{bmatrix} -4 & 1 & 3 \\ 1 & -2 & 1 \\ 3 & 1 & -4 \end{bmatrix} \begin{bmatrix} a_1 \\ b_1 \\ c_1 \end{bmatrix} = \begin{bmatrix} 0 \\ 0 \\ 0 \end{bmatrix}, \quad \begin{bmatrix} 5 & 0 & -6 \\ 2 & -1 & -2 \\ 4 & -2 & -4 \end{bmatrix} \begin{bmatrix} a_2 \\ b_2 \\ c_2 \end{bmatrix} = \begin{bmatrix} 0 \\ 0 \\ 0 \end{bmatrix}, \quad \begin{bmatrix} 6 & 0 & -6 \\ 2 & 0 & -2 \\ 4 & -2 & -3 \end{bmatrix} \begin{bmatrix} a_3 \\ b_3 \\ c_3 \end{bmatrix} = \begin{bmatrix} 0 \\ 0 \\ 0 \end{bmatrix}$$

Eigenvectors $\mathbf{v}_1 = [6 \ \ 2 \ \ 5]^T, \quad \mathbf{v}_2 = [3 \ \ 1 \ \ 2]^T, \quad \mathbf{v}_3 = [2 \ \ 1 \ \ 2]^T$

$$x_1(t) = 6c_1 + 3c_2 e^t + 2c_3 e^{-t}$$
$$x_2(t) = 2c_1 + c_2 e^t + c_3 e^{-t}$$
$$x_3(t) = 5c_1 + 2c_2 e^t + 2c_3 e^{-t}$$

23. Characteristic equation $-\lambda^3 + 3\lambda^2 + 4\lambda - 12 = 0$

Eigenvalues $\lambda_1 = 2, \ \lambda_2 = -2, \ \lambda_3 = 3$

Eigenvector equations

$$\begin{bmatrix} 1 & 1 & 1 \\ -5 & -5 & -1 \\ 5 & 5 & 1 \end{bmatrix} \begin{bmatrix} a_1 \\ b_1 \\ c_1 \end{bmatrix} = \begin{bmatrix} 0 \\ 0 \\ 0 \end{bmatrix}, \quad \begin{bmatrix} 5 & 1 & 1 \\ -5 & -1 & -1 \\ 5 & 5 & 5 \end{bmatrix} \begin{bmatrix} a_2 \\ b_2 \\ c_2 \end{bmatrix} = \begin{bmatrix} 0 \\ 0 \\ 0 \end{bmatrix}, \quad \begin{bmatrix} 0 & 1 & 1 \\ -5 & -6 & -1 \\ 5 & 5 & 0 \end{bmatrix} \begin{bmatrix} a_3 \\ b_3 \\ c_3 \end{bmatrix} = \begin{bmatrix} 0 \\ 0 \\ 0 \end{bmatrix}$$

Eigenvectors $\mathbf{v}_1 = [1 \ -1 \ \ 0]^T, \quad \mathbf{v}_2 = [0 \ \ 1 \ -1]^T, \quad \mathbf{v}_3 = [1 \ -1 \ \ 1]^T$

$$x_1(t) = c_1 e^{2t} \qquad\qquad + c_3 e^{3t}$$
$$x_2(t) = -c_1 e^{2t} + c_2 e^{-2t} - c_3 e^{3t}$$
$$x_3(t) = \qquad\qquad - c_2 e^{-2t} + c_3 e^{3t}$$

25. Characteristic equation $-\lambda^3 + 4\lambda^2 - 13\lambda = 0$

Eigenvalues $\lambda = 0$ and $2 \pm 3i$

With $\lambda = 1$ the eigenvector equation

$$\begin{bmatrix} 5 & 5 & 2 \\ -6 & -6 & -5 \\ 6 & 6 & 5 \end{bmatrix} \begin{bmatrix} a_1 \\ b_1 \\ c_1 \end{bmatrix} = \begin{bmatrix} 0 \\ 0 \\ 0 \end{bmatrix} \quad \text{gives eigenvector } \mathbf{v}_1 = [1 \ -1 \ 0]^T.$$

With $\lambda = 2 + 3i$ we solve the eigenvector equation

$$\begin{bmatrix} 3-3i & 5 & 2 \\ -6 & -8-3i & -5 \\ 6 & 6 & 3-3i \end{bmatrix} \begin{bmatrix} a \\ b \\ c \end{bmatrix} = \begin{bmatrix} 0 \\ 0 \\ 0 \end{bmatrix}$$

to find the complex-valued eigenvector $\mathbf{v} = [1+i \ -2 \ 2]^T$. The corresponding complex-valued solution is

$$\mathbf{x}(t) = \mathbf{v}e^{(2+3i)t} = e^{2t} \begin{bmatrix} (\cos 3t - \sin 3t) + i(\cos 3t + \sin 3t) \\ -2\cos 3t - 2i\sin 3t \\ 2\cos 3t + 2i\sin 3t \end{bmatrix}.$$

The scalar components of the resulting general solution are

$$x_1(t) = c_1 + e^{2t}[(c_2 + c_3)\cos 3t + (-c_2 + c_3)\sin 3t]$$
$$x_2(t) = -c_1 + 2e^{2t}(-c_2\cos 3t - c_3\sin 3t)$$
$$x_3(t) = 2e^{2t}(c_2\cos 3t + c_3\sin 3t)$$

27. The coefficient matrix

$$\mathbf{A} = \begin{bmatrix} -0.2 & 0 \\ 0.2 & -0.4 \end{bmatrix}$$

has characteristic equation $\lambda^2 + 0.6\lambda + 0.08 = 0$ with eigenvalues $\lambda_1 = -0.2$ and $\lambda_2 = -0.4$. We find easily that the associated eigenvectors are $\mathbf{v}_1 = [1 \ \ 1]^T$ and $\mathbf{v}_2 = [0 \ \ 1]^T$, so we get the general solution

$$x_1(t) = c_1 e^{-0.2t}, \quad x_2(t) = c_1 e^{-0.2t} + c_2 e^{-0.4t}.$$

The initial conditions $x_1(0) = 15$, $x_2(0) = 0$ give $c_1 = 15$ and $c_2 = -15$, so we get

$$x_1(t) = 15e^{-0.2t}, \quad x_2(t) = 15e^{-0.2t} - 15e^{-0.4t}.$$

To find the maximum value of $x_2(t)$, we solve the equation $x_2'(t) = 0$ for $t = 5\ln 2$, which gives the maximum value $x_2(5\ln 2) = 3.75$ lb.

29. The coefficient matrix

$$A = \begin{bmatrix} -0.2 & 0.4 \\ 0.2 & -0.4 \end{bmatrix}$$

has eigenvalues $\lambda_1 = 0$ and $\lambda_2 = -0.6$, with eigenvectors $\mathbf{v}_1 = [2\ \ 1]^T$ and $\mathbf{v}_2 = [1\ \ -1]^T$ that yield the general solution

$$x_1(t) = 2c_1 + c_2 e^{-0.6t}, \quad x_2(t) = c_1 - c_2 e^{-0.6t}.$$

The initial conditions $x_1(0) = 15$, $x_2(0) = 0$ give $c_1 = c_2 = 5$, so we get

$$x_1(t) = 10 + 5e^{-0.6t}, \quad x_2(t) = 5 - 5e^{-0.6t}.$$

31. The coefficient matrix

$$A = \begin{bmatrix} -1 & 0 & 0 \\ 1 & -2 & 0 \\ 0 & 2 & -3 \end{bmatrix}$$

has as eigenvalues its diagonal elements $\lambda_1 = -1$, $\lambda_2 = -2$, and $\lambda_3 = -3$. We find readily that the associated eigenvectors are $\mathbf{v}_1 = [1\ \ 1\ \ 1]^T$, $\mathbf{v}_2 = [0\ \ 1\ \ 2]^T$, and $\mathbf{v}_3 = [0\ \ 0\ \ 1]^T$. The resulting general solution is solution is given by

$$x_1(t) = c_1 e^{-t}$$
$$x_2(t) = c_1 e^{-t} + c_2 e^{-2t}$$
$$x_3(t) = c_1 e^{-t} + 2c_2 e^{-2t} + c_3 e^{-3t}.$$

The initial conditions $x_1(0) = 27$, $x_2(0) = x_2(0) = 0$ give $c_1 = c_3 = 27$, $c_2 = -27$, so we get

$$x_1(t) = 27 e^{-t}$$
$$x_2(t) = 27 e^{-t} - 27 e^{-2t}$$
$$x_3(t) = 27 e^{-t} - 54 e^{-2t} + 27 e^{-3t}.$$

The equation $x_3'(t) = 0$ simplifies to the equation

$$3e^{-2t} - 4e^{-t} + 1 = (3e^{-t} - 1)(e^{-t} - 1) = 0$$

with positive solution $t_m = \ln 3$. Thus the maximum amount of salt ever in tank 3 is $x_3(\ln 3) = 4$ pounds.

154　　　　　　　　　　　　　　　　　Chapter 7

33. The coefficient matrix

$$A = \begin{bmatrix} -4 & 0 & 0 \\ 4 & -6 & 0 \\ 0 & 6 & -2 \end{bmatrix}$$

has as eigenvalues its diagonal elements $\lambda_1 = -4$, $\lambda_2 = -6$, and $\lambda_3 = -2$. We find readily that the associated eigenvectors are $\mathbf{v}_1 = [-1 \ -2 \ 6]^T$, $\mathbf{v}_2 = [0 \ -2 \ 3]^T$, and $\mathbf{v}_3 = [0 \ 0 \ 1]^T$. The resulting general solution is solution is given by

$$x_1(t) = -c_1 e^{-4t}$$
$$x_2(t) = -2c_1 e^{-4t} - 2c_2 e^{-6t}$$
$$x_3(t) = 6c_1 e^{-4t} + 3c_2 e^{-6t} + c_3 e^{-2t}.$$

The initial conditions $x_1(0) = 45$, $x_2(0) = x_2(0) = 0$ give $c_1 = -45$, $c_2 = 45$, $c_3 = 135$, so we get

$$x_1(t) = 45 e^{-4t}$$
$$x_2(t) = 90 e^{-4t} - 90 e^{-6t}$$
$$x_3(t) = -270 e^{-4t} + 135 e^{-6t} + 135 e^{-2t}.$$

The equation $x_3'(t) = 0$ simplifies to the equation

$$3e^{-4t} - 4e^{-2t} + 1 = (3e^{-2t} - 1)(e^{-2t} - 1) = 0$$

with positive solution $t_m = \frac{1}{2}\ln 3$. Thus the maximum amount of salt ever in tank 3 is $x_3(\frac{1}{2}\ln 3) = 20$ pounds.

35. The coefficient matrix

$$A = \begin{bmatrix} -6 & 0 & 3 \\ 6 & -20 & 0 \\ 0 & 20 & -3 \end{bmatrix}$$

has characteristic equation $-\lambda^3 - 29\lambda^2 - 198\lambda = -\lambda(\lambda - 18)(\lambda - 11) = 0$ with eigenvalues $\lambda_0 = 0$, $\lambda_1 = -18$, and $\lambda_2 = -11$. We find that associated eigenvectors are $\mathbf{v}_0 = [10 \ 3 \ 20]^T$, $\mathbf{v}_1 = [-1 \ -3 \ 4]^T$, and $\mathbf{v}_2 = [-3 \ -2 \ 5]^T$. The resulting general solution is solution is given by

$$x_1(t) = 10c_0 - c_1 e^{-18t} - 3c_2 e^{-11t}$$
$$x_2(t) = 3c_0 - 3c_1 e^{-18t} - 2c_2 e^{-11t}$$
$$x_3(t) = 20c_0 + 4c_1 e^{-18t} + 5c_2 e^{-11t}.$$

The initial conditions $x_1(0) = 33$, $x_2(0) = x_2(0) = 0$ give $c_1 = 1$, $c_2 = 55/7$, $c_3 = -72/7$, so we get

$$x_1(t) = 10 - \tfrac{1}{7}\left(55e^{-18t} - 216e^{-11t}\right)$$
$$x_2(t) = 3 - \tfrac{1}{7}\left(165e^{-18t} - 144e^{-11t}\right)$$
$$x_3(t) = 20 + \tfrac{1}{7}\left(220e^{-18t} - 360e^{-11t}\right).$$

Thus the limiting amounts of salt in tanks 1, 2, and 3 are 10 lb, 3 lb, and 20 lb.

37. The coefficient matrix

$$\mathbf{A} = \begin{bmatrix} -1 & 0 & 2 \\ 1 & -3 & 0 \\ 0 & 3 & -2 \end{bmatrix}$$

has characteristic equation $-\lambda^3 - 6\lambda^2 - 11\lambda = 0$ with eigenvalues $\lambda_0 = 0$, $\lambda_1 = -3 - i\sqrt{2}$, and $\lambda_2 = -3 + i\sqrt{2}$. The eigenvector equation

$$\begin{bmatrix} -1 & 0 & 2 \\ 1 & -3 & 0 \\ 0 & 3 & -2 \end{bmatrix} \begin{bmatrix} a \\ b \\ c \end{bmatrix} = \begin{bmatrix} 0 \\ 0 \\ 0 \end{bmatrix}$$

associated with the eigenvalue $\lambda_0 = 0$ yields the associated eigenvector $\mathbf{v}_0 = [6 \ 2 \ 3]^T$ and consequently the constant solution $\mathbf{x}_0(t) \equiv \mathbf{v}_0$. Then the eigenvector equation

$$\begin{bmatrix} 2+i\sqrt{2} & 0 & 2 \\ 1 & i\sqrt{2} & 0 \\ 0 & 3 & 1+i\sqrt{2} \end{bmatrix} \begin{bmatrix} a \\ b \\ c \end{bmatrix} = \begin{bmatrix} 0 \\ 0 \\ 0 \end{bmatrix}$$

associated with $\lambda_1 = -3 - i\sqrt{2}$ yields the complex-valued eigenvector $\mathbf{v}_1 = \left[(-2+i\sqrt{2})/3 \ \ (-1-i\sqrt{2})/3 \ \ 1\right]^T$. The corresponding complex-valued solution is

$$\mathbf{x}_1(t) = \mathbf{v}_1 e^{(-3-i\sqrt{2})t}$$

$$= \frac{1}{3}e^{-3t}\begin{bmatrix} \left(-2\cos(t\sqrt{2})+\sqrt{2}\sin(t\sqrt{2})\right)+i\left(\sqrt{2}\cos(t\sqrt{2})+2\sin(t\sqrt{2})\right) \\ \left(-\cos(t\sqrt{2})-\sqrt{2}\sin(t\sqrt{2})\right)+i\left(-\sqrt{2}\cos(t\sqrt{2})+\sin(t\sqrt{2})\right) \\ 3\cos(t\sqrt{2})-3i\sin(t\sqrt{2}) \end{bmatrix}.$$

The scalar components of resulting general solution $\mathbf{x} = c_0\mathbf{x}_0 + c_1\,\mathrm{Re}(\mathbf{x}_1) + c_2\,\mathrm{Im}(\mathbf{x}_1)$ are given by

$$x_1(t) = 6c_0 + \tfrac{1}{3}e^{-3t}\left[\left(-2c_1+\sqrt{2}c_2\right)\cos(t\sqrt{2})+\left(\sqrt{2}c_1+2c_2\right)\sin(t\sqrt{2})\right]$$

$$x_2(t) = 2c_0 + \tfrac{1}{3}e^{-3t}\left[\left(-c_1-\sqrt{2}c_2\right)\cos(t\sqrt{2})+\left(-\sqrt{2}c_1+c_2\right)\sin(t\sqrt{2})\right]$$

$$x_3(t) = 3c_0 + e^{-3t}\left[c_1\cos(t\sqrt{2})-c_2\sin(t\sqrt{2})\right].$$

When we impose the initial conditions $x_1(0) = 55$, $x_2(0) = x_2(0) = 0$ we find that $c_0 = 5$, $c_1 = -15$, and $c_2 = 45/\sqrt{2}$. This finally gives the particular solution

$$x_1(t) = 30 + e^{-3t}\left[25\cos(t\sqrt{2})+10\sqrt{2}\sin(t\sqrt{2})\right]$$

$$x_2(t) = 10 - e^{-3t}\left[10\cos(t\sqrt{2})-\tfrac{25}{2}\sqrt{2}\sin(t\sqrt{2})\right]$$

$$x_3(t) = 15 - e^{-3t}\left[15\cos(t\sqrt{2})+\tfrac{45}{2}\sqrt{2}\sin(t\sqrt{2})\right].$$

Thus the limiting amounts of salt in tanks 1, 2, and 3 are 30 lb, 10 lb, and 15 lb.

In Problems 38-41 the Maple command `with(linalg):eigenvects(A)`, the Mathematica command `Eigensystem[A]`, or the MATLAB command `[V,D] = eig(A)` can be used to find the eigenvalues and associated eigenvectors of the given coefficient matrix **A**.

39. Characteristic equation: $(\lambda^2 - 1)(\lambda^2 - 4) = 0$

Eigenvalues and associated eigenvectors:

$\lambda = 1$, $\mathbf{v} = [3\ \ -2\ \ 4\ \ 1]^T$
$\lambda = -1$, $\mathbf{v} = [0\ \ 0\ \ 1\ \ 0]^T$
$\lambda = 2$, $\mathbf{v} = [0\ \ 1\ \ 0\ \ 0]^T$
$\lambda = -2$, $\mathbf{v} = [1\ \ -1\ \ 0\ \ 0]^T$

Section 7.3

Scalar solution equations:

$$x_1(t) = 3c_1e^t \qquad\qquad + c_4e^{-2t}$$
$$x_2(t) = -2c_1e^t \quad + c_3e^{2t} - c_4e^{-2t}$$
$$x_3(t) = 4c_1e^t + c_2e^{-t}$$
$$x_4(t) = \quad c_1e^t$$

41. The eigenvectors associated with the respective eigenvalues $\lambda_1 = -3$, $\lambda_2 = -6$, $\lambda_3 = 10$, and $\lambda_4 = 15$ are

$$\mathbf{v}_1 = [\ 1 \quad 0 \quad 0 \quad -1\ ]^T$$
$$\mathbf{v}_2 = [\ 0 \quad 1 \quad -1 \quad 0\ ]^T$$
$$\mathbf{v}_3 = [-2 \quad 1 \quad 1 \quad -2\ ]^T$$
$$\mathbf{v}_4 = [\ 1 \quad 2 \quad 2 \quad 1\ ]^T.$$

Hence the general solution has scalar component functions

$$x_1(t) = c_1e^{-3t} \qquad\qquad - 2c_3e^{10t} + c_4e^{15t}$$
$$x_2(t) = \qquad\quad c_2e^{-6t} + c_3e^{10t} + 2c_4e^{15t}$$
$$x_3(t) = \qquad\quad -c_2e^{-6t} + c_3e^{10t} + 2c_4e^{15t}$$
$$x_4(t) = -c_1e^{-3t} \qquad\qquad - 2c_3e^{10t} + c_4e^{15t}.$$

The given initial conditions are satisfied by choosing $c_1 = c_2 = 0$, $c_3 = -1$, and $c_4 = 1$, so the desired particular solution is given by

$$x_1(t) = 2e^{10t} + e^{15t} = x_4(t)$$
$$x_2(t) = -e^{10t} + 2e^{15t} = x_3(t).$$

SECTION 7.4

SECOND-ORDER SYSTEMS AND MECHANICAL APPLICATIONS

This section uses the eigenvalue method to exhibit realistic applications of linear systems. If a computer system like Maple, Mathematica, MATLAB, or even a TI-85/86/89/92 calculator is available, then a system of more than three railway cars, or a multistory building with four or more floors (as in the project), can be investigated. However, the problems in the text are intended for manual solution.

Problems 1-7 involve the system

$$m_1 x_1'' = -(k_1+k_2)x_1 + k_2 x_2$$
$$m_2 x_2'' = k_2 x_1 - (k_2+k_3)x_2$$

with various values of m_1, m_2 and k_1, k_2, k_3. In each problem we divide the first equation by m_1 and the second one by m_2 to obtain a second-order linear system $\mathbf{x}'' = \mathbf{A}\mathbf{x}$ in the standard form of Theorem 1 in this section. If the eigenvalues λ_1 and λ_2 are both negative, then the natural (circular) frequencies of the system are $\omega_1 = \sqrt{-\lambda_1}$ and $\omega_2 = \sqrt{-\lambda_2}$, and — according to Eq. (11) in Theorem 1 of this section — the eigenvalues $\mathbf{v}_1$ and $\mathbf{v}_2$ associated with λ_1 and λ_2 determine the natural modes of oscillations at these frequencies.

1. The matrix $\mathbf{A} = \begin{bmatrix} -2 & 2 \\ 2 & -2 \end{bmatrix}$ has eigenvalues $\lambda_0 = 0$ and $\lambda_1 = -4$ with associated eigenvalues $\mathbf{v}_0 = [1\ \ 1]^T$ and $\mathbf{v}_1 = [1\ \ -1]^T$. Thus we have the special case described in Eq. (12) of Theorem 1, and a general solution is given by

$$x_1(t) = a_1 + a_2 t + b_1 \cos 2t + b_2 \sin 2t,$$
$$x_2(t) = a_1 + a_2 t - b_1 \cos 2t - b_2 \sin 2t.$$

The natural frequencies are $\omega_1 = 0$ and $\omega_2 = 2$. In the degenerate natural mode with "frequency" $\omega_1 = 0$ the two masses move by translation without oscillating. At frequency $\omega_2 = 2$ they oscillate in opposite directions with equal amplitudes.

3. The matrix $\mathbf{A} = \begin{bmatrix} -3 & 2 \\ 1 & -2 \end{bmatrix}$ has eigenvalues $\lambda_1 = -1$ and $\lambda_2 = -4$ with associated eigenvalues $\mathbf{v}_1 = [1\ \ 1]^T$ and $\mathbf{v}_2 = [2\ \ -1]^T$. Hence a general solution is given by

$$x_1(t) = a_1 \cos t + a_2 \sin t + 2b_1 \cos 2t + 2b_2 \sin 2t,$$
$$x_2(t) = a_1 \cos t + a_2 \sin t - b_1 \cos 2t - b_2 \sin 2t.$$

The natural frequencies are $\omega_1 = 1$ and $\omega_2 = 2$. In the natural mode with frequency ω_1, the two masses m_1 and m_2 move in the same direction with equal amplitudes of oscillation. In the natural mode with frequency ω_2 they move in opposite directions with the amplitude of oscillation of m_1 twice that of m_2.

5. The matrix $\mathbf{A} = \begin{bmatrix} -3 & 1 \\ 1 & -3 \end{bmatrix}$ has eigenvalues $\lambda_1 = -2$ and $\lambda_2 = -4$ with associated eigenvalues $\mathbf{v}_1 = [1\ \ 1]^T$ and $\mathbf{v}_2 = [1\ \ -1]^T$. Hence a general solution is given by

$$x_1(t) = a_1 \cos t\sqrt{2} + a_2 \sin t\sqrt{2} + b_1 \cos 2t + b_2 \sin 2t,$$
$$x_2(t) = a_1 \cos t\sqrt{2} + a_2 \sin t\sqrt{2} - b_1 \cos 2t - b_2 \sin 2t.$$

The natural frequencies are $\omega_1 = \sqrt{2}$ and $\omega_2 = 2$. In the natural mode with frequency ω_1, the two masses m_1 and m_2 move in the same direction with equal amplitudes of oscillation. At frequency ω_2 they move in opposite directions with equal amplitudes.

7. The matrix $\mathbf{A} = \begin{bmatrix} -10 & 6 \\ 6 & -10 \end{bmatrix}$ has eigenvalues $\lambda_1 = -4$ and $\lambda_2 = -16$ with associated eigenvalues $\mathbf{v}_1 = [1 \quad 1]^T$ and $\mathbf{v}_2 = [1 \quad -1]^T$. Hence a general solution is given by

$$x_1(t) = a_1 \cos 2t + a_2 \sin 2t + b_1 \cos 4t + b_2 \sin 4t,$$
$$x_2(t) = a_1 \cos 2t + a_2 \sin 2t - b_1 \cos 4t - b_2 \sin 4t.$$

The natural frequencies are $\omega_1 = 2$ and $\omega_2 = 4$. In the natural mode with frequency ω_1, the two masses m_1 and m_2 move in the same direction with equal amplitudes of oscillation. At frequency ω_2 they move in opposite directions with equal amplitudes.

9. Substitution of the trial solution $x_1 = c_1 \cos 3t$, $x_2 = c_2 \cos 3t$ in the system

$$x_1'' = -3x_1 + 2x_2, \quad 2x_2'' = 2x_1 - 4x_2 + 120\cos 3t$$

yields $c_1 = 3$, $c_2 = -9$, so a general solution is given by

$$x_1(t) = a_1 \cos t + a_2 \sin t + 2b_1 \cos 2t + 2b_2 \sin 2t + 3\cos 3t,$$
$$x_2(t) = a_1 \cos t + a_2 \sin t - b_1 \cos 2t - b_2 \sin 2t - 9\cos 3t.$$

Imposition of the initial conditions $x_1(0) = x_2(0) = x_1'(0) = x_2'(0) = 0$ now yields $a_1 = 5$, $a_2 = 0$, $b_1 = -4$, $b_2 = 0$. The resulting particular solution is

$$x_1(t) = 5\cos t - 8\cos 2t + 3\cos 3t,$$
$$x_2(t) = 5\cos t + 4\cos 2t - 9\cos 3t.$$

We have a superposition of three oscillations, in which the two masses move

- in the same direction with frequency $\omega_1 = 1$ and equal amplitudes;
- in opposite directions with frequency $\omega_2 = 2$ and with the amplitude of motion of m_1 being twice that of m_2;
- in opposite directions with frequency $\omega_3 = 3$ and with the amplitude of motion of m_2 being 3 times that of m_1.

11. (a) The matrix $\mathbf{A} = \begin{bmatrix} -40 & 8 \\ 12 & -60 \end{bmatrix}$ has eigenvalues $\lambda_1 = -36$ and $\lambda_2 = -64$ with associated eigenvalues $\mathbf{v}_2 = [2 \quad 1]^T$ and $\mathbf{v}_2 = [1 \quad -3]^T$. Hence a general solution is given by

$$x(t) = 2a_1 \cos 6t + 2a_2 \sin 6t + b_1 \cos 8t + b_2 \sin 8t,$$
$$y(t) = a_1 \cos 6t + a_2 \sin 6t - 3b_1 \cos 8t - 3b_2 \sin 8t.$$

The natural frequencies are $\omega_1 = 6$ and $\omega_2 = 8$. In mode 1 the two masses oscillate in the same direction with frequency $\omega_1 = 6$ and with the amplitude of motion of m_1 being twice that of m_2. In mode 2 the two masses oscillate in opposite directions with frequency $\omega_2 = 8$ and with the amplitude of motion of m_2 being 3 times that of m_1.

(b) Substitution of the trial solution $x = c_1 \cos 7t$, $y = c_2 \cos 7t$ in the system

$$x'' = -40x + 8y - 195\cos 7t, \qquad y'' = 12x - 60y - 195\cos 7t$$

yields $c_1 = 19$, $c_2 = 3$, so a general solution is given by

$$x(t) = 2a_1 \cos 6t + 2a_2 \sin 6t + b_1 \cos 8t + b_2 \sin 8t + 19\cos 7t,$$
$$y(t) = a_1 \cos 6t + a_2 \sin 6t - 3b_1 \cos 8t - 3b_2 \sin 8t + 3\cos 7t.$$

Imposition of the initial conditions $x(0) = 19$, $x'(0) = 12$, $y(0) = 3$, $y'(0) = 6$ now yields $a_1 = 0$, $a_2 = 1$, $b_1 = 0$, $b_2 = 0$. The resulting particular solution is

$$x(t) = 2\sin 6t + 19\cos 7t,$$
$$y(t) = \sin 6t + 3\cos 7t.$$

Thus the expected oscillation with frequency $\omega_2 = 8$ is missing, and we have a superposition of (only two) oscillations, in which the two masses move

- in the same direction with frequency $\omega_1 = 6$ and with the amplitude of motion of m_1 being twice that of m_2;
- in the same direction with frequency $\omega_3 = 7$ and with the amplitude of motion of m_1 being 19/7 times that of m_2.

13. The coefficient matrix $\mathbf{A} = \begin{bmatrix} -4 & 2 & 0 \\ 2 & -4 & 2 \\ 0 & 2 & -4 \end{bmatrix}$ has characteristic polynomial

$$-\lambda^3 - 12\lambda^2 - 40\lambda - 32 = -(\lambda + 4)(\lambda^2 + 8\lambda + 8).$$

Its eigenvalues $\lambda_1 = -4$, $\lambda_2 = -4-2\sqrt{2}$, $\lambda_3 = -4+2\sqrt{2}$ have associated eigenvectors $\mathbf{v}_1 = [1 \ \ 0 \ \ -1]^T$, $\mathbf{v}_2 = [1 \ \ -\sqrt{2} \ \ 1]^T$, $\mathbf{v}_3 = [1 \ \ \sqrt{2} \ \ 1]^T$. Hence the system's three natural modes of oscillation have

- Natural frequency $\omega_1 = 2$ with amplitude ratios $1 : 0 : -1$.
- Natural frequency $\omega_2 = \sqrt{4+2\sqrt{2}}$ with amplitude ratios $1 : -\sqrt{2} : 1$.
- Natural frequency $\omega_2 = \sqrt{4-2\sqrt{2}}$ with amplitude ratios $1 : \sqrt{2} : 1$.

15. First we need the general solution of the homogeneous system $\mathbf{x}'' = \mathbf{A}\mathbf{x}$ with

$$\mathbf{A} = \begin{bmatrix} -50 & 25/2 \\ 50 & -50 \end{bmatrix}$$

The eigenvalues of $\mathbf{A}$ are $\lambda_1 = -25$ and $\lambda_2 = -75$, so the natural frequencies of the system are $\omega_1 = 5$ and $\omega_2 = 5\sqrt{3}$. The associated eigenvectors are $\mathbf{v}_1 = [1 \ \ 2]^T$ and $\mathbf{v}_2 = [1 \ \ -2]^T$, so the complementary solution $\mathbf{x}_c(t)$ is given by

$$x_1(t) = a_1 \cos 5t + a_2 \sin 5t + b_1 \cos 5\sqrt{3}t + b_2 \sin 5\sqrt{3}t,$$
$$x_2(t) = 2a_1 \cos 5t + 2a_2 \sin 5t - 2b_1 \cos 5\sqrt{3}t - 2b_2 \sin 5\sqrt{3}t.$$

When we substitute the trial solution $\mathbf{x}_p(t) = [c_1 \ \ c_2]^T \cos 10t$ in the nonhomogeneous system, we find that $c_1 = 4/3$ and $c_2 = -16/3$, so a particular solution $\mathbf{x}_p(t)$ is described by

$$x_1(t) = (4/3)\cos 10t, \qquad x_2(t) = -(16/3)\cos 10t.$$

Finally, when we impose the zero initial conditions on the solution $\mathbf{x}(t) = \mathbf{x}_c(t) + \mathbf{x}_p(t)$ we find that $a_1 = 2/3$, $a_2 = 0$, $b_1 = -2$, and $b_2 = 0$. Thus the solution we seek is described by

$$x_1(t) = (2/3)\cos 5t - 2\cos 5\sqrt{3}t + (4/3)\cos 10t$$
$$x_2(t) = (4/3)\cos 5t + 4\cos 5\sqrt{3}t + (16/3)\cos 10t.$$

We have a superposition of two oscillations with the natural frequencies $\omega_1 = 5$ and $\omega_2 = 5\sqrt{3}$ and a forced oscillation with frequency $\omega = 10$. In each of the two natural oscillations the amplitude of motion of m_2 is twice that of m_1, while in the forced oscillation the amplitude of motion of m_2 is four times that of m_1.

17. With $c_1 = c_2 = 2$, it follows from Problem 16 that the natural frequencies and associated eigenvectors are $\omega_1 = 0$, $\mathbf{v}_1 = [1 \ \ 1]^T$ and $\omega_2 = 2$, $\mathbf{v}_2 = [1 \ \ -1]^T$.

Hence Theorem 1 gives the general solution

$$x_1(t) = a_1 + b_1 t + a_2 \cos 2t + b_2 \sin 2t$$
$$x_2(t) = a_1 + b_1 t - a_2 \cos 2t - b_2 \sin 2t.$$

The initial conditions $x_1'(0) = v_0$, $x_1(0) = x_2(0) = x_2'(0) = 0$ yield $a_1 = a_2 = 0$ and $b_1 = v_0/2$, $b_2 = v_0/4$, so

$$x_1(t) = (v_0/4)(2t + \sin 2t)$$
$$x_2(t) = (v_0/4)(2t - \sin 2t)$$

while $x_2 - x_1 = (v_0/4)(-2 \sin 2t) < 0$, that is, until $t = \pi/2$. Finally, $x_1'(\pi/2) = 0$ and $x_2'(\pi/2) = v_0$.

19. With $c_1 = 1$ and $c_2 = 3$, it follows from Problem 16 that the natural frequencies and associated eigenvectors are $\omega_1 = 0$, $\mathbf{v}_1 = [1 \quad 1]^T$ and $\omega_2 = 2$, $\mathbf{v}_2 = [1 \quad -3]^T$. Hence Theorem 1 gives the general solution

$$x_1(t) = a_1 + b_1 t + a_2 \cos 2t + b_2 \sin 2t$$
$$x_2(t) = a_1 + b_1 t - 3a_2 \cos 2t - 3b_2 \sin 2t.$$

The initial conditions $x_1'(0) = v_0$, $x_1(0) = x_2(0) = x_2'(0) = 0$ yield $a_1 = a_2 = 0$ and $b_1 = 3v_0/4$, $b_2 = v_0/8$, so

$$x_1(t) = (v_0/8)(6t + \sin 2t)$$
$$x_2(t) = (v_0/8)(6t - 3 \sin 2t)$$

while $x_2 - x_1 = (v_0/8)(-4 \sin 2t) < 0$; that is, until $t = \pi/2$. Finally, $x_1'(\pi/2) = v_0/2$ and $x_2'(\pi/2) = 3v_0/2$.

21. (a) The matrix

$$\mathbf{A} = \begin{bmatrix} -160/3 & 320/3 \\ 8 & -116 \end{bmatrix}$$

has eigenvalues $\lambda_1 \approx -41.8285$ and $\lambda_2 \approx -127.5049$, so the natural frequencies are

$$\omega_1 \approx 6.4675 \text{ rad/sec} \approx 1.0293 \text{ Hz}$$
$$\omega_2 \approx 11.2918 \text{ rad/sec} \approx 1.7971 \text{ Hz}.$$

(b) Resonance occurs at the two critical speeds

$$v_1 = 20\omega_1/\pi \approx 41 \text{ ft/sec} \approx 28 \text{ mi/h}$$
$$v_2 = 20\omega_2/\pi \approx 72 \text{ ft/sec} \approx 49 \text{ mi/h}.$$

In Problems 23–25 we substitute the given physical parameters into the equations in (42):

$$mx'' = -(k_1 + k_2)x + (k_1L_1 - k_2L_2)\theta$$

$$I\theta'' = (k_1L_1 - k_2L_2)x - (k_1L_1^2 + k_2L_2^2)\theta$$

As in Problem 21, a critical frequency of ω rad/sec yields a critical velocity of $v = 20\omega/\pi$ ft/sec.

23. $100x'' = -4000x, \quad 800\theta'' = 100000\theta$

Obviously the matrix $\mathbf{A} = \begin{bmatrix} -40 & 0 \\ 0 & -125 \end{bmatrix}$ has eigenvalues $\lambda_1 = -40$ and $\lambda_2 = -125$.

Up-and-down: $\omega_1 = \sqrt{40}, \qquad v_1 \approx 40.26 \text{ ft/sec} \approx 27 \text{ mph}$
Angular: $\omega_2 = \sqrt{125}, \qquad v_2 \approx 71.18 \text{ ft/sec} \approx 49 \text{ mph}$

25. $100x'' = -3000x - 5000\theta$
$800\theta'' = -5000x - 75000\theta$

The matrix $\mathbf{A} = \begin{bmatrix} -30 & -50 \\ -25/4 & -375/4 \end{bmatrix}$ has eigenvalues $\lambda_1, \lambda_2 = \dfrac{5}{8}\left(-99 \pm \sqrt{3401}\right)$.

$\omega_1 \approx 5.0424, \qquad v_1 \approx 32.10 \text{ ft/sec} \approx 22 \text{ mph}$
$\omega_2 \approx 9.9158, \qquad v_2 \approx 63.13 \text{ ft/sec} \approx 43 \text{ mph}$

SECTION 7.5

MULTIPLE EIGENVALUE SOLUTIONS

In each of Problems 1–6 we give first the characteristic equation with repeated (multiplicity 2) eigenvalue λ. In each case we find that $(\mathbf{A} - \lambda\mathbf{I})^2 = \mathbf{0}$. Then $\mathbf{w} = [1 \quad 0]^T$ is a generalized eigenvector and $\mathbf{v} = (\mathbf{A} - \lambda\mathbf{I})\mathbf{w} \neq \mathbf{0}$ is an ordinary eigenvector associated with λ. We give finally the scalar component functions $x_1(t), x_2(t)$ of the general solution

$$\mathbf{x}(t) = c_1\mathbf{v}e^{\lambda t} + c_2(\mathbf{v}t + \mathbf{w})e^{\lambda t}$$

of the given system $\mathbf{x}' = \mathbf{A}\mathbf{x}$.

1. Characteristic equation $\quad \lambda^2 + 6\lambda + 9 = 0$
Repeated eigenvalue $\quad \lambda = -3$
Generalized eigenvector $\quad \mathbf{w} = [1 \quad 0]^T$

$$\mathbf{v} = (\mathbf{A} - \lambda\mathbf{I})\mathbf{w} = \begin{bmatrix} 1 & 1 \\ -1 & -1 \end{bmatrix} \begin{bmatrix} 1 \\ 0 \end{bmatrix} = \begin{bmatrix} 1 \\ -1 \end{bmatrix}$$

$x_1(t) = (c_1 + c_2 + c_2 t)e^{-3t}$
$x_2(t) = (-c_1 \quad - c_2 t)e^{-3t}$.

3. Characteristic equation $\quad \lambda^2 - 6\lambda + 9 = 0$
Repeated eigenvalue $\quad \lambda = 3$
Generalized eigenvector $\quad \mathbf{w} = [1 \quad 0]^T$

$$\mathbf{v} = (\mathbf{A} - \lambda\mathbf{I})\mathbf{w} = \begin{bmatrix} -2 & -2 \\ 2 & 2 \end{bmatrix} \begin{bmatrix} 1 \\ 0 \end{bmatrix} = \begin{bmatrix} -2 \\ 2 \end{bmatrix}$$

$x_1(t) = (-2c_1 + c_2 - 2c_2 t)e^{3t}$
$x_2(t) = (2c_1 + \quad 2c_2 t)e^{3t}$.

5. Characteristic equation $\quad \lambda^2 - 10\lambda + 25 = 0$
Repeated eigenvalue $\quad \lambda = 5$
Generalized eigenvector $\quad \mathbf{w} = [1 \quad 0]^T$

$$\mathbf{v} = (\mathbf{A} - \lambda\mathbf{I})\mathbf{w} = \begin{bmatrix} 2 & 1 \\ -4 & -2 \end{bmatrix} \begin{bmatrix} 1 \\ 0 \end{bmatrix} = \begin{bmatrix} 2 \\ -4 \end{bmatrix}$$

$x_1(t) = (2c_1 + c_2 + 2c_2 t)e^{5t}$
$x_2(t) = (-4c_1 \quad - 4c_2 t)e^{5t}$.

In each of Problems 7-10 the characteristic polynomial is easily calculated by expansion along the row or column of $\mathbf{A}$ that contains two zeros. The matrix $\mathbf{A}$ has only two distinct eigenvalues, so we write $\lambda_1, \lambda_2, \lambda_3$ with either $\lambda_1 = \lambda_2$ or $\lambda_2 = \lambda_3$. Nevertheless, we find that it has 3 linearly independent eigenvectors $\mathbf{v}_1, \mathbf{v}_2$, and $\mathbf{v}_3$. We list also the scalar components $x_1(t), x_2(t), x_3(t)$ of the general solution $\mathbf{x}(t) = c_1\mathbf{v}_1 e^{\lambda_1 t} + c_2\mathbf{v}_2 e^{\lambda_2 t} + c_3\mathbf{v}_3 e^{\lambda_3 t}$ of the system.

7. Characteristic equation $\quad -\lambda^3 + 13\lambda^2 - 40\lambda + 36 = -(\lambda-2)^2(\lambda-9)$
Eigenvalues $\quad \lambda = 2, 2, 9$
Eigenvectors $\quad [1 \quad 1 \quad 0]^T, [1 \quad 0 \quad 1]^T, [0 \quad 1 \quad 0]^T$

$x_1(t) = c_1 e^{2t} + c_2 e^{2t}$
$x_2(t) = c_1 e^{2t} \quad\quad + c_3 e^{9t}$
$x_3(t) = \quad\quad c_1 e^{2t}$

Section 7.5

9. Characteristic equation $\quad -\lambda^3 +19\lambda^2 -115\lambda +225 = -(\lambda-5)^2(\lambda-9)$
Eigenvalues $\quad \lambda = 5, 5, 9$
Eigenvectors $\quad [1\ 2\ 0]^T,\ [7\ 0\ 2]^T,\ [3\ 0\ 1]^T$

$x_1(t) = c_1 e^{5t} + 7c_2 e^{5t} + 3c_3 e^{9t}$

$x_2(t) = 2c_1 e^{5t}$

$x_3(t) = \quad\quad 2c_2 e^{5t} + c_3 e^{9t}$

In each of Problems 11–14, the characteristic equation is $-\lambda^3 -3\lambda^2 -3\lambda -1 = -(\lambda+1)^3$. Hence $\lambda = -1$ is a triple eigenvalue of defect 2, and we find that $(\mathbf{A}-\lambda\mathbf{I})^3 = \mathbf{0}$. In each problem we start with $\mathbf{v}_3 = [1\ 0\ 0]^T$ and then calculate $\mathbf{v}_2 = (\mathbf{A}-\lambda\mathbf{I})\mathbf{v}_3$ and $\mathbf{v}_1 = (\mathbf{A}-\lambda\mathbf{I})\mathbf{v}_2 \neq \mathbf{0}$. It follows that $(\mathbf{A}-\lambda\mathbf{I})\mathbf{v}_1 = (\mathbf{A}-\lambda\mathbf{I})^2 \mathbf{v}_2 = (\mathbf{A}-\lambda\mathbf{I})^3 \mathbf{v}_3 = \mathbf{0}$, so $\mathbf{v}_1$ is an ordinary eigenvector associated with the triple eigenvalue λ. Hence $\{\mathbf{v}_1, \mathbf{v}_2, \mathbf{v}_3\}$ is a length 3 chain of generalized eigenvectors, and the corresponding general solution is described by

$$\mathbf{x}(t) = e^{-t}[c_1 \mathbf{v}_1 + c_2(\mathbf{v}_1 t + \mathbf{v}_2) + c_3(\mathbf{v}_1 t^2/2 + \mathbf{v}_2 t + \mathbf{v}_3)].$$

We give the scalar components $x_1(t), x_2(t), x_3(t)$ of $\mathbf{x}(t)$.

11. $\mathbf{v}_1 = [0\ 1\ 0]^T,\quad \mathbf{v}_2 = [-2\ -1\ 1]^T,\quad \mathbf{v}_3 = [1\ 0\ 0]^T$

$x_1(t) = e^{-t}(-2c_2 + c_3 - 2c_3 t)$

$x_2(t) = e^{-t}(c_1 - c_2 + c_2 t - c_3 t + c_3 t^2/2)$

$x_3(t) = e^{-t}(c_2 + c_3 t)$

13. $\mathbf{v}_1 = [1\ 0\ 0]^T,\quad \mathbf{v}_2 = [0\ 2\ 1]^T,\quad \mathbf{v}_3 = [1\ 0\ 0]^T$

$x_1(t) = e^{-t}(c_1 + c_2 t + c_3 t^2/2)$

$x_2(t) = e^{-t}(2c_2 + c_3 + 2c_3 t)$

$x_3(t) = e^{-t}(c_2 + c_3 t)$

In each of Problems 15-18, the characteristic equation is $-\lambda^3 + 3\lambda^2 - 3\lambda + 1 = -(\lambda-1)^3$. Hence $\lambda = 1$ is a triple eigenvalue of defect 1, and we find that $(\mathbf{A}-\lambda\mathbf{I})^2 = \mathbf{0}$. First we find the two linearly independent (ordinary) eigenvectors $\mathbf{u}_1$ and $\mathbf{u}_2$ associated with λ. Then we start with $\mathbf{v}_2 = [1\ 0\ 0]^T$ and calculate $\mathbf{v}_1 = (\mathbf{A}-\lambda\mathbf{I})\mathbf{v}_2 \neq \mathbf{0}$. It follows that $(\mathbf{A}-\lambda\mathbf{I})\mathbf{v}_1 = (\mathbf{A}-\lambda\mathbf{I})^2 \mathbf{v}_2 = \mathbf{0}$, so $\mathbf{v}_1$ is an ordinary eigenvector associated with λ. However,

$\mathbf{v}_1$ is a linear combination of $\mathbf{u}_1$ and $\mathbf{u}_2$, so $\mathbf{v}_1 e^t$ is a linear combination of the independent solutions $\mathbf{u}_1 e^t$ and $\mathbf{u}_2 e^t$. But $\{\mathbf{v}_1, \mathbf{v}_2\}$ is a length 2 chain of generalized eigenvectors associated with λ, so $(\mathbf{v}_1 t + \mathbf{v}_2) e^t$ is the desired third independent solution. The corresponding general solution is described by

$$\mathbf{x}(t) = e^t [c_1 \mathbf{u}_1 + c_2 \mathbf{u}_2 + c_3 (\mathbf{v}_1 t + \mathbf{v}_2)]$$

We give the scalar components $x_1(t), x_2(t), x_3(t)$ of $\mathbf{x}(t)$.

15. $\mathbf{u}_1 = [3\ -1\ 0]^T \quad \mathbf{u}_2 = [0\ 0\ 1]^T$
$\mathbf{v}_1 = [-3\ 1\ 1]^T \quad \mathbf{v}_2 = [1\ 0\ 0]^T$

$x_1(t) = e^t (3c_1 + c_3 - 3c_3 t)$
$x_2(t) = e^t (-c_1 + c_3 t)$
$x_3(t) = e^t (c_2 + c_3 t)$

17. $\mathbf{u}_1 = [2\ 0\ -9]^T \quad \mathbf{u}_2 = [1\ -3\ 0]^T$
$\mathbf{v}_1 = [0\ 6\ -9]^T \quad \mathbf{v}_2 = [0\ 1\ 0]^T$
(Either $\mathbf{v}_2 = [1\ 0\ 0]^T$ or $\mathbf{v}_2 = [0\ 0\ 1]^T$ can be used also, but they yield different forms of the solution than given in the book's answer section.)
$x_1(t) = e^t (2c_1 + c_2)$
$x_2(t) = e^t (-3c_2 + c_3 + 6c_3 t)$
$x_3(t) = e^t (-9c_1 - 9c_3 t)$

19. Characteristic equation $\quad \lambda^4 - 2\lambda^2 + 1 = 0$
Double eigenvalue $\lambda = -1$ with eigenvectors

$$\mathbf{v}_1 = [1\ 0\ 0\ 1]^T \text{ and } \mathbf{v}_2 = [0\ 0\ 1\ 0]^T.$$

Double eigenvalue $\lambda = +1$ with eigenvectors

$$\mathbf{v}_3 = [0\ 1\ 0\ -2]^T \text{ and } \mathbf{v}_4 = [1\ 0\ 3\ 0]^T.$$

General solution

$$\mathbf{x}(t) = e^{-t}(c_1 \mathbf{v}_1 + c_2 \mathbf{v}_2) + e^t (c_3 \mathbf{v}_3 + c_4 \mathbf{v}_4)$$

Scalar components

$$x_1(t) = c_1 e^{-t} + c_4 e^{t}$$
$$x_2(t) = c_3 e^{t}$$
$$x_3(t) = c_2 e^{-t} + 3 c_4 e^{t}$$
$$x_4(t) = c_1 e^{-t} - 2 c_3 e^{t}$$

21. Characteristic equation $\quad \lambda^4 - 4\lambda^3 + 6\lambda^2 - 4\lambda + 1 = (\lambda - 1)^4 = 0$
Eigenvalue $\lambda = 1$ with multiplicity 4 and defect 2.

We find that $(\mathbf{A} - \lambda \mathbf{I})^2 \neq 0$ but $(\mathbf{A} - \lambda \mathbf{I})^3 = 0$. We therefore start with $\mathbf{v}_3 = [1 \ 0 \ 0 \ 0]^T$ and define $\mathbf{v}_2 = (\mathbf{A} - \lambda \mathbf{I})\mathbf{v}_3$ and $\mathbf{v}_1 = (\mathbf{A} - \lambda \mathbf{I})\mathbf{v}_2 \neq 0$, thereby obtaining the length 3 chain $\{\mathbf{v}_1, \mathbf{v}_2, \mathbf{v}_3\}$ with

$$\mathbf{v}_1 = [0 \ 0 \ 0 \ 1]^T, \quad \mathbf{v}_2 = [-2 \ 1 \ 1 \ 0]^T, \quad \mathbf{v}_3 = [1 \ 0 \ 0 \ 0]^T.$$

Then we find the second ordinary eigenvector $\mathbf{v}_4 = [0 \ 0 \ 1 \ 0]^T$. The corresponding general solution

$$\mathbf{x}(t) = e^{t} [c_1 \mathbf{v}_1 + c_2(\mathbf{v}_1 t + \mathbf{v}_2) + c_3(\mathbf{v}_1 t^2/2 + \mathbf{v}_2 t + \mathbf{v}_3) + c_4 \mathbf{v}_4]$$

has scalar components

$$x_1(t) = e^{t}(-2c_2 + c_3 - 2c_3 t)$$
$$x_2(t) = e^{t}(c_2 + c_3 t)$$
$$x_3(t) = e^{t}(c_2 + c_4 + c_3 t).$$
$$x_4(t) = e^{t}(c_1 + c_2 t + c_3 t^2/2.)$$

In Problems 23 and 24 there are only two distinct eigenvalues λ_1 and λ_2. However, the eigenvector equation $(\mathbf{A} - \lambda \mathbf{I})\mathbf{v} = 0$ yields the three linearly independent eigenvectors $\mathbf{v}_1, \mathbf{v}_2,$ and $\mathbf{v}_3$ that are given. We list the scalar components of the corresponding general solution $\mathbf{x}(t) = c_1 \mathbf{v}_1 e^{\lambda_1 t} + c_2 \mathbf{v}_2 e^{\lambda_2 t} + c_3 \mathbf{v}_3 e^{\lambda_2 t}$.

23. $\lambda_1 = -1:$ $\quad \{\mathbf{v}_1\}$ with $\mathbf{v}_1 = [1 \ -1 \ 2]^T$
$\lambda_2 = 3:$ $\quad \{\mathbf{v}_2\}$ with $\mathbf{v}_2 = [4 \ 0 \ 9]^T$ and
$\quad \{\mathbf{v}_3\}$ with $\mathbf{v}_3 = [0 \ 2 \ 1]^T$

Scalar components

$$x_1(t) = c_1 e^{-t} + 4 c_2 e^{3t}$$
$$x_2(t) = -c_1 e^{-t} \quad\quad\quad + 2 c_3 e^{3t}$$
$$x_3(t) = 2 c_1 e^{-t} + 9 c_2 e^{3t} + c_3 e^{3t}$$

In Problems 25, 26, and 28 there is given a single eigenvalue λ of multiplicity 3. We find that $(\mathbf{A} - \lambda\mathbf{I})^2 \neq 0$ but $(\mathbf{A} - \lambda\mathbf{I})^3 = 0$. We therefore start with $\mathbf{v}_3 = [1 \ 0 \ 0]^T$ and define $\mathbf{v}_2 = (\mathbf{A} - \lambda\mathbf{I})\mathbf{v}_3$ and $\mathbf{v}_1 = (\mathbf{A} - \lambda\mathbf{I})\mathbf{v}_2 \neq 0$, thereby obtaining the length 3 chain $\{\mathbf{v}_1, \mathbf{v}_2, \mathbf{v}_3\}$ of generalized eigenvectors based on the ordinary eigenvector $\mathbf{v}_1$. We list the scalar components of the corresponding general solution

$$\mathbf{x}(t) = c_1\mathbf{v}_1 e^{\lambda t} + c_2(\mathbf{v}_1 t + \mathbf{v}_2) e^{\lambda t} + c_3(\mathbf{v}_1 t^2/2 + \mathbf{v}_2 t + \mathbf{v}_3) e^{\lambda t}.$$

25. $\{\mathbf{v}_1, \mathbf{v}_2, \mathbf{v}_3\}$ with

$$\mathbf{v}_1 = [-1 \ 0 \ -1]^T, \quad \mathbf{v}_2 = [-4 \ -1 \ 0]^T, \quad \mathbf{v}_3 = [1 \ 0 \ 0]^T$$

Scalar components

$$x_1(t) = e^{2t}(-c_1 - 4c_2 + c_3 - c_2 t - 4c_3 t - c_3 t^2/2)$$
$$x_2(t) = e^{2t}(-c_2 - c_3 t)$$
$$x_3(t) = e^{2t}(-c_1 - c_2 t - c_3 t^2/2)$$

27. We find that the triple eigenvalue $\lambda = 2$ has the two linearly independent eigenvectors $[1 \ 1 \ 0]^T$ and $[-1 \ 0 \ 1]^T$. Next we find that $(\mathbf{A} - \lambda\mathbf{I}) \neq 0$ but $(\mathbf{A} - \lambda\mathbf{I})^2 = 0$. We therefore start with $\mathbf{v}_2 = [1 \ 0 \ 0]^T$ and define

$$\mathbf{v}_1 = (\mathbf{A} - \lambda\mathbf{I})\mathbf{v}_2 = [-5 \ 3 \ 8]^T \neq \mathbf{0},$$

thereby obtaining the length 2 chain $\{\mathbf{v}_1, \mathbf{v}_2\}$ of generalized eigenvectors based on the ordinary eigenvector $\mathbf{v}_1$. If we take $\mathbf{v}_3 = [1 \ 1 \ 0]^T$, then the general solution $\mathbf{x}(t) = e^{2t}[c_1\mathbf{v}_1 + c_2(\mathbf{v}_1 t + \mathbf{v}_2) + c_3\mathbf{v}_3]$ has scalar components

$$x_1(t) = e^{2t}(-5c_1 + c_2 + c_3 - 5c_2 t)$$
$$x_2(t) = e^{2t}(3c_1 + 3c_2 t)$$
$$x_3(t) = e^{2t}(8c_1 + 8c_2 t).$$

In Problems 29 and 30 the matrix $\mathbf{A}$ has two distinct eigenvalues λ_1 and λ_2 each having multiplicity 2 and defect 1. First, we select $\mathbf{v}_2$ so that $\mathbf{v}_1 = (\mathbf{A} - \lambda_1\mathbf{I})\mathbf{v}_2 \neq \mathbf{0}$ but $(\mathbf{A} - \lambda_1\mathbf{I})\mathbf{v}_1 = \mathbf{0}$, so $\{\mathbf{v}_1, \mathbf{v}_2\}$ is a length 2 chain based on $\mathbf{v}_1$. Next, we select $\mathbf{u}_2$ so that $\mathbf{u}_1 = (\mathbf{A} - \lambda_1\mathbf{I})\mathbf{u}_2 \neq \mathbf{0}$ but $(\mathbf{A} - \lambda_1\mathbf{I})\mathbf{u}_1 = \mathbf{0}$, so $\{\mathbf{u}_1, \mathbf{u}_2\}$ is a length 2 chain based on $\mathbf{u}_1$. We give the scalar components of the corresponding general solution

$$\mathbf{x}(t) = e^{\lambda_1 t}[c_1\mathbf{v}_1 + c_2(\mathbf{v}_1 t + \mathbf{v}_2)] + e^{\lambda_2 t}[c_3\mathbf{u}_1 + c_4(\mathbf{u}_1 t + \mathbf{u}_2)].$$

Section 7.5

29. $\lambda = -1$: $\{\mathbf{v}_1, \mathbf{v}_2\}$ with $\mathbf{v}_1 = \begin{bmatrix} 1 & -3 & -1 & -2 \end{bmatrix}^T$ and $\mathbf{v}_2 = \begin{bmatrix} 0 & 1 & 0 & 0 \end{bmatrix}^T$,
$\lambda = 2$: $\{\mathbf{u}_1, \mathbf{u}_2\}$ with $\mathbf{u}_1 = \begin{bmatrix} 0 & -1 & 1 & 0 \end{bmatrix}^T$ and $\mathbf{u}_2 = \begin{bmatrix} 0 & 0 & 2 & 1 \end{bmatrix}^T$

Scalar components

$$x_1(t) = e^{-t}(c_1 + c_2 t)$$
$$x_2(t) = e^{-t}(-3c_1 + c_2 - 3c_2 t) + e^{2t}(-c_3 - c_4 t)$$
$$x_3(t) = e^{-t}(-c_1 - c_2 t) + e^{2t}(c_3 + 2c_4 + c_4 t)$$
$$x_4(t) = e^{-t}(-2c_1 - 2c_2 t) + e^{2t}(c_4)$$

31. We have the single eigenvalue $\lambda = 1$ of multiplicity 4. Starting with $\mathbf{v}_3 = \begin{bmatrix} 1 & 0 & 0 & 0 \end{bmatrix}^T$, we calculate $\mathbf{v}_2 = (\mathbf{A} - \lambda\mathbf{I})\mathbf{v}_3$ and $\mathbf{v}_1 = (\mathbf{A} - \lambda\mathbf{I})\mathbf{v}_2 \neq \mathbf{0}$, and find that $(\mathbf{A} - \lambda\mathbf{I})\mathbf{v}_1 = \mathbf{0}$. Therefore $\{\mathbf{v}_1, \mathbf{v}_2, \mathbf{v}_3\}$ is a length 3 chain based on the ordinary eigenvector $\mathbf{v}_1$. Next, the eigenvector equation $(\mathbf{A} - \lambda\mathbf{I})\mathbf{v} = \mathbf{0}$ yields the second linearly independent eigenvector $\mathbf{v}_4 = \begin{bmatrix} 0 & 1 & 3 & 0 \end{bmatrix}^T$. With

$$\mathbf{v}_1 = \begin{bmatrix} 42 & 7 & -21 & -42 \end{bmatrix}^T, \quad \mathbf{v}_2 = \begin{bmatrix} 34 & 22 & -10 & -27 \end{bmatrix}^T,$$
$$\mathbf{v}_3 = \begin{bmatrix} 1 & 0 & 0 & 0 \end{bmatrix}^T \quad \text{and} \quad \mathbf{v}_4 = \begin{bmatrix} 0 & 1 & 3 & 0 \end{bmatrix}$$

the general solution

$$\mathbf{x}(t) = e^t[c_1\mathbf{v}_1 + c_2(\mathbf{v}_1 t + \mathbf{v}_2) + c_3(\mathbf{v}_1 t^2/2 + \mathbf{v}_2 t + \mathbf{v}_3) + c_4\mathbf{v}_4]$$

has scalar components

$$x_1(t) = e^t(42c_1 + 34c_2 + c_3 + 42c_2 t + 34c_3 t + 21c_3 t^2)$$
$$x_2(t) = e^t(7c_1 + 22c_2 + c_4 + 7c_2 t + 22c_3 t + 7c_3 t^2/2)$$
$$x_3(t) = e^t(-21c_1 - 10c_2 + 3c_4 - 21c_2 t - 10c_3 t - 21c_3 t^2/2)$$
$$x_4(t) = e^t(-42c_1 - 27c_2 - 42c_2 t - 27c_3 t - 21c_3 t^2).$$

33. The chain $\{\mathbf{v}_1, \mathbf{v}_2\}$ was found using the matrices

$$\mathbf{A} - \lambda\mathbf{I} = \begin{bmatrix} 4i & -4 & 1 & 0 \\ 4 & 4i & 0 & 1 \\ 0 & 0 & 4i & -4 \\ 0 & 0 & 4 & 4i \end{bmatrix} \rightarrow \begin{bmatrix} 1 & i & 0 & 0 \\ 0 & 0 & 1 & 0 \\ 0 & 0 & 0 & 1 \\ 0 & 0 & 0 & 0 \end{bmatrix}$$

and

$$(\mathbf{A} - \lambda\mathbf{I})^2 = \begin{bmatrix} -32 & -32i & 8i & -8 \\ 32i & -32 & 8 & 8i \\ 0 & 0 & -32 & -32i \\ 0 & 0 & 32i & -32 \end{bmatrix} \rightarrow \begin{bmatrix} 1 & i & 0 & 0 \\ 0 & 0 & 1 & i \\ 0 & 0 & 0 & 0 \\ 0 & 0 & 0 & 0 \end{bmatrix}$$

where $\rightarrow$ signifies reduction to row-echelon form. The resulting real-valued solution vectors are

$$\mathbf{x}_1(t) = e^{3t} [\ \cos 4t \quad \sin 4t \quad 0 \quad 0\ ]^T$$
$$\mathbf{x}_2(t) = e^{3t} [-\sin 4t \quad \cos 4t \quad 0 \quad 0\ ]^T$$
$$\mathbf{x}_3(t) = e^{3t} [\ t\cos 4t \quad t\sin 4t \quad \cos 4t \quad \sin 4t\ ]^T$$
$$\mathbf{x}_4(t) = e^{3t} [-t\sin 4t \quad t\cos 4t \quad -\sin 4t \quad \cos 4t\ ]^T.$$

35. The coefficient matrix

$$\mathbf{A} = \begin{bmatrix} 0 & 0 & 1 & 0 \\ 0 & 0 & 0 & 1 \\ -1 & 1 & -2 & 1 \\ 1 & -1 & 1 & -2 \end{bmatrix}$$

has eigenvalues

$\lambda = 0$ with eigenvector $\mathbf{v}_1 = [1 \quad 1 \quad 0 \quad 0]^T$
$\lambda = -1$ with eigenvectors $\mathbf{v}_2 = [1 \quad 0 \quad -1 \quad 0]^T$ and $\mathbf{v}_3 = [0 \quad 1 \quad 0 \quad -1]^T$,
$\lambda = -2$ with eigenvector $\mathbf{v}_4 = [1 \quad -1 \quad -2 \quad 2]^T.$

When we impose the given initial conditions on the general solution

$$\mathbf{x}(t) = c_1\mathbf{v}_1 + c_2\mathbf{v}_2 e^{-t} + c_3\mathbf{v}_3 e^{-t} + c_4\mathbf{v}_4 e^{-2t}$$

we find that $c_1 = v_0$, $c_2 = c_3 = -v_0$, $c_4 = 0$. Hence the position functions of the two masses are given by

$$x_1(t) = x_2(t) = v_0(1 - e^{-t}).$$

Each mass travels a distance v_0 before stopping.

In Problems 37–46 we use the eigenvectors and generalized eigenvectors found in Problems 23–32 to construct a matrix $\mathbf{Q}$ such that $\mathbf{J} = \mathbf{Q}^{-1}\mathbf{A}\mathbf{Q}$ is a Jordan normal form of the given matrix $\mathbf{A}$.

37. $\mathbf{v}_1 = [1 \quad -1 \quad 2]^T,\quad \mathbf{v}_2 = [4 \quad 0 \quad 9]^T,\quad \mathbf{v}_3 = [0 \quad 2 \quad 1]^T$

$$\mathbf{Q} = [\mathbf{v}_1 \quad \mathbf{v}_2 \quad \mathbf{v}_3] = \begin{bmatrix} 1 & 4 & 0 \\ -1 & 0 & 2 \\ 2 & 9 & 1 \end{bmatrix}$$

Section 7.5

$$J = Q^{-1}AQ = \frac{1}{2}\begin{bmatrix} -18 & -4 & 8 \\ 5 & 1 & -2 \\ -9 & -1 & 4 \end{bmatrix}\begin{bmatrix} 39 & 8 & -16 \\ -36 & -5 & 16 \\ 72 & 16 & -29 \end{bmatrix}\begin{bmatrix} 1 & 4 & 0 \\ -1 & 0 & 2 \\ 2 & 9 & 1 \end{bmatrix} = \begin{bmatrix} -1 & 0 & 0 \\ 0 & 3 & 0 \\ 0 & 0 & 3 \end{bmatrix}$$

39. $\mathbf{v}_1 = [-1 \ 0 \ -1]^T, \quad \mathbf{v}_2 = [-4 \ -1 \ 0]^T, \quad \mathbf{v}_3 = [1 \ 0 \ 0]^T$

$$Q = [\mathbf{v}_1 \ \mathbf{v}_2 \ \mathbf{v}_3] = \begin{bmatrix} -1 & -4 & 1 \\ 0 & -1 & 0 \\ -1 & 0 & 0 \end{bmatrix}$$

$$J = Q^{-1}AQ = \begin{bmatrix} 0 & 0 & -1 \\ 0 & -1 & 0 \\ 1 & -4 & -1 \end{bmatrix}\begin{bmatrix} -2 & 17 & 4 \\ -1 & 6 & 1 \\ 0 & 1 & 2 \end{bmatrix}\begin{bmatrix} -1 & -4 & 1 \\ 0 & -1 & 0 \\ -1 & 0 & 0 \end{bmatrix} = \begin{bmatrix} 2 & 1 & 0 \\ 0 & 2 & 1 \\ 0 & 0 & 2 \end{bmatrix}$$

41. $\mathbf{v}_1 = [-5 \ 3 \ 8]^T, \quad \mathbf{v}_2 = [1 \ 0 \ 0]^T, \quad \mathbf{v}_3 = [1 \ 1 \ 0]^T$

$$Q = [\mathbf{v}_1 \ \mathbf{v}_2 \ \mathbf{v}_3] = \begin{bmatrix} -5 & 1 & 1 \\ 3 & 0 & 1 \\ 8 & 0 & 0 \end{bmatrix}$$

$$J = Q^{-1}AQ = \frac{1}{8}\begin{bmatrix} 0 & 0 & 1 \\ 8 & -8 & 8 \\ 0 & 8 & -3 \end{bmatrix}\begin{bmatrix} -3 & 5 & -5 \\ 3 & -1 & 3 \\ 8 & -8 & 10 \end{bmatrix}\begin{bmatrix} -5 & 1 & 1 \\ 3 & 0 & 1 \\ 8 & 0 & 0 \end{bmatrix} = \begin{bmatrix} 2 & 1 & 0 \\ 0 & 2 & 0 \\ 0 & 0 & 2 \end{bmatrix}$$

43. $\mathbf{v}_1 = [1 \ -3 \ -1 \ -2]^T, \quad \mathbf{v}_2 = [0 \ 1 \ 0 \ 0]^T,$
$\mathbf{u}_1 = [0 \ -1 \ 1 \ 0]^T, \quad \mathbf{u}_2 = [0 \ 0 \ 2 \ 1]^T$

$$Q = [\mathbf{v}_1 \ \mathbf{v}_2 \ \mathbf{u}_1 \ \mathbf{u}_2] = \begin{bmatrix} 1 & 0 & 0 & 0 \\ -3 & 1 & -1 & 0 \\ -1 & 0 & 1 & 2 \\ -2 & 0 & 0 & 1 \end{bmatrix}$$

$$J = Q^{-1}AQ$$

$$= \begin{bmatrix} 1 & 0 & 0 & 0 \\ 0 & 1 & 1 & -2 \\ -3 & 0 & 1 & -2 \\ 2 & 0 & 0 & 1 \end{bmatrix}\begin{bmatrix} -1 & 1 & 1 & -2 \\ 7 & -4 & -6 & 11 \\ 5 & -1 & 1 & 3 \\ 6 & -2 & -2 & 6 \end{bmatrix}\begin{bmatrix} 1 & 0 & 0 & 0 \\ -3 & 1 & -1 & 0 \\ -1 & 0 & 1 & 2 \\ -2 & 0 & 0 & 1 \end{bmatrix} = \begin{bmatrix} -1 & 1 & 0 & 0 \\ 0 & -1 & 0 & 0 \\ 0 & 0 & 2 & 1 \\ 0 & 0 & 0 & 2 \end{bmatrix}$$

45. $\mathbf{v}_1 = [42 \ 7 \ -21 \ -42]^T, \quad \mathbf{v}_2 = [34 \ 22 \ -10 \ -27]^T,$
$\mathbf{v}_3 = [1 \ 0 \ 0 \ 0]^T, \quad \mathbf{v}_4 = [0 \ 1 \ 3 \ 0]^T$

$$\mathbf{Q} = \begin{bmatrix} \mathbf{v}_1 & \mathbf{v}_2 & \mathbf{v}_3 & \mathbf{v}_4 \end{bmatrix} = \begin{bmatrix} 42 & 34 & 1 & 0 \\ 7 & 22 & 0 & 1 \\ -21 & -10 & 0 & 3 \\ -42 & -27 & 0 & 0 \end{bmatrix}$$

$$\mathbf{J} = \mathbf{Q}^{-1}\mathbf{A}\mathbf{Q}$$

$$= \frac{1}{2058} \begin{bmatrix} 0 & -81 & 27 & -76 \\ 0 & 126 & -42 & 42 \\ 2058 & -882 & 294 & 1764 \\ 0 & -147 & 735 & -392 \end{bmatrix} \begin{bmatrix} 2 & 1 & -2 & 1 \\ 0 & 3 & -5 & 3 \\ 0 & -13 & 22 & -12 \\ 0 & -27 & 45 & -25 \end{bmatrix} \begin{bmatrix} 42 & 34 & 1 & 0 \\ 7 & 22 & 0 & 1 \\ -21 & -10 & 0 & 3 \\ -42 & -27 & 0 & 0 \end{bmatrix}$$

$$= \begin{bmatrix} 1 & 1 & 0 & 0 \\ 0 & 1 & 1 & 0 \\ 0 & 0 & 1 & 0 \\ 0 & 0 & 0 & 1 \end{bmatrix}$$

SECTION 7.6

NUMERICAL METHODS FOR SYSTEMS

In Problems 1–8 we first write the given system in the form $x' = f(t, x, y)$, $y' = g(t, x, y)$. Then we use the template

$$h = 0.1; \quad t_1 = t_0 + h$$
$$x_1 = x_0 + h f(t_0, x_0, y_0); \quad y_1 = y_0 + h g(t_0, x_0, y_0)$$
$$x_2 = x_1 + h f(t_1, x_1, y_1); \quad y_2 = y_1 + h g(t_1, x_1, y_1)$$

(with the given values of t_0, x_0, and y_0) to calculate the Euler approximations $x_1 \approx x(0.1)$, $y_1 \approx y(0.1)$ and $x_2 \approx x(0.2)$, $y_2 \approx y(0.2)$ in part (a). We give these approximations and the actual values $x_{act} = x(0.2)$, $y_{act} = y(0.2)$ in tabular form. We use the template

$$h = 0.2; \quad t_1 = t_0 + h$$
$$u_1 = x_0 + h f(t_0, x_0, y_0); \quad v_1 = y_0 + h g(t_0, x_0, y_0)$$
$$x_1 = x_0 + \tfrac{1}{2} h \big[f(t_0, x_0, y_0) + f(t_1, u_1, v_1) \big]$$
$$y_1 = y_0 + \tfrac{1}{2} h \big[g(t_0, x_0, y_0) + g(t_1, u_1, v_1) \big]$$

to calculate the improved Euler approximations $u_1 \approx x(0.2)$, $u_1 \approx y(0.2)$ and $x_1 \approx x(0.2)$, $y_1 \approx y(0.2)$ in part (b). We give these approximations and the actual values $x_{act} = x(0.2)$, $y_{act} = y(0.2)$ in tabular form. We use the template

$$h = 0.2;$$
$$F_1 = f(t_0, x_0, y_0); \quad G_1 = g(t_0, x_0, y_0)$$
$$F_2 = f(t_0 + \tfrac{1}{2}h, x_0 + \tfrac{1}{2}hF_1, y_0 + \tfrac{1}{2}hG_1); \quad G_2 = g(t_0 + \tfrac{1}{2}h, x_0 + \tfrac{1}{2}hF_1, y_0 + \tfrac{1}{2}hG_1)$$
$$F_3 = f(t_0 + \tfrac{1}{2}h, x_0 + \tfrac{1}{2}hF_2, y_0 + \tfrac{1}{2}hG_2); \quad G_3 = g(t_0 + \tfrac{1}{2}h, x_0 + \tfrac{1}{2}hF_2, y_0 + \tfrac{1}{2}hG_2)$$
$$F_4 = f(t_0 + h, x_0 + hF_3, y_0 + hG_3); \quad G_4 = g(t_0 + h, x_0 + hF_3, y_0 + hG_3)$$
$$x_1 = x_0 + \tfrac{h}{6}(F_1 + 2F_2 + 2F_3 + F_4); \quad y_1 = y_0 + \tfrac{h}{6}(G_1 + 2G_2 + 2G_3 + G_4)$$

to calculate the intermediate slopes and Runge-Kutta approximations $x_1 \approx x(0.2)$, $y_1 \approx y(0.2)$ for part (c). Again, we give the results in tabular form.

1. **(a)**

x_1	y_1	x_2	y_2	x_{act}	y_{act}
0.4	2.2	0.88	2.5	1.0034	2.6408

(b)

u_1	v_1	x_1	y_1	x_{act}	y_{act}
0.8	2.4	0.96	2.6	1.0034	2.6408

(c)

F_1	G_1	F_2	G_2	F_3	G_3	F_4	G_4
4	2	4.8	3	5.08	3.26	6.32	4.684

x_1	y_1	x_{act}	y_{act}
1.0027	2.6401	1.0034	2.6408

3. **(a)**

x_1	y_1	x_2	y_2	x_{act}	y_{act}
1.7	1.5	2.81	2.31	3.6775	2.9628

(b)

u_1	v_1	x_1	y_1	x_{act}	y_{act}
2.4	2	3.22	2.62	3.6775	2.9628

(c)

F_1	G_1	F_2	G_2	F_3	G_3	F_4	G_4
7	5	11.1	8.1	13.57	9.95	23.102	17.122

x_1	y_1	x_{act}	y_{act}
3.6481	2.9407	3.6775	2.9628

5. **(a)**

x_1	y_1	x_2	y_2	x_{act}	y_{act}
0.9	3.2	−0.52	2.92	−0.5793	2.4488

(b)

u_1	v_1	x_1	y_1	x_{act}	y_{act}
−0.2	3.4	−0.84	2.44	−0.5793	2.4488

(c)

F_1	G_1	F_2	G_2	F_3	G_3	F_4	G_4
−11	2	−14.2	−2.8	−12.44	−3.12	−12.856	−6.704
x_1	y_1	x_{act}	y_{act}				
−0.5712	2.4485	−0.5793	2.4488				

7. **(a)**

x_1	y_1	x_2	y_2	x_{act}	y_{act}
2.5	1.3	3.12	1.68	3.2820	1.7902

(b)

u_1	v_1	x_1	y_1	x_{act}	y_{act}
3	1.6	3.24	1.76	3.2820	1.7902

(c)

F_1	G_1	F_2	G_2	F_3	G_3	F_4	G_4
5	3	6.2	3.8	6.48	4	8.088	5.096
x_1	y_1	x_{act}	y_{act}				
3.2816	1.7899	3.2820	1.7902				

In Problems 9–11 we use the same Runge-Kutta template as in part (c) of Problems 1–8 above, and give both the Runge-Kutta approximate values with step sizes $h = 0.1$ and $h = 0.05$, and also the actual values.

9. With $h = 0.1$: $x(1) \approx 3.99261$, $y(1) \approx 6.21770$
    With $h = 0.05$: $x(1) \approx 3.99234$, $y(1) \approx 6.21768$
    Actual values: $x(1) \approx 3.99232$, $y(1) \approx 6.21768$

11. With $h = 0.1$: $x(1) \approx -0.05832$, $y(1) \approx 0.56664$
    With $h = 0.05$: $x(1) \approx -0.05832$, $y(1) \approx 0.56665$
    Actual values: $x(1) \approx -0.05832$, $y(1) \approx 0.56665$

13. With $y = x'$ we want to solve numerically the initial value problem

$$x' = y, \qquad x(0) = 0$$
$$y' = -32 - 0.04y, \qquad y(0) = 288.$$

When we run Program RK2DIM with step size $h = 0.1$ we find that the change of sign in the velocity v occurs as follows:

t	x	v
7.6	1050.2	+2.8
7.7	1050.3	-0.4

Thus the bolt attains a maximum height of about 1050 feet in about 7.7 seconds.

15. With $y = x'$, and with x in miles and t in seconds, we want to solve numerically the initial value problem

$$x' = y$$
$$y' = -95485.5/(x^2 + 7920x + 15681600)$$
$$x(0) = 0, \qquad y(0) = 1.$$

We find (running RK2DIM with $h = 1$) that the projectile reaches a maximum height of about 83.83 miles in about $168 \text{ sec} = 2 \text{ min } 48 \text{ sec}$.

16. We first defined the MATLAB function

```
function xp  =  fnball(t,x)
%   Defines the baseball system
%       x1'  =  x'  =  x3,   x3'  =  -cvx'
%       x2'  =  y'  =  x4,   x4'  =  -cvy'- g
%   with air resistance coefficient c.

g    =  32;
c    =  0.0025;
xp   =  x;
v    =  sqrt(x(3).^2) + x(4).^2);
xp(1)  =  x(3);
xp(2)  =  x(4);
xp(3)  =  -c*v*x(3);
xp(4)  =  -c*v*x(4) - g;
```

Then, using the n-dimensional program **rkn** with step size 0.1 and initial data corresponding to the indicated initial inclination angles, we got the following results:

Angle	Time	Range
40	5.0	352.9
45	5.4	347.2
50	5.8	334.2

We have listed the time to the nearest tenth of a second, but have interpolated to find the range in feet.

17. The data in Problem 16 indicate that the range increases when the initial angle is decreased below 45°. The further data

Angle	Range
41.0	352.1
40.5	352.6
40.0	352.9
39.5	352.8
39.0	352.7
35.0	350.8

indicate that a maximum range of about 353 ft is attained with $\alpha \approx 40°$.

19. First we run program **rkn** (with $h = 0.1$) with $v_0 = 250$ ft/sec and obtain the following results:

t	x	y
5.0	457.43	103.90
6.0	503.73	36.36

Interpolation gives $x = 494.4$ when $y = 50$. Then a run with $v_0 = 255$ ft/sec gives the following results:

t	x	y
5.5	486.75	77.46
6.0	508.86	41.62

Finally a run with $v_0 = 253$ ft/sec gives these results:

t	x	y
5.5	484.77	75.44
6.0	506.82	39.53

Now $x \approx 500$ ft when $y = 50$ ft. Thus Babe Ruth's home run ball had an initial velocity of 253 ft/sec.

21.   A run of program **rkn** with $h = 0.1$ indicates that the projectile has a range of about 21,400 ft $\approx$ 4.05 mi and a flight time of about 46 sec. It attains a maximum height of about 8970 ft in about 17.5 sec. At time $t \approx 23$ sec it has its minimum velocity of about 368 ft/sec. It hits the ground ($t \approx 46$ sec) at an angle of about 77° with a velocity of about 518 ft/sec.

CHAPTER 8

MATRIX EXPONENTIAL METHODS

SECTION 8.1

MATRIX EXPONENTIALS AND LINEAR SYSTEMS

In Problems 1–8 we first use the eigenvalues and eigenvectors of the coefficient matrix $\mathbf{A}$ to find first a fundamental matrix $\Phi(t)$ for the homogeneous system $\mathbf{x}' = \mathbf{A}\mathbf{x}$. Then we apply the formula

$$\mathbf{x}(t) = \Phi(t)\Phi(0)^{-1}\mathbf{x}_0,$$

to find the solution vector $\mathbf{x}(t)$ that satisfies the initial condition $\mathbf{x}(0) = \mathbf{x}_0$. Formulas (11) and (12) in the text provide inverses of 2-by-2 and 3-by-3 matrices.

1. Eigensystem: $\lambda_1 = 1$, $\mathbf{v}_1 = [1\ \ -1]^T$; $\ \ \lambda_2 = 3$, $\mathbf{v}_2 = [1\ \ 1]^T$

$$\Phi(t) = \begin{bmatrix} e^{\lambda_1 t}\mathbf{v}_1 & e^{\lambda_2 t}\mathbf{v}_2 \end{bmatrix} = \begin{bmatrix} e^t & e^{3t} \\ -e^t & e^{3t} \end{bmatrix}$$

$$\mathbf{x}(t) = \begin{bmatrix} e^t & e^{3t} \\ -e^t & e^{3t} \end{bmatrix} \cdot \frac{1}{2}\begin{bmatrix} 1 & -1 \\ 1 & 1 \end{bmatrix} \cdot \begin{bmatrix} 3 \\ -2 \end{bmatrix} = \frac{1}{2}\begin{bmatrix} 5e^t + e^{3t} \\ -5e^t + e^{3t} \end{bmatrix}$$

3. Eigensystem: $\lambda = 4i$, $\mathbf{v} = [1+2i\ \ 2]^T$

$$\Phi(t) = \begin{bmatrix} \text{Re}(\mathbf{v}e^{\lambda t}) & \text{Im}(\mathbf{v}e^{\lambda t}) \end{bmatrix} = \begin{bmatrix} \cos 4t - 2\sin 4t & 2\cos 4t + \sin 4t \\ 2\cos 4t & 2\sin 4t \end{bmatrix}$$

$$\mathbf{x}(t) = \begin{bmatrix} \cos 4t - 2\sin 4t & 2\cos 4t + \sin 4t \\ 2\cos 4t & 2\sin 4t \end{bmatrix} \cdot \frac{1}{4}\begin{bmatrix} 0 & 2 \\ 2 & -1 \end{bmatrix} \cdot \begin{bmatrix} 0 \\ 1 \end{bmatrix} = \frac{1}{4}\begin{bmatrix} -5\sin 4t \\ 4\cos 4t - 2\sin 4t \end{bmatrix}$$

5. Eigensystem: $\lambda = 3i$, $\mathbf{v} = [-1+i\ \ 3]^T$

$$\Phi(t) = \begin{bmatrix} \text{Re}(\mathbf{v}e^{\lambda t}) & \text{Im}(\mathbf{v}e^{\lambda t}) \end{bmatrix} = \begin{bmatrix} -\cos 3t - \sin 3t & \cos 3t - \sin 3t \\ 3\cos 3t & 3\sin 3t \end{bmatrix}$$

$$\mathbf{x}(t) = \begin{bmatrix} -\cos 3t - \sin 3t & \cos 3t - \sin 3t \\ 3\cos 3t & 3\sin 3t \end{bmatrix} \cdot \frac{1}{3}\begin{bmatrix} 0 & 1 \\ 3 & 1 \end{bmatrix} \cdot \begin{bmatrix} 1 \\ -1 \end{bmatrix} = \frac{1}{3}\begin{bmatrix} 3\cos 3t - \sin 3t \\ -3\cos 3t + 6\sin 3t \end{bmatrix}$$

7. Eigensystem:
$$\lambda_1 = 0, \quad \mathbf{v}_1 = [6 \; 2 \; 5]^T; \quad \lambda_2 = 1, \quad \mathbf{v}_2 = [3 \; 1 \; 2]^T; \quad \lambda_3 = -1, \quad \mathbf{v}_3 = [2 \; 1 \; 2]^T$$

$$\Phi(t) = \begin{bmatrix} e^{\lambda_1 t}\mathbf{v}_1 & e^{\lambda_2 t}\mathbf{v}_2 & e^{\lambda_3 t}\mathbf{v}_3 \end{bmatrix} = \begin{bmatrix} 6 & 3e^t & 2e^{-t} \\ 2 & e^t & e^{-t} \\ 5 & 2e^t & 2e^{-t} \end{bmatrix}$$

$$\mathbf{x}(t) = \begin{bmatrix} 6 & 3e^t & 2e^{-t} \\ 2 & e^t & e^{-t} \\ 5 & 2e^t & 2e^{-t} \end{bmatrix} \begin{bmatrix} 0 & -2 & 1 \\ 1 & 2 & -2 \\ -1 & 3 & 0 \end{bmatrix} \begin{bmatrix} 2 \\ 1 \\ 0 \end{bmatrix} = \begin{bmatrix} -12 + 12e^t + 2e^{-t} \\ -4 + 4e^t + e^{-t} \\ -10 + 8e^t + 2e^{-t} \end{bmatrix}$$

In each of Problems 9–20 we first solve the given linear system to find two linearly independent solutions $\mathbf{x}_1$ and $\mathbf{x}_2$, then set up the fundamental matrix $\Phi(t) = [\mathbf{x}_1(t) \; \mathbf{x}_2(t)]$, and finally calculate the matrix exponential $e^{At} = \Phi(t)\Phi(0)^{-1}$.

9. Eigensystem: $\lambda_1 = 1, \quad \mathbf{v}_1 = [1 \; 1]^T; \quad \lambda_2 = 3, \quad \mathbf{v}_2 = [2 \; 1]^T$

$$\Phi(t) = \begin{bmatrix} e^{\lambda_1 t}\mathbf{v}_1 & e^{\lambda_2 t}\mathbf{v}_2 \end{bmatrix} = \begin{bmatrix} e^t & 2e^{3t} \\ e^t & e^{3t} \end{bmatrix}$$

$$e^{At} = \begin{bmatrix} e^t & 2e^{3t} \\ e^t & e^{3t} \end{bmatrix} \begin{bmatrix} -1 & 2 \\ 1 & -1 \end{bmatrix} = \begin{bmatrix} -e^t + 2e^{3t} & 2e^t - 2e^{3t} \\ -e^t + e^{3t} & 2e^t - e^{3t} \end{bmatrix}$$

11. Eigensystem: $\lambda_1 = 2, \quad \mathbf{v}_1 = [1 \; 1]^T; \quad \lambda_2 = 3, \quad \mathbf{v}_2 = [3 \; 2]^T$

$$\Phi(t) = \begin{bmatrix} e^{\lambda_1 t}\mathbf{v}_1 & e^{\lambda_2 t}\mathbf{v}_2 \end{bmatrix} = \begin{bmatrix} e^{2t} & 3e^{3t} \\ e^{2t} & 2e^{3t} \end{bmatrix}$$

$$e^{At} = \begin{bmatrix} e^{2t} & 3e^{3t} \\ e^{2t} & 2e^{3t} \end{bmatrix} \begin{bmatrix} -2 & 3 \\ 1 & -1 \end{bmatrix} = \begin{bmatrix} -2e^{2t} + 3e^{3t} & 3e^{2t} - 3e^{3t} \\ -2e^{2t} + 2e^{3t} & 3e^{2t} - 2e^{3t} \end{bmatrix}$$

13. Eigensystem: $\lambda_1 = 1, \quad \mathbf{v}_1 = [1 \; 1]^T; \quad \lambda_2 = 3, \quad \mathbf{v}_2 = [4 \; 3]^T$

$$\Phi(t) = \begin{bmatrix} e^{\lambda_1 t}\mathbf{v}_1 & e^{\lambda_2 t}\mathbf{v}_2 \end{bmatrix} = \begin{bmatrix} e^t & 4e^{3t} \\ e^t & 3e^{3t} \end{bmatrix}$$

$$e^{At} = \begin{bmatrix} e^t & 4e^{3t} \\ e^t & 3e^{3t} \end{bmatrix} \begin{bmatrix} -3 & 4 \\ 1 & -1 \end{bmatrix} = \begin{bmatrix} -3e^t + 4e^{3t} & 4e^t - 4e^{3t} \\ -3e^t + 3e^{3t} & 4e^t - 3e^{3t} \end{bmatrix}$$

15. Eigensystem: $\lambda_1 = 1$, $\mathbf{v}_1 = [2 \ \ 1]^T$; $\qquad \lambda_2 = 2$, $\mathbf{v}_2 = [5 \ \ 2]^T$

$$\Phi(t) = \begin{bmatrix} e^{\lambda_1 t}\mathbf{v}_1 & e^{\lambda_2 t}\mathbf{v}_2 \end{bmatrix} = \begin{bmatrix} 2e^t & 5e^{2t} \\ e^t & 2e^{2t} \end{bmatrix}$$

$$e^{\mathbf{A}t} = \begin{bmatrix} 2e^t & 5e^{2t} \\ e^t & 2e^{2t} \end{bmatrix} \begin{bmatrix} -2 & 5 \\ 1 & -2 \end{bmatrix} = \begin{bmatrix} -4e^t + 5e^{2t} & 10e^t - 10e^{2t} \\ -2e^t + 2e^{2t} & 5e^t - 4e^{2t} \end{bmatrix}$$

17. Eigensystem: $\lambda_1 = 2$, $\mathbf{v}_1 = [1 \ \ -1]^T$; $\qquad \lambda_2 = 4$, $\mathbf{v}_2 = [1 \ \ 1]^T$

$$\Phi(t) = \begin{bmatrix} e^{\lambda_1 t}\mathbf{v}_1 & e^{\lambda_2 t}\mathbf{v}_2 \end{bmatrix} = \begin{bmatrix} e^{2t} & e^{4t} \\ -e^{2t} & e^{4t} \end{bmatrix}$$

$$e^{\mathbf{A}t} = \begin{bmatrix} e^{2t} & e^{4t} \\ -e^{2t} & e^{4t} \end{bmatrix} \cdot \frac{1}{2}\begin{bmatrix} 1 & -1 \\ 1 & 1 \end{bmatrix} = \frac{1}{2}\begin{bmatrix} e^{2t} + e^{4t} & -e^{2t} + e^{4t} \\ -e^{2t} + e^{4t} & e^{2t} + e^{4t} \end{bmatrix}$$

19. Eigensystem: $\lambda_1 = 5$, $\mathbf{v}_1 = [1 \ \ -2]^T$; $\qquad \lambda_2 = 10$, $\mathbf{v}_2 = [2 \ \ 1]^T$

$$\Phi(t) = \begin{bmatrix} e^{\lambda_1 t}\mathbf{v}_1 & e^{\lambda_2 t}\mathbf{v}_2 \end{bmatrix} = \begin{bmatrix} e^{5t} & 2e^{10t} \\ -2e^{5t} & e^{10t} \end{bmatrix}$$

$$e^{\mathbf{A}t} = \begin{bmatrix} e^{5t} & 2e^{10t} \\ -2e^{5t} & e^{10t} \end{bmatrix} \cdot \frac{1}{5}\begin{bmatrix} 1 & -2 \\ 2 & 1 \end{bmatrix} = \frac{1}{5}\begin{bmatrix} e^{5t} + 4e^{10t} & -2e^{5t} + 2e^{10t} \\ -2e^{5t} + 2e^{10t} & 4e^{5t} + e^{10t} \end{bmatrix}$$

21. $\mathbf{A}^2 = \mathbf{0}$ so $e^{\mathbf{A}t} = \mathbf{I} + \mathbf{A}t = \begin{bmatrix} 1+t & -t \\ t & 1-t \end{bmatrix}$

23. $\mathbf{A}^3 = \mathbf{0}$ so $e^{\mathbf{A}t} = \mathbf{I} + \mathbf{A}t + \frac{1}{2}\mathbf{A}^2 t^2 = \begin{bmatrix} 1+t & -t & -t-t^2 \\ t & 1-t & t-t^2 \\ 0 & 0 & 1 \end{bmatrix}$

25. $\mathbf{A} = 2\mathbf{I} + \mathbf{D}$ where $\mathbf{D}^2 = \mathbf{0}$, so $e^{\mathbf{A}t} = e^{2\mathbf{I}t}e^{\mathbf{D}t} = (e^{2t}\mathbf{I})(\mathbf{I} + \mathbf{D}t)$. Hence

$$e^{\mathbf{A}t} = \begin{bmatrix} e^{2t} & 5t\,e^{2t} \\ 0 & e^{2t} \end{bmatrix}, \qquad \mathbf{x}(t) = e^{\mathbf{A}t}\begin{bmatrix} 4 \\ 7 \end{bmatrix} = e^{2t}\begin{bmatrix} 4 + 35t \\ 7 \end{bmatrix}$$

27. $A = I+D$ where $D^3 = 0$, so $e^{At} = e^{It}e^{Dt} = (e^t I)(I+Dt+\frac{1}{2}D^2t^2)$. Hence

$$e^{At} = \begin{bmatrix} e^t & 2te^t & (3t+2t^2)e^t \\ 0 & e^t & 2te^t \\ 0 & 0 & e^t \end{bmatrix}, \quad x(t) = e^{At}\begin{bmatrix} 4 \\ 5 \\ 6 \end{bmatrix} = e^t\begin{bmatrix} 4+28t+12t^2 \\ 5+12t \\ 6 \end{bmatrix}$$

29. $A = I+D$ where $D^4 = 0$, so $e^{At} = e^{It}e^{Dt} = (e^t I)(I+Dt+\frac{1}{2}D^2t^2+\frac{1}{6}D^3t^3)$. Hence

$$e^{At} = e^t\begin{bmatrix} 1 & 2t & 3t+6t^2 & 4t+6t^2+4t^3 \\ 0 & 1 & 6t & 3t+6t^2 \\ 0 & 0 & 1 & 2t \\ 0 & 0 & 0 & 1 \end{bmatrix}, \quad x(t) = e^{At}\begin{bmatrix} 1 \\ 1 \\ 1 \\ 1 \end{bmatrix} = e^t\begin{bmatrix} 1+9t+12t^2+4t^3 \\ 1+9t+6t^2 \\ 1+2t \\ 1 \end{bmatrix}$$

33. $e^{At} = I\cosh t + A\sinh t = \begin{bmatrix} \cosh t & \sinh t \\ \sinh t & \cosh t \end{bmatrix}$, so the general solution of $x' = Ax$ is

$$x(t) = e^{At}c = \begin{bmatrix} c_1\cosh t + c_2\sinh t \\ c_1\sinh t + c_2\cosh t \end{bmatrix}.$$

In Problems 35–40 we give first the linearly independent generalized eigenvectors $u_1, u_2, \cdots, u_n$ of the matrix A and the corresponding solution vectors $x_1(t), x_2(t), \cdots, x_n(t)$ defined by Eq. (34) in the text, then the fundamental matrix $\Phi(t) = [x_1(t) \ x_2(t) \ \cdots \ x_n(t)]$. Finally we calculate the exponential matrix $e^{At} = \Phi(t)\Phi(0)^{-1}$.

35. $\lambda = 3$: $u_1 = [4 \ 0]^T$, $u_2 = [0 \ 1]^T$

$\{u_1, u_2\}$ is a length 2 chain based on the ordinary (rank 1) eigenvector u_1, so u_2 is a generalized eigenvector of rank 2.

$x_1(t) = e^{\lambda t}u_1, \quad x_2(t) = e^{\lambda t}(u_2 + (A-\lambda I)u_2 t)$

$\Phi(t) = [x_1(t) \ x_2(t)] = e^{3t}\begin{bmatrix} 4 & 4t \\ 0 & 1 \end{bmatrix}$

$e^{At} = e^{3t}\begin{bmatrix} 4 & 4t \\ 0 & 1 \end{bmatrix} \cdot \frac{1}{4}\begin{bmatrix} 1 & 0 \\ 0 & 4 \end{bmatrix} = e^{3t}\begin{bmatrix} 1 & 4t \\ 0 & 1 \end{bmatrix}$

37. $\lambda_1 = 2$: $u_1 = [1 \ 0 \ 0]^T$, $x_1(t) = e^{\lambda_1 t}u_1$

$\lambda_2 = 1$: $u_2 = [9 \ -3 \ 0]^T$, $u_3 = [10 \ 1 \ -1]^T$

$\{\mathbf{u}_2, \mathbf{u}_3\}$ is a length 2 chain based on the ordinary (rank 1) eigenvector $\mathbf{u}_2$, so $\mathbf{u}_3$ is a generalized eigenvector of rank 2.

$$\mathbf{x}_2(t) = e^{\lambda_2 t}\mathbf{u}_2, \quad \mathbf{x}_3(t) = e^{\lambda_2 t}\left(\mathbf{u}_3 + (\mathbf{A} - \lambda_2\mathbf{I})\mathbf{u}_3 t\right)$$

$$\Phi(t) = [\mathbf{x}_1(t) \ \mathbf{x}_2(t) \ \mathbf{x}_3(t)] = \begin{bmatrix} e^{2t} & 9e^t & (10+9t)e^t \\ 0 & -3e^t & (1-3t)e^t \\ 0 & 0 & -e^t \end{bmatrix}$$

$$e^{\mathbf{A}t} = \begin{bmatrix} e^{2t} & 9e^t & (10+9t)e^t \\ 0 & -3e^t & (1-3t)e^t \\ 0 & 0 & -e^t \end{bmatrix} \cdot \frac{1}{3}\begin{bmatrix} 3 & 9 & 13 \\ 0 & -1 & -1 \\ 0 & 0 & -3 \end{bmatrix}$$

$$= \begin{bmatrix} e^{2t} & -3e^t + 3e^{2t} & (-13-9t)e^t + 13e^{2t} \\ 0 & e^t & 3t\,e^t \\ 0 & 0 & e^t \end{bmatrix}$$

39. $\lambda_2 = 1$: $\quad \mathbf{u}_1 = [3 \ 0 \ 0 \ 0]^T, \quad \mathbf{u}_2 = [0 \ 1 \ 0 \ 0]^T$

$\{\mathbf{u}_1, \mathbf{u}_2\}$ is a length 2 chain based on the ordinary (rank 1) eigenvector $\mathbf{u}_1$, so $\mathbf{u}_2$ is a generalized eigenvector of rank 2.

$$\mathbf{x}_1(t) = e^{\lambda_1 t}\mathbf{u}_1, \quad \mathbf{x}_2(t) = e^{\lambda_1 t}\left(\mathbf{u}_2 + (\mathbf{A} - \lambda_1\mathbf{I})\mathbf{u}_2 t\right)$$

$\lambda_2 = 2$: $\quad \mathbf{u}_3 = [144 \ 36 \ 12 \ 0]^T, \quad \mathbf{u}_4 = [0 \ 27 \ 17 \ 4]^T$

$\{\mathbf{u}_3, \mathbf{u}_4\}$ is a length 2 chain based on the ordinary (rank 1) eigenvector $\mathbf{u}_3$, so $\mathbf{u}_4$ is a generalized eigenvector of rank 2.

$$\mathbf{x}_3(t) = e^{\lambda_2 t}\mathbf{u}_3, \quad \mathbf{x}_4(t) = e^{\lambda_2 t}\left(\mathbf{u}_4 + (\mathbf{A} - \lambda_2\mathbf{I})\mathbf{u}_4 t\right)$$

$$\Phi(t) = [\mathbf{x}_1(t) \ \mathbf{x}_2(t) \ \mathbf{x}_3(t) \ \mathbf{x}_4(t)] = \begin{bmatrix} 3e^t & 3t\,e^t & 144e^{2t} & 144t\,e^{2t} \\ 0 & e^t & 36e^{2t} & (27+36t)e^{2t} \\ 0 & 0 & 12e^{2t} & (17+12t)e^{2t} \\ 0 & 0 & 0 & 4e^{2t} \end{bmatrix}$$

$$e^{\mathbf{A}t} = \begin{bmatrix} 3e^t & 3t\,e^t & 144e^{2t} & 144t\,e^{2t} \\ 0 & e^t & 36e^{2t} & (27+36t)e^{2t} \\ 0 & 0 & 12e^{2t} & (17+12t)e^{2t} \\ 0 & 0 & 0 & 4e^{2t} \end{bmatrix} \cdot \frac{1}{48}\begin{bmatrix} 16 & 0 & -192 & 816 \\ 0 & 48 & -144 & 288 \\ 0 & 0 & 4 & -17 \\ 0 & 0 & 0 & 12 \end{bmatrix}$$

$$= \begin{bmatrix} e^t & 3te^t & (-12-9t)e^t+12te^{2t} & (51+18t)e^t+(-51+36t)e^{2t} \\ 0 & e^t & -3e^t+3e^{2t} & 6e^t+(-6+9t)e^{2t} \\ 0 & 0 & e^{2t} & 3te^{2t} \\ 0 & 0 & 0 & e^{2t} \end{bmatrix}$$

SECTION 8.2

NONHOMOGENEOUS LINEAR SYSTEMS

1. Substitution of the trial solution $x_p(t) = a$, $y_p(t) = b$ yields the equations $a + 2b + 3 = 0$, $2a + b - 2 = 0$ with solution $a = 7/3$, $b = -8/3$. Thus we obtain the particular solution $x(t) = 7/3$, $y(t) = -8/3$.

3. When we substitute the trial solution

$$x_p = a_1 + b_1 t + c_1 t^2, \quad y_p = a_2 + b_2 t + c_2 t^2$$

$\begin{bmatrix} b_1 + 2c_1 t \\ b_2 + 2c_2 t \end{bmatrix} = \begin{bmatrix} 3 & 4 \\ 3 & 2 \end{bmatrix}$

and collect coefficients, we get the equations

$$3a_1 + 4a_2 = b_1 \qquad 3b_1 + 4b_2 = 2c_1 \qquad 3c_1 + 4c_2 = 0$$
$$3a_1 + 2a_2 = b_2 \qquad 3b_1 + 2b_2 = 2c_2 \qquad 3c_1 + 2c_2 + 1 = 0.$$

Working backwards, we solve first for $c_1 = -2/3$, $c_2 = 1/2$, then for $b_1 = 10/9$, $b_2 = -7/6$, and finally for $a_1 = -31/27$, $a_2 = 41/36$. This determines the particular solution $x_p(t)$, $y_p(t)$. Next, the coefficient matrix of the associated homogeneous system has eigenvalues $\lambda_1 = -1$ and $\lambda_2 = 6$ with eigenvectors $\mathbf{v}_1 = [1 \ -1]^T$ and $\mathbf{v}_2 = [4 \ 3]^T$, respectively, so the complementary solution is given by

$$x_c(t) = c_1 e^{-t} + 4c_2 e^{6t}, \qquad x_c(t) = -c_1 e^{-t} + 3c_2 e^{6t}.$$

When we impose the initial conditions $x(0) = 0$, $y(0) = 0$ on the general solution $x(t) = x_c(t) + x_p(t)$, $y(t) = y_c(t) + y_p(t)$ we find that $c_1 = 8/7$, $c_2 = 1/756$. This finally gives the desired particular solution

$$x(t) = \frac{1}{756}(864e^{-t} + 4e^{6t} - 868 + 840t - 504t^2)$$

$$y(t) = \frac{1}{756}(-864e^{-t} + 3e^{6t} + 861 - 882t + 378t^2).$$

5. The coefficient matrix of the associated homogeneous system has eigenvalues $\lambda_1 = -1$ and $\lambda_2 = 5$, so the nonhomogeneous term e^{-t} duplicates part of the complementary solution. We therefore try the particular solution

$$x_p(t) = a_1 + b_1 e^{-t} + c_1 t e^{-t}, \quad y_p(t) = a_2 + b_2 e^{-t} + c_2 t e^{-t}.$$

Upon solving the six linear equations we get by collecting coefficients after substitution of this trial solution into the given nonhomogeneous system, we obtain the particular solution

$$x(t) = \frac{1}{3}(-12 - e^{-t} - 7te^{-t}), \quad y(t) = \frac{1}{3}(-6 - 7te^{-t}).$$

7. First we try the particular solution

$$x_p(t) = a_1 \sin t + b_1 \cos t, \quad y_p(t) = a_2 \sin t + b_2 \cos t.$$

Upon solving the four linear equations we get by collecting coefficients after substitution of this trial solution into the given nonhomogeneous system, we find that $a_1 = -21/82$, $b_1 = -25/82$, $a_2 = -15/41$, $b_2 = -12/41$. The coefficient matrix of the associated homogeneous system has eigenvalues $\lambda_1 = 1$ and $\lambda_2 = -9$ with eigenvectors $\mathbf{v}_1 = [1 \quad 1]^T$ and $\mathbf{v}_2 = [2 \quad -3]^T$, respectively, so the complementary solution is given by

$$x_c(t) = c_1 e^t + 2c_2 e^{-9t}, \quad y_c(t) = c_1 e^t - 3c_2 e^{-9t}.$$

When we impose the initial conditions $x(0) = 1$, $y(0) = 0$, we find that $c_1 = 9/10$ and $c_2 = 83/410$. It follows that the desired particular solution $x = x_c + x_p$, $y = y_c + y_p$ is given by

$$x(t) = \frac{1}{410}(369e^t + 166e^{-9t} - 125\cos t - 105\sin t)$$

$$y(t) = \frac{1}{410}(369e^t - 249e^{-9t} - 120\cos t - 150\sin t).$$

9. Here the associated homogeneous system is the same as in Problem 8, so the nonhomogeneous term $\cos 2t$ term duplicates the complementary function. We therefore substitute the trial solution

$$x_p(t) = a_1 \sin 2t + b_1 \cos 2t + c_1 t \sin 2t + d_1 t \cos 2t$$
$$y_p(t) = a_2 \sin 2t + b_2 \cos 2t + c_2 t \sin 2t + d_2 t \cos 2t$$

Section 8.2

and use a computer algebra system to solve the system of 8 linear equations that results when we collect coefficients in the usual way. This gives the particular solution

$$x(t) = \frac{1}{4}(\sin 2t + 2t\cos 2t + t\sin 2t), \quad y(t) = \frac{1}{4}t\sin 2t.$$

11. The coefficient matrix of the associated homogeneous system has eigenvalues $\lambda_1 = 0$ and $\lambda_2 = 4$, so there is duplication of constant terms. We therefore substitute the particular solution

$$x_p(t) = a_1 + b_1 t, \quad y_p(t) = a_2 + b_2 t$$

and solve the resulting equations for $a_1 = -2, a_2 = 0, b_1 = -2, b_2 = 1$. The eigenvectors of the coefficient matrix associated with the eigenvalues $\lambda_1 = 0$ and $\lambda_2 = 4$ are $\mathbf{v}_1 = [2 \quad -1]^T$ and $\mathbf{v}_2 = [2 \quad 1]^T$, respectively, so the general solution of the given nonhomogeneous system is given by

$$x(t) = 2c_1 + 2c_2 e^{4t} - 2 - 2t, \quad y(t) = -c_1 + c_2 e^{4t} + t.$$

When we impose the initial conditions $x(0) = 1, y(0) = -1$ we find readily that $c_1 = 5/4, c_2 = 1/4$. This gives the desired particular solution

$$x(t) = \tfrac{1}{2}(1 - 4t + e^{4t}), \quad y(t) = \tfrac{1}{4}(-5 + 4t + e^{4t}).$$

13. The coefficient matrix of the associated homogeneous system has eigenvalues $\lambda_1 = 1$ and $\lambda_2 = 3$, so there is duplication of e^t terms. We therefore substitute the trial solution

$$x_p(t) = (a_1 + b_1 t)e^t, \quad y_p(t) = (a_2 + b_2 t)e^t.$$

This leads readily to the particular solution

$$x(t) = \frac{1}{2}(1 + 5t)e^t, \quad y(t) = -\frac{5}{2}t e^t.$$

In Problems 15 and 16 the amounts $x_1(t)$ and $x_2(t)$ in the two tanks satisfy the equations

$$x_1' = rc_0 - k_1 x_1, \quad x_2' = k_1 x_1 - k_2 x_2$$

where $k_i = r/V_i$ in terms of the flow rate r, the inflowing concentration c_0, and the volumes V_1 and V_2 of the two tanks.

15. **(a)** We solve the initial value problem

$$x_1' = 20 - x_1/10, \qquad x_1(0) = 0$$
$$x_2' = x_1/10 - x_2/20, \qquad x_2(0) = 0$$

for $x_1(t) = 200(1-e^{-t/10})$, $x_2(t) = 400(1+e^{-t/10} - 2e^{-t/20})$.

(b) Evidently $x_1(t) \to 200$ gal and $x_2(t) \to 400$ gal as $t \to \infty$.

(c) It takes about 6 min 56 sec for tank 1 to reach a salt concentration of 1 lb/gal, and about 24 min 34 sec for tank 2 to reach this concentration.

In Problems 17–34 we apply the variation of parameters formula in Eq. (28) of Section 8.2. This is a highly formal process best carried out using a computer algebra system. The answers shown below were actually calculated using the Mathematica code listed in the computing project for Section 8.2. For instance, for Problem 17 we first enter the coefficient matrix

```
A = {{6,  -7},
     {1,  -2}};
```

the initial vector

```
x0 = {{0},
      {0}};
```

and the vector

```
f[t_] := {{60},
          {90}};
```

of nonhomogeneous terms. It simplifies the notation to rename Mathematica's exponential matrix function by defining

```
exp[A_] := MatrixExp[A] // FullSimplify
```

Then the integral in the variation of parameters formula is given by

```
integral =
Integrate[exp[-A*s] . f[s], {s, 0, t}] // Simplify
```

$$\begin{bmatrix} -102 + 7e^{-5t} + 95e^t \\ -96 + e^{-5t} + 95e^t \end{bmatrix}.$$

Finally the desired particular solution is given by

```
solution =
exp[A*t] . (x0 + integral) // Simplify
```

$$\begin{bmatrix} 102 - 7e^{-5t} - 95e^t \\ 96 - e^{-5t} - 95e^t \end{bmatrix}.$$

(Maple and MATLAB versions of this computation are provided in the computing projects manual that accompanies the textbook.)

In each succeeding problem, we need only substitute the given coefficient matrix **A**, initial vector **x0**, and the vector **f** of nonhomogeneous terms in the above commands, and then re-execute them in turn. This formal process is illustrated with intermediate results in Problems 17 and 25. In the remaining problems we give below only the component functions of the final results.

17. $\quad e^{\mathbf{A}t} = \dfrac{1}{6}\begin{bmatrix} -e^{-t} + 7e^{5t} & 7e^{-t} - 7e^{5t} \\ -e^{-t} + e^{5t} & 7e^{-t} - e^{5t} \end{bmatrix}$

$e^{-\mathbf{A}s}\mathbf{f}(s) = \dfrac{1}{6}\begin{bmatrix} -e^s + 7e^{-5s} & 7e^s - 7e^{-5s} \\ -e^s + e^{-5s} & 7e^s - e^{-5s} \end{bmatrix}\begin{bmatrix} 60 \\ 90 \end{bmatrix} = \begin{bmatrix} 95e^s - 35e^{-5s} \\ 95e^s - 5e^{-5s} \end{bmatrix}$

$\mathbf{x}(0) + \displaystyle\int_0^t e^{-\mathbf{A}s}\mathbf{f}(s)\,ds = \begin{bmatrix} 0 \\ 0 \end{bmatrix} + \int_0^t \begin{bmatrix} 95e^s - 35e^{-5s} \\ 95e^s - 5e^{-5s} \end{bmatrix}ds = \begin{bmatrix} -102 + 95e^t + 7e^{-5t} \\ -96 + 95e^t + e^{-5t} \end{bmatrix}$

$\mathbf{x}(t) = e^{\mathbf{A}t}\left(\mathbf{x}(0) + \displaystyle\int_0^t e^{-\mathbf{A}s}\mathbf{f}(s)\,ds\right) = \dfrac{1}{6}\begin{bmatrix} -e^{-t} + 7e^{5t} & 7e^{-t} - 7e^{5t} \\ -e^{-t} + e^{5t} & 7e^{-t} - e^{5t} \end{bmatrix}\begin{bmatrix} -102 + 95e^t + 7e^{-5t} \\ -96 + 95e^t + e^{-5t} \end{bmatrix}$

$= \begin{bmatrix} 102 - 95e^{-t} - 7e^{5t} \\ 96 - 95e^{-t} - e^{5t} \end{bmatrix}$

Thus the solution vector $\mathbf{x}(t)$ has scalar components

$$x_1(t) = 102 - 95e^{-t} - 7e^{5t}, \quad x_2(t) = 96 - 95e^{-t} - e^{5t}.$$

19. $\quad x_1(t) = -70 - 60t + 16e^{-3t} + 54e^{2t}, \quad x_2(t) = 5 - 60t - 32e^{-3t} + 27e^{2t}$

21. See the solution to Problem 31 in Section 8.3 for the variation of parameters calculation.

$x_1(t) = -e^{-t} - 14e^{2t} + 15e^{3t}, \quad x_2(t) = -5e^{-t} - 10e^{2t} + 15e^{3t}$

23. $\quad x_1(t) = 3 + 11t + 8t^2, \quad x_2(t) = 5 + 17t + 24t^2$

25. $\quad e^{\mathbf{A}t} = \begin{bmatrix} \cos t + 2\sin t & -5\sin t \\ \sin t & \cos t - 2\sin t \end{bmatrix}$

$$e^{-As}\mathbf{f}(s) = \begin{bmatrix} \cos s - 2\sin s & 5\sin s \\ -\sin s & \cos t + 2\sin s \end{bmatrix} \begin{bmatrix} 4t \\ 1 \end{bmatrix}$$

$$= \begin{bmatrix} 5\sin s + 4s\cos s - 8s\sin s \\ \cos s + 2\sin s - 4s\sin s \end{bmatrix}$$

$$\mathbf{x}(0) + \int_0^t e^{-As}\mathbf{f}(s)\,ds = \begin{bmatrix} 0 \\ 0 \end{bmatrix} + \int_0^t \begin{bmatrix} 5\sin s + 4s\cos s - 8s\sin s \\ \cos s + 2\sin s - 4s\sin s \end{bmatrix} ds =$$

$$= \begin{bmatrix} 1 - \cos t - 8\sin t + 8t\cos t + 4t\sin t \\ 2 - 2\cos t - 3\sin t + 4t\cos t \end{bmatrix}$$

$$\mathbf{x}(t) = e^{At}\left(\mathbf{x}(0) + \int_0^t e^{-As}\mathbf{f}(s)\,ds\right)$$

$$= \begin{bmatrix} \cos t + 2\sin t & -5\sin t \\ \sin t & \cos t - 2\sin t \end{bmatrix} \begin{bmatrix} 1 - \cos t - 8\sin t + 8t\cos t + 4t\sin t \\ 2 - 2\cos t - 3\sin t + 4t\cos t \end{bmatrix}$$

$$= \begin{bmatrix} -1 + 8t + \cos t - 8\sin t \\ -2 + 4t + 2\cos t - 3\sin t \end{bmatrix}$$

Thus the solution vector $\mathbf{x}(t)$ has scalar components

$$x_1(t) = -1 + 8t + \cos t - 8\sin t, \quad x_2(t) = -2 + 4t + 2\cos t - 3\sin t.$$

27. $x_1(t) = 8t^3 + 6t^4, \quad x_2(t) = 3t^2 - 2t^3 + 3t^4$

29. See the solution to Problem 31 in Section 8.3 for the variation of parameters calculation.
$x_1(t) = t\cos t - \ln(\cos t)\sin t, \quad x_2(t) = t\sin t + \ln(\cos t)\cos t$

31. $x_1(t) = (9t^2 + 4t^3)e^t, \quad x_2(t) = 6t^2 e^t, \quad x_3(t) = 6te^t$

33. $x_1(t) = 15t^2 + 60t^3 + 95t^4 + 12t^5, \quad x_2(t) = 15t^2 + 55t^3 + 15t^4$,
$x_3(t) = 15t^2 + 20t^3, \quad x_4(t) = 15t^2$

SECTION 8.3

SPECTRAL DECOMPOSITION METHODS

In Problems 1–20 here we want to use projection matrices to find fundamental matrix solutions of the linear systems given in Problems 1-20 of Section 7.3. In each of Problems 1–16, the coefficient matrix $\mathbf{A}$ is 2×2 with distinct eigenvalues λ_1 and λ_2. We can therefore use the method of Example 1 in Section 8.3. That is, if we define the projection matrices

$$P_1 = \frac{A - \lambda_2 I}{\lambda_1 - \lambda_2} \quad \text{and} \quad P_2 = \frac{A - \lambda_1 I}{\lambda_2 - \lambda_1}, \tag{1}$$

then the desired fundamental matrix solution of the system $x' = Ax$ is the exponential matrix

$$e^{At} = e^{\lambda_1 t} P_1 + e^{\lambda_2 t} P_2. \tag{2}$$

We use the eigenvalues λ_1 and λ_2 given in the Section 7.3 solutions for these problems.

1. $A = \begin{bmatrix} 1 & 2 \\ 2 & 1 \end{bmatrix}; \quad \lambda_1 = -1, \quad \lambda_2 = 3$

$$P_1 = \frac{1}{-4}(A - 3I) = \frac{1}{2}\begin{bmatrix} 1 & -1 \\ -1 & 1 \end{bmatrix}, \quad P_2 = \frac{1}{4}(A + I) = \frac{1}{2}\begin{bmatrix} 1 & 1 \\ 1 & 1 \end{bmatrix}$$

$$e^{At} = e^{-t} P_1 + e^{3t} P_2 = \frac{1}{2}\begin{bmatrix} e^{-t} + e^{3t} & -e^{-t} + e^{3t} \\ -e^{-t} + e^{3t} & e^{-t} + e^{3t} \end{bmatrix}$$

3. $A = \begin{bmatrix} 3 & 4 \\ 3 & 2 \end{bmatrix}; \quad \lambda_1 = -1, \quad \lambda_2 = 6$

$$P_1 = \frac{1}{-7}(A - 6I) = \frac{1}{7}\begin{bmatrix} 3 & -4 \\ -3 & 4 \end{bmatrix}, \quad P_2 = \frac{1}{7}(A + I) = \frac{1}{7}\begin{bmatrix} 4 & 4 \\ 3 & 3 \end{bmatrix}$$

$$e^{At} = e^{-t} P_1 + e^{6t} P_2 = \frac{1}{7}\begin{bmatrix} 3e^{-t} + 4e^{6t} & -4e^{-t} + 4e^{6t} \\ -3e^{-t} + 3e^{6t} & 4e^{-t} + 3e^{6t} \end{bmatrix}$$

5. $A = \begin{bmatrix} 6 & -7 \\ 1 & -2 \end{bmatrix}; \quad \lambda_1 = -1, \quad \lambda_2 = 5$

$$P_1 = \frac{1}{-6}(A - 5I) = \frac{1}{6}\begin{bmatrix} -1 & 7 \\ -1 & 7 \end{bmatrix}, \quad P_2 = \frac{1}{6}(A + I) = \frac{1}{6}\begin{bmatrix} 7 & -7 \\ 1 & -1 \end{bmatrix}$$

$$e^{At} = e^{-t} P_1 + e^{5t} P_2 = \frac{1}{6}\begin{bmatrix} -e^{-t} + 7e^{5t} & 7e^{-t} - 7e^{5t} \\ -e^{-t} + e^{5t} & 7e^{-t} - e^{5t} \end{bmatrix}$$

7. $A = \begin{bmatrix} -3 & 4 \\ 6 & -5 \end{bmatrix}; \quad \lambda_1 = -9, \quad \lambda_2 = 1$

$$P_1 = \frac{1}{-10}(A - I) = \frac{1}{5}\begin{bmatrix} 2 & -2 \\ -3 & 3 \end{bmatrix}, \quad P_2 = \frac{1}{10}(A + 9I) = \frac{1}{10}\begin{bmatrix} 3 & 2 \\ 3 & 2 \end{bmatrix}$$

$$e^{\mathbf{A}t} = e^{-9t}\mathbf{P}_1 + e^t\mathbf{P}_2 = \frac{1}{5}\begin{bmatrix} 2e^{-9t}+3e^t & -2e^{-9t}+2e^t \\ -3e^{-9t}+3e^t & 3e^{-9t}+2e^t \end{bmatrix}$$

9. $\mathbf{A} = \begin{bmatrix} 2 & -5 \\ 4 & 2 \end{bmatrix}; \quad \lambda_1 = -4i, \quad \lambda_2 = 4i$

$$\mathbf{P}_1 = \frac{1}{-8i}(\mathbf{A} - 4i\mathbf{I}) = \frac{1}{8}\begin{bmatrix} 4+2i & -5i \\ 4i & 4-2i \end{bmatrix}, \quad \mathbf{P}_2 = \frac{1}{8i}(\mathbf{A} + 4i\mathbf{I}) = \frac{1}{8}\begin{bmatrix} 4-2i & 5i \\ -4i & 4+2i \end{bmatrix}$$

$$e^{\mathbf{A}t} = e^{-4it}\mathbf{P}_1 + e^{4it}\mathbf{P}_2 = (\cos 4t - i\sin 4t)\mathbf{P}_1 + (\cos 4t + i\sin 4t)\mathbf{P}_2$$

$$= \frac{1}{4}\begin{bmatrix} 4\cos 4t + 2\sin 4t & -5\sin 4t \\ 4\sin 4t & 4\cos 4t - 2\sin 4t \end{bmatrix}$$

11. $\mathbf{A} = \begin{bmatrix} 1 & -2 \\ 2 & 1 \end{bmatrix}; \quad \lambda_1 = 1-2i, \quad \lambda_2 = 1+2i$

$$\mathbf{P}_1 = \frac{1}{-4i}(\mathbf{A} - (1+2i)\mathbf{I}) = \frac{1}{2}\begin{bmatrix} 1 & -i \\ i & 1 \end{bmatrix}, \quad \mathbf{P}_2 = \frac{1}{4i}(\mathbf{A} - (1-2i)\mathbf{I}) = \frac{1}{2}\begin{bmatrix} 1 & i \\ -i & 1 \end{bmatrix}$$

$$e^{\mathbf{A}t} = e^{(1-2i)t}\mathbf{P}_1 + e^{(1+2i)t}\mathbf{P}_2 = e^t(\cos 2t - i\sin 2t)\mathbf{P}_1 + e^t(\cos 2t + i\sin 2t)\mathbf{P}_2$$

$$= e^t \begin{bmatrix} \cos 2t & -\sin 2t \\ \sin 2t & \cos 2t \end{bmatrix}$$

13. $\mathbf{A} = \begin{bmatrix} 5 & -9 \\ 2 & -1 \end{bmatrix}; \quad \lambda_1 = 2-3i, \quad \lambda_2 = 2+3i$

$$\mathbf{P}_1 = \frac{1}{-6i}(\mathbf{A} - (2+3i)\mathbf{I}) = \frac{1}{6}\begin{bmatrix} 3+3i & -9i \\ 2i & 3-3i \end{bmatrix}$$

$$\mathbf{P}_2 = \frac{1}{6i}(\mathbf{A} - (2-3i)\mathbf{I}) = \frac{1}{6}\begin{bmatrix} 3-3i & 9i \\ -2i & 3+3i \end{bmatrix}$$

$$e^{\mathbf{A}t} = e^{(2-3i)t}\mathbf{P}_1 + e^{(2+3i)t}\mathbf{P}_2 = e^{2t}(\cos 3t - i\sin 3t)\mathbf{P}_1 + e^{2t}(\cos 3t + i\sin 3t)\mathbf{P}_2$$

$$= \frac{e^{2t}}{3}\begin{bmatrix} 3\cos 3t + \sin 3t & -9\sin 3t \\ 2\sin 3t & 3\cos 3t - \sin 3t \end{bmatrix}$$

15. $\mathbf{A} = \begin{bmatrix} 7 & -5 \\ 4 & 3 \end{bmatrix}; \quad \lambda_1 = 5-4i, \quad \lambda_2 = 5+4i$

$$\mathbf{P}_1 = \frac{1}{-8i}(\mathbf{A} - (5+4i)\mathbf{I}) = \frac{1}{8}\begin{bmatrix} 4+2i & -5i \\ 4i & 4-2i \end{bmatrix}$$

$$\mathbf{P}_2 = \frac{1}{8i}(\mathbf{A} - (5-4i)\mathbf{I}) = \frac{1}{8}\begin{bmatrix} 4-2i & 5i \\ -4i & 4+2i \end{bmatrix}$$

$$e^{\mathbf{A}t} = e^{(5-4i)t}\mathbf{P}_1 + e^{(5+4i)t}\mathbf{P}_2 = e^{5t}(\cos 4t - i\sin 4t)\mathbf{P}_1 + e^{5t}(\cos 4t + i\sin 4t)\mathbf{P}_2$$

$$= \frac{e^{5t}}{4}\begin{bmatrix} 4\cos 4t + 2\sin 2t & -5\sin 4t \\ 4\sin 4t & 4\cos 4t - 2\sin 2t \end{bmatrix}$$

In Problems 17, 18, and 20 the coefficient matrix $\mathbf{A}$ is 3×3 with distinct eigenvalues $\lambda_1, \lambda_2, \lambda_3$. Looking at Equations (7) and (3) in Section 8.3, we see that

$$e^{\mathbf{A}t} = e^{\lambda_1 t}\mathbf{P}_1 + e^{\lambda_2 t}\mathbf{P}_2 + e^{\lambda_3 t}\mathbf{P}_3 \tag{3}$$

where the projection matrices are defined by

$$\mathbf{P}_1 = \frac{(\mathbf{A}-\lambda_2\mathbf{I})(\mathbf{A}-\lambda_3\mathbf{I})}{(\lambda_1-\lambda_2)(\lambda_1-\lambda_3)}, \quad \mathbf{P}_2 = \frac{(\mathbf{A}-\lambda_1\mathbf{I})(\mathbf{A}-\lambda_3\mathbf{I})}{(\lambda_2-\lambda_1)(\lambda_2-\lambda_3)}, \quad \mathbf{P}_3 = \frac{(\mathbf{A}-\lambda_1\mathbf{I})(\mathbf{A}-\lambda_2\mathbf{I})}{(\lambda_3-\lambda_1)(\lambda_3-\lambda_2)}. \tag{4}$$

17. $\mathbf{A} = \begin{bmatrix} 4 & 1 & 4 \\ 1 & 7 & 1 \\ 4 & 1 & 4 \end{bmatrix}$; $\lambda_1 = 0$, $\lambda_2 = 6$, $\lambda_3 = 9$

$$\mathbf{P}_1 = \frac{1}{54}(\mathbf{A}-6\mathbf{I})(\mathbf{A}-9\mathbf{I}) = \frac{1}{2}\begin{bmatrix} 1 & 0 & -1 \\ 0 & 0 & 0 \\ -1 & 0 & 1 \end{bmatrix}$$

$$\mathbf{P}_2 = \frac{1}{-18}(\mathbf{A}-0\mathbf{I})(\mathbf{A}-9\mathbf{I}) = \frac{1}{6}\begin{bmatrix} 1 & -2 & 1 \\ -2 & 4 & -2 \\ 1 & -2 & 1 \end{bmatrix}$$

$$\mathbf{P}_3 = \frac{1}{27}(\mathbf{A}-0\mathbf{I})(\mathbf{A}-6\mathbf{I}) = \frac{1}{3}\begin{bmatrix} 1 & 1 & 1 \\ 1 & 1 & 1 \\ 1 & 1 & 1 \end{bmatrix}$$

$$e^{\mathbf{A}t} = \mathbf{P}_1 + e^{6t}\mathbf{P}_2 + e^{9t}\mathbf{P}_3 = \frac{1}{6}\begin{bmatrix} 3+e^{6t}+2e^{9t} & -2e^{6t}+2e^{9t} & -3+e^{6t}+2e^{9t} \\ -2e^{6t}+2e^{9t} & 4e^{6t}+2e^{9t} & -2e^{6t}+2e^{9t} \\ -3+e^{6t}+2e^{9t} & -2e^{6t}+2e^{9t} & 3+e^{6t}+2e^{9t} \end{bmatrix}$$

19. $\mathbf{A} = \begin{bmatrix} 4 & 1 & 1 \\ 1 & 4 & 1 \\ 1 & 1 & 4 \end{bmatrix}$; $\lambda_1 = 6$, $\lambda_2 = 3$, $\lambda_3 = 3$

Here we have the eigenvalue $\lambda_1 = 6$ of multiplicity 1 and the eigenvalue $\lambda_2 = 3$ of multiplicity 2. By Example 2 in Section 8.3, the desired matrix exponential is given by

$$e^{\mathbf{A}t} = e^{\lambda_1 t}\mathbf{P}_1 + e^{\lambda_2 t}\mathbf{P}_2\left[\mathbf{I}+(\mathbf{A}-\lambda_2\mathbf{I})t\right] \tag{5}$$

192   Chapter 8

where $\mathbf{P}_1$ and $\mathbf{P}_2$ are the projection matrices of $\mathbf{A}$ corresponding to the eigenvalues λ_1 and λ_2. The reciprocal of the characteristic polynomial $p(\lambda) = (\lambda-6)(\lambda-3)^2$ has the partial fractions decomposition

$$\frac{1}{p(\lambda)} = \frac{1}{9(\lambda-6)} - \frac{\lambda}{9(\lambda-3)^2},$$

so $a_1(\lambda) = 1/9$ and $a_2(\lambda) = -\lambda/9$ in the notation of Equation (25) in the text. Therefore Equation (26) there gives

$$\mathbf{P}_1 = a_1(\mathbf{A}) \cdot (\mathbf{A}-\lambda_2\mathbf{I})^2 = \frac{1}{9}(\mathbf{A}-3\mathbf{I})^2 = \frac{1}{3}\begin{bmatrix} 1 & 1 & 1 \\ 1 & 1 & 1 \\ 1 & 1 & 1 \end{bmatrix},$$

$$\mathbf{P}_2 = a_2(\mathbf{A}) \cdot (\mathbf{A}-\lambda_1\mathbf{I}) = -\frac{1}{9}\mathbf{A}(\mathbf{A}-6\mathbf{I}) = \frac{1}{3}\begin{bmatrix} 2 & -1 & -1 \\ -1 & 2 & -1 \\ -1 & -1 & 2 \end{bmatrix}.$$

Finally, Equation (5) above gives

$$e^{\mathbf{A}t} = e^{6t}\mathbf{P}_1 + e^{3t}\mathbf{P}_2\left(\mathbf{I}+(\mathbf{A}-3\mathbf{I})t\right) = \frac{1}{3}\begin{bmatrix} 3e^{3t}+e^{6t} & -e^{3t}+e^{6t} & -e^{3t}+e^{6t} \\ -e^{3t}+e^{6t} & 2e^{3t}+e^{6t} & -e^{3t}+e^{6t} \\ -e^{3t}+e^{6t} & -e^{3t}+e^{6t} & 2e^{3t}+e^{6t} \end{bmatrix}.$$

In Problems 21–30 here we want to use projection matrices to find fundamental matrix solutions of the linear systems given in Problems 1–10 of Section 7.5. In each of Problems 21–26, the 2×2 coefficient matrix $\mathbf{A}$ has characteristic polynomial of the form $p(\lambda) = (\lambda-\lambda_1)^2$ and thus a single eigenvalue λ_1 of multiplicity 2. Consequently, Example 5 in Section 8.3 gives the desired fundamental matrix

$$e^{\mathbf{A}t} = e^{\lambda_1 t}\left[\mathbf{I}+(\mathbf{A}-\lambda_1\mathbf{I})t\right]. \tag{6}$$

21. $\mathbf{A} = \begin{bmatrix} -2 & 1 \\ -1 & -4 \end{bmatrix}; \quad \lambda_1 = -3$

$$e^{\mathbf{A}t} = e^{-3t}\left[\mathbf{I}+(\mathbf{A}+3\mathbf{I})t\right] = e^{-3t}\begin{bmatrix} 1+t & t \\ -t & 1-t \end{bmatrix}$$

23. $\mathbf{A} = \begin{bmatrix} 1 & -2 \\ 2 & 5 \end{bmatrix}$; $\quad \lambda_1 = 3$

$$e^{\mathbf{A}t} = e^{3t}[\mathbf{I} + (\mathbf{A}-3\mathbf{I})t] = e^{3t}\begin{bmatrix} 1-2t & -2t \\ 2t & 1+2t \end{bmatrix}$$

25. $\mathbf{A} = \begin{bmatrix} 7 & 1 \\ -4 & 3 \end{bmatrix}$; $\quad \lambda_1 = 5$

$$e^{\mathbf{A}t} = e^{5t}[\mathbf{I} + (\mathbf{A}-5\mathbf{I})t] = e^{5t}\begin{bmatrix} 1-4t & -4t \\ 4t & 1+4t \end{bmatrix}$$

Each of the 3×3 coefficient matrices in Problems 27–30 has a characteristic polynomial of the form $p(\lambda) = (\lambda - \lambda_1)((\lambda - \lambda_2)^2$ yielding an eigenvalue λ_1 of multiplicity 1 and an eigenvalue λ_2 of multiplicity 2. We therefore use the method explained in Problem 19 above, and list here the results of the principal steps in the calculation of the fundamental matrix $e^{\mathbf{A}t}$.

27. $\mathbf{A} = \begin{bmatrix} 2 & 0 & 0 \\ -7 & 9 & 7 \\ 0 & 0 & 2 \end{bmatrix}$; $\quad p(\lambda) = (\lambda-9)(\lambda-2)^2$; $\quad \lambda_1 = 9, \ \lambda_2 = 2$

$$\frac{1}{p(\lambda)} = \frac{1}{49(\lambda-9)} - \frac{\lambda+5}{49(\lambda-3)^2}; \quad a_1(\lambda) = \frac{1}{49}, \quad a_2(\lambda) = -\frac{\lambda+5}{49}$$

$$\mathbf{P}_1 = a_1(\mathbf{A})\cdot(\mathbf{A}-\lambda_2\mathbf{I})^2 = \frac{1}{49}(\mathbf{A}-2\mathbf{I})^2 = \begin{bmatrix} 0 & 0 & 0 \\ -1 & 1 & 1 \\ 0 & 0 & 0 \end{bmatrix}$$

$$\mathbf{P}_2 = a_2(\mathbf{A})\cdot(\mathbf{A}-\lambda_1\mathbf{I}) = -\frac{1}{49}(\mathbf{A}+5\mathbf{I})(\mathbf{A}-9\mathbf{I}) = \begin{bmatrix} 1 & 0 & 0 \\ 1 & 0 & -1 \\ 0 & 0 & 1 \end{bmatrix}$$

$$e^{\mathbf{A}t} = e^{9t}\mathbf{P}_1 + e^{2t}\mathbf{P}_2(\mathbf{I}+(\mathbf{A}-2\mathbf{I})t) = \begin{bmatrix} e^{2t} & 0 & 0 \\ e^{2t}-e^{9t} & e^{9t} & -e^{2t}+e^{9t} \\ 0 & 0 & e^{2t} \end{bmatrix}$$

29. $\mathbf{A} = \begin{bmatrix} -19 & 12 & 84 \\ 0 & 5 & 0 \\ -8 & 4 & 33 \end{bmatrix}$; $\quad p(\lambda) = (\lambda-9)(\lambda-5)^2$; $\quad \lambda_1 = 9, \ \lambda_2 = 5$

$$\frac{1}{p(\lambda)} = \frac{1}{16(\lambda-9)} + \frac{1-\lambda}{16(\lambda-5)^2}; \quad a_1(\lambda) = \frac{1}{16}, \quad a_2(\lambda) = \frac{1-\lambda}{16}$$

$$\mathbf{P}_1 = a_1(\mathbf{A}) \cdot (\mathbf{A} - \lambda_2 \mathbf{I})^2 = \frac{1}{16}(\mathbf{A} - 5\mathbf{I})^2 = \begin{bmatrix} -6 & 3 & 21 \\ 0 & 0 & 0 \\ -2 & 1 & 7 \end{bmatrix}$$

$$\mathbf{P}_2 = a_2(\mathbf{A}) \cdot (\mathbf{A} - \lambda_1 \mathbf{I}) = \frac{1}{16}(\mathbf{I} - \mathbf{A})(\mathbf{A} - 9\mathbf{I}) = \begin{bmatrix} 7 & -3 & -21 \\ 0 & 1 & 0 \\ 2 & -1 & -6 \end{bmatrix}$$

$$e^{\mathbf{A}t} = e^{9t}\mathbf{P}_1 + e^{5t}\mathbf{P}_2\left(\mathbf{I} + (\mathbf{A} - 5\mathbf{I})t\right) = \begin{bmatrix} 7e^{5t} - 6e^{9t} & -3e^{5t} + 3e^{9t} & -21e^{5t} + 21e^{9t} \\ 0 & e^{5t} & 0 \\ 2e^{5t} - 2e^{9t} & -e^{5t} + e^{9t} & -6e^{5t} + 7e^{9t} \end{bmatrix}$$

In Problems 31–40 we use the methods of this section to find the matrix exponentials that were *given* in in the statements of Problems 21–30 of Section 8.2. Once $e^{\mathbf{A}t}$ is known, the desired particular solution $\mathbf{x}(t)$ is provided by the variation of parameters formula

$$\mathbf{x}(t) = e^{\mathbf{A}t}\left(\mathbf{x}(0) + \int_0^t e^{-\mathbf{A}s}\mathbf{f}(s)\,ds\right) \qquad (7)$$

of Section 8.2. We give first the calculation of the matrix exponential using projection matrices and then the final result, which we obtained in each case using a computer algebra system to evaluate the right-hand side in Equation (7) — as described in the remarks preceding Problems 21–30 of Section 8.2. We illustrate this highly formal process by giving intermediate results in Problems 31, 34, and 39.

31. $\mathbf{A} = \begin{bmatrix} 4 & -1 \\ 5 & 2 \end{bmatrix}$, $\lambda_1 = -1$, $\lambda_2 = 3$

$$\mathbf{P}_1 = \frac{1}{-4}(\mathbf{A} - 3\mathbf{I}) = \frac{1}{4}\begin{bmatrix} -1 & 1 \\ -5 & 5 \end{bmatrix}, \quad \mathbf{P}_2 = \frac{1}{4}(\mathbf{A} + \mathbf{I}) = \frac{1}{4}\begin{bmatrix} 5 & -1 \\ 5 & -1 \end{bmatrix}$$

$$e^{\mathbf{A}t} = e^{-t}\mathbf{P}_1 + e^{3t}\mathbf{P}_2 = \frac{1}{4}\begin{bmatrix} -e^{-t} + 5e^{3t} & e^{-t} - e^{3t} \\ -5e^{-t} + 5e^{3t} & 5e^{-t} - e^{3t} \end{bmatrix}$$

$$\mathbf{x}(0) = \begin{bmatrix} 0 \\ 0 \end{bmatrix}, \quad \mathbf{f}(t) = \begin{bmatrix} 18e^{2t} \\ 30e^{2t} \end{bmatrix}$$

$$e^{-\mathbf{A}s}\mathbf{f}(s) = \frac{1}{4}\begin{bmatrix} -e^s + 5e^{-3s} & e^s - e^{-3s} \\ -5e^s + 5e^{-3s} & 5e^s - e^{-3s} \end{bmatrix}\begin{bmatrix} 18e^{2s} \\ 30e^{2s} \end{bmatrix} = \begin{bmatrix} 15e^{-s} + 3e^{3s} \\ 15e^{-s} + 15e^{3s} \end{bmatrix}$$

$$\mathbf{x}(0) + \int_0^t e^{-\mathbf{A}s}\mathbf{f}(s)\,ds = \begin{bmatrix} 0 \\ 0 \end{bmatrix} + \int_0^t \begin{bmatrix} 15e^{-s} + 3e^{3s} \\ 15e^{-s} + 15e^{3s} \end{bmatrix} ds = \begin{bmatrix} 14 - 15e^{-t} + e^{3t} \\ 10 - 15e^{-t} + 5e^{3t} \end{bmatrix}$$

Section 8.3

$$\mathbf{x}(t) = e^{\mathbf{A}t}\left(\mathbf{x}(0) + \int_0^t e^{-\mathbf{A}s}\mathbf{f}(s)\,ds\right) = \frac{1}{4}\begin{bmatrix} -e^{-t}+5e^{3t} & e^{-t}-e^{3t} \\ -5e^{-t}+5e^{3t} & 5e^{-t}-e^{3t} \end{bmatrix}\begin{bmatrix} 14-15e^{-t}+e^{3t} \\ 10-15e^{-t}+5e^{3t} \end{bmatrix}$$

$$\mathbf{x}(t) = \begin{bmatrix} -e^{-t}-14e^{2t}+15e^{3t} \\ -5e^{-t}-10e^{2t}+15e^{3t} \end{bmatrix}$$

33. $\mathbf{A} = \begin{bmatrix} 3 & -1 \\ 9 & -3 \end{bmatrix}$, $\lambda_1 = 0$, $\lambda_2 = 0$

$e^{\mathbf{A}t} = e^{-0t}[\mathbf{I}+(\mathbf{A}+0\mathbf{I})t] = \mathbf{I}+\mathbf{A}t = \begin{bmatrix} 1+3t & -t \\ 9t & 1-3t \end{bmatrix}$ (as in Problems 21-26)

$\mathbf{x}(0) = \begin{bmatrix} 3 \\ 5 \end{bmatrix}$, $\mathbf{f}(t) = \begin{bmatrix} 7 \\ 5 \end{bmatrix}$

$\mathbf{x}(t) = \begin{bmatrix} 3+11t+8t^2 \\ 5+17t+24t^2 \end{bmatrix}$

35. $\mathbf{A} = \begin{bmatrix} 2 & -5 \\ 1 & -2 \end{bmatrix}$, $\lambda_1 = -i$, $\lambda_2 = i$

$\mathbf{P}_1 = \frac{1}{-2i}(\mathbf{A}-i\mathbf{I}) = \frac{1}{2}\begin{bmatrix} 1+2i & -5i \\ i & 1-2i \end{bmatrix}$, $\mathbf{P}_2 = \frac{1}{2i}(\mathbf{A}+i\mathbf{I}) = \frac{1}{2}\begin{bmatrix} 1-2i & 5i \\ -i & 1+2i \end{bmatrix}$

$e^{\mathbf{A}t} = e^{-it}\mathbf{P}_1 + e^{it}\mathbf{P}_2 = (\cos t - i\sin t)\mathbf{P}_1 + (\cos t + i\sin t)\mathbf{P}_2$

$= \begin{bmatrix} \cos t + 2\sin t & -5\sin t \\ \sin t & \cos t - 2\sin t \end{bmatrix}$

$\mathbf{x}(0) = \begin{bmatrix} 0 \\ 0 \end{bmatrix}$, $\mathbf{f}(t) = \begin{bmatrix} 4t \\ 1 \end{bmatrix}$

The solution vector $\mathbf{x}(t) = \begin{bmatrix} -1+8t+\cos t-8\sin t \\ -2+4t+2\cos t-3\sin t \end{bmatrix}$ is derived by variation of parameters in the solution to Problem 25 of Section 8.2.

37. $\mathbf{A} = \begin{bmatrix} 3 & -1 \\ 9 & -3 \end{bmatrix}$, $\lambda_1 = 0$, $\lambda_2 = 0$

$e^{\mathbf{A}t} = e^{-0t}[\mathbf{I}+(\mathbf{A}+0\mathbf{I})t] = \mathbf{I}+\mathbf{A}t = \begin{bmatrix} 1+2t & -4t \\ t & 1-2t \end{bmatrix}$ (as in Problems 21-26)

$\mathbf{x}(0) = \begin{bmatrix} 0 \\ 0 \end{bmatrix}$, $\mathbf{f}(t) = \begin{bmatrix} 36t^2 \\ 6t \end{bmatrix}$

$$\mathbf{x}(t) = \begin{bmatrix} 8t^3 + 6t^4 \\ 3t^2 - 2t^3 + 3t^4 \end{bmatrix}$$

39. $\mathbf{A} = \begin{bmatrix} 0 & -1 \\ 1 & 0 \end{bmatrix}, \quad \lambda_1 = -i, \quad \lambda_2 = i$

$$\mathbf{P}_1 = \frac{1}{-2i}(\mathbf{A} - i\mathbf{I}) = \frac{1}{2}\begin{bmatrix} 1 & -i \\ i & 1 \end{bmatrix}, \quad \mathbf{P}_2 = \frac{1}{2i}(\mathbf{A} + i\mathbf{I}) = \frac{1}{2}\begin{bmatrix} 1 & i \\ -i & 1 \end{bmatrix}$$

$$e^{\mathbf{A}t} = e^{-it}\mathbf{P}_1 + e^{it}\mathbf{P}_2$$

$$= (\cos t - i\sin t)\mathbf{P}_1 + (\cos t + i\sin t)\mathbf{P}_2 = \begin{bmatrix} \cos t & -\sin t \\ \sin t & \cos t \end{bmatrix}$$

$$\mathbf{x}(0) = \begin{bmatrix} 0 \\ 0 \end{bmatrix}, \quad \mathbf{f}(t) = \begin{bmatrix} \sec t \\ 0 \end{bmatrix}$$

$$e^{-\mathbf{A}s}\mathbf{f}(s) = \begin{bmatrix} \cos s & \sin s \\ -\sin s & \cos s \end{bmatrix}\begin{bmatrix} \sec s \\ 0 \end{bmatrix} = \begin{bmatrix} 1 \\ -\tan s \end{bmatrix}$$

$$\mathbf{x}(0) + \int_0^t e^{-\mathbf{A}s}\mathbf{f}(s)\,ds = \begin{bmatrix} 0 \\ 0 \end{bmatrix} + \int_0^t \begin{bmatrix} 1 \\ -\tan s \end{bmatrix} ds = \begin{bmatrix} t \\ \ln(\cos t) \end{bmatrix}$$

$$\mathbf{x}(t) = e^{\mathbf{A}t}\left(\mathbf{x}(0) + \int_0^t e^{-\mathbf{A}s}\mathbf{f}(s)\,ds\right) = \begin{bmatrix} \cos t & -\sin t \\ \sin t & \cos t \end{bmatrix}\begin{bmatrix} t \\ \ln(\cos t) \end{bmatrix}$$

$$\mathbf{x}(t) = \begin{bmatrix} t\cos t - \ln(\cos t)\sin t \\ t\sin t + \ln(\cos t)\cos t \end{bmatrix}$$

Each of the 3×3 coefficient matrices in Problems 41 and 42 has a characteristic polynomial of the form $p(\lambda) = (\lambda - \lambda_1)((\lambda - \lambda_2)^2$ yielding an eigenvalue λ_1 of multiplicity 1 and an eigenvalue λ_2 of multiplicity 2. We therefore use the method explained in Problem 19 above, and list here the results of the principal steps in the calculation of the fundamental matrix $e^{\mathbf{A}t}$.

41. $\mathbf{A} = \begin{bmatrix} 39 & 8 & -16 \\ -36 & -5 & 16 \\ 72 & 16 & -29 \end{bmatrix}; \quad p(\lambda) = (\lambda + 1)(\lambda - 3)^2, \quad \lambda_1 = -1, \quad \lambda_3 = 3$

$$\frac{1}{p(\lambda)} = \frac{1}{16(\lambda + 1)} + \frac{7 - \lambda}{16(\lambda - 3)^2}; \quad a_1(\lambda) = \frac{1}{16}, \quad a_2(\lambda) = \frac{7 - \lambda}{16}$$

$$\mathbf{P}_1 = a_1(\mathbf{A})\cdot(\mathbf{A} - \lambda_2\mathbf{I})^2 = \frac{1}{16}(\mathbf{A} - 3\mathbf{I})^2 = \begin{bmatrix} -9 & -2 & 4 \\ 9 & 2 & -4 \\ -18 & -4 & 8 \end{bmatrix}$$

Section 8.3

$$\mathbf{P}_2 = a_2(\mathbf{A})\cdot(\mathbf{A}-\lambda_1\mathbf{I}) = \frac{1}{16}(7\mathbf{I}-\mathbf{A})(\mathbf{A}+\mathbf{I}) = \begin{bmatrix} 10 & 2 & -4 \\ -9 & -1 & 4 \\ 18 & 4 & -7 \end{bmatrix}$$

$$e^{\mathbf{A}t} = e^{-t}\mathbf{P}_1 + e^{3t}\mathbf{P}_2\left(\mathbf{I}+(\mathbf{A}-3\mathbf{I})t\right) = \begin{bmatrix} -9e^{-t}+10e^{3t} & -2e^{-t}+2e^{3t} & 4e^{-t}-4e^{3t} \\ 9e^{-t}-9e^{3t} & 2e^{-t}-e^{3t} & -4e^{-t}+4e^{3t} \\ -18e^{-t}+18e^{3t} & -4e^{-t}+4e^{3t} & 8e^{-t}-7e^{3t} \end{bmatrix}$$

In each of Problems 43 and 44, the given 3×3 coefficient matrix $\mathbf{A}$ has characteristic polynomial of the form $p(\lambda) = (\lambda - \lambda_1)^3$ and thus a single eigenvalue λ_1 of multiplicity 3. Consequently, Equations (25) and (26) in Section 8.3 of the text imply that $a_1(\lambda) = b_1(\lambda) = 1$ and hence that the associated projection matrix $\mathbf{P}_1 = \mathbf{I}$. Therefore Equation (35) in the text reduces (with $q = 1$ and $m_1 = 3$) to

$$e^{\mathbf{A}t} = e^{\lambda_1 t}\left[\mathbf{I} + (\mathbf{A}-\lambda_1\mathbf{I})t + \tfrac{1}{2}(\mathbf{A}-\lambda_1\mathbf{I})^2 t^2\right]. \tag{8}$$

43. $\mathbf{A} = \begin{bmatrix} -2 & 17 & 4 \\ -1 & 6 & 1 \\ 0 & 1 & 2 \end{bmatrix}$; $p(\lambda) = (\lambda-2)^3$, $\lambda_1 = 2$

Substitution of $\mathbf{A}$ and $\lambda_1 = 2$ into the formula in (8) above yields

$$e^{\mathbf{A}t} = \frac{1}{2}e^{2t}\begin{bmatrix} -t^2-8t+2 & 4t^2+34t & t^2+8t \\ -2t & 8t+2 & 2t \\ -t^2 & 4t^2+2t & t^2+2 \end{bmatrix}.$$

45. $\mathbf{A} = \begin{bmatrix} -1 & 1 & 1 & -2 \\ 7 & -4 & -6 & 11 \\ 5 & -1 & 1 & 3 \\ 6 & -2 & -2 & 6 \end{bmatrix}$; $p(\lambda) = (\lambda+1)^2(\lambda-2)^2$, $\lambda_1 = -1$, $\lambda_2 = 2$

Since $\mathbf{A}$ has two eigenvalues $\lambda_1 = -1$ and $\lambda_2 = 2$ each of multiplicity 2, Equation (35) in the text reduces (with $q = m_1 = m_2 = 2$) to

$$e^{\mathbf{A}t} = e^{\lambda_1 t}\mathbf{P}_1\left[\mathbf{I}+(\mathbf{A}-\lambda_1\mathbf{I})t\right] + e^{\lambda_2 t}\mathbf{P}_2\left[\mathbf{I}+(\mathbf{A}-\lambda_2\mathbf{I})t\right]. \tag{9}$$

To calculate the projection matrices $\mathbf{P}_1$ and $\mathbf{P}_2$, we start with the partial fraction decomposition

$$\frac{1}{p(\lambda)} = \frac{2\lambda+5}{27(\lambda+1)^2} + \frac{7-2\lambda}{27(\lambda-2)^2}.$$

Then Equation (25) in the text implies that $a_1(\lambda) = (2\lambda+5)/27$ and $a_2(\lambda) = (7-2\lambda)/27$. Hence Equation (26) yields

$$\mathbf{P}_1 = \frac{1}{27}(2\mathbf{A}+5\mathbf{I})(\mathbf{A}+\mathbf{I})^2 = \begin{bmatrix} 1 & 0 & 0 & 0 \\ -3 & 1 & 1 & -2 \\ -1 & 0 & 0 & 0 \\ -2 & 0 & 0 & 0 \end{bmatrix}$$

and

$$\mathbf{P}_2 = \frac{1}{27}(7\mathbf{I}-2\mathbf{A})(\mathbf{A}-2\mathbf{I})^2 = \begin{bmatrix} 0 & 0 & 0 & 0 \\ 3 & 0 & -1 & 2 \\ 1 & 0 & 1 & 0 \\ 2 & 0 & 0 & 1 \end{bmatrix}.$$

When we substitute these eigenvalues and projection matrices in Eq. (9) above we get

$$e^{\mathbf{A}t} = \begin{bmatrix} e^{-t} & te^{-t} & te^{-t} & -2te^{-t} \\ -3e^{-t}+(3-2t)e^{2t} & (1-3t)e^{-t} & (1-3t)e^{-t}-e^{2t} & 2(3t-1)e^{-t}+(2-t)e^{2t} \\ -e^{-t}+(1+2t)e^{2t} & -te^{-t} & -te^{-t}+e^{2t} & 2te^{-t}+te^{2t} \\ -2e^{-t}+2e^{2t} & -2te^{-t} & -2te^{-t} & 4te^{-t}+e^{2t} \end{bmatrix}.$$

In each of problems 47–50 we use the given values of $k_1, k_2,$ and k_3 set up the second-order linear system $\mathbf{x}'' = \mathbf{Ax}$ with coefficient matrix

$$\mathbf{A} = \begin{bmatrix} -k_1-k_2 & k_2 \\ k_2 & -k_2-k_3 \end{bmatrix}.$$

We then calculate the particular solution

$$\mathbf{x}(t) = e^{\sqrt{\mathbf{A}}\,t}\mathbf{c}_1 + e^{-\sqrt{\mathbf{A}}\,t}\mathbf{c}_2 \tag{10}$$

given by Theorem 3 with $\mathbf{c}_1 = \begin{bmatrix} 1 & 0 \end{bmatrix}^T$ and $\mathbf{c}_2 = \begin{bmatrix} 0 & 1 \end{bmatrix}^T$. Given the distinct eigenvalues λ_1, λ_2 and the projection matrices $\mathbf{P}_1, \mathbf{P}_2$ of $\mathbf{A}$, the matrix exponential needed in (10) is given by

$$e^{\sqrt{\mathbf{A}}\,t} = e^{\sqrt{\lambda_1}\,t}\mathbf{P}_1 + e^{\sqrt{\lambda_2}\,t}\mathbf{P}_2. \tag{11}$$

47. $\mathbf{A} = \begin{bmatrix} -5 & 4 \\ 4 & -5 \end{bmatrix}$; $p(\lambda) = \lambda^2 - 10\lambda + 9$, $\lambda_1 = -1$, $\lambda_2 = -9$

$\mathbf{P}_1 = \dfrac{1}{-10}(\mathbf{A}+9\mathbf{I}) = \dfrac{1}{2}\begin{bmatrix} 1 & 1 \\ 1 & 1 \end{bmatrix}$, $\mathbf{P}_2 = \dfrac{1}{10}(\mathbf{A}+\mathbf{I}) = \dfrac{1}{2}\begin{bmatrix} 1 & -1 \\ -1 & 1 \end{bmatrix}$

$e^{\sqrt{\mathbf{A}}t} = e^{\sqrt{\lambda_1}t}\mathbf{P}_1 + e^{\sqrt{\lambda_2}t}\mathbf{P}_2 = e^{it}\mathbf{P}_1 + e^{3it}\mathbf{P}_2 = \dfrac{1}{2}\begin{bmatrix} e^{it}+e^{3it} & e^{it}-e^{3it} \\ e^{it}-e^{3it} & e^{it}+e^{3it} \end{bmatrix}$

$\mathbf{x}(t) = e^{\sqrt{\mathbf{A}}t}\mathbf{c}_1 + e^{-\sqrt{\mathbf{A}}t}\mathbf{c}_2 = \dfrac{1}{2}\begin{bmatrix} (e^{it}+e^{-it})-(e^{3it}-e^{3it}) \\ (e^{it}+e^{-it})+(e^{3it}-e^{3it}) \end{bmatrix} = \begin{bmatrix} \cos t \\ \cos t \end{bmatrix} + i\begin{bmatrix} -\sin 3t \\ \sin 3t \end{bmatrix}$

Note that $\mathbf{x}(t)$ is a linear combination of two motions — one in which the two masses move in the same direction with frequency $\omega_1 = 1$ and with equal amplitudes, and one in which they move in opposite directions with frequency $\omega_2 = 3$ and with equal amplitudes.

49. $\mathbf{A} = \begin{bmatrix} -3 & 1 \\ 1 & -3 \end{bmatrix}$; $p(\lambda) = \lambda^2 - 6\lambda + 8$, $\lambda_1 = -2$, $\lambda_2 = -4$

$\mathbf{P}_1 = \dfrac{1}{-6}(\mathbf{A}+4\mathbf{I}) = \dfrac{1}{2}\begin{bmatrix} 1 & 1 \\ 1 & 1 \end{bmatrix}$, $\mathbf{P}_2 = \dfrac{1}{6}(\mathbf{A}+2\mathbf{I}) = \dfrac{1}{2}\begin{bmatrix} 1 & -1 \\ -1 & 1 \end{bmatrix}$

$e^{\sqrt{\mathbf{A}}t} = e^{\sqrt{\lambda_1}t}\mathbf{P}_1 + e^{\sqrt{\lambda_2}t}\mathbf{P}_2 = e^{\sqrt{2}it}\mathbf{P}_1 + e^{2it}\mathbf{P}_2 = \dfrac{1}{2}\begin{bmatrix} e^{\sqrt{2}it}+e^{2it} & e^{\sqrt{2}it}-e^{2it} \\ e^{\sqrt{2}it}-e^{2it} & e^{\sqrt{2}it}+e^{2it} \end{bmatrix}$

$\mathbf{x}(t) = e^{\sqrt{\mathbf{A}}t}\mathbf{c}_1 + e^{-\sqrt{\mathbf{A}}t}\mathbf{c}_2 = \dfrac{1}{2}\begin{bmatrix} (e^{\sqrt{2}it}+e^{-\sqrt{2}it})-(e^{2it}-e^{2it}) \\ (e^{\sqrt{2}it}+e^{-\sqrt{2}it})+(e^{2it}-e^{2it}) \end{bmatrix} = \begin{bmatrix} \cos\sqrt{2}\,t \\ \cos\sqrt{2}\,t \end{bmatrix} + i\begin{bmatrix} -\sin 2t \\ \sin 2t \end{bmatrix}$

Note that $\mathbf{x}(t)$ is a linear combination of two motions — one in which the two masses move in the same direction with frequency $\omega_1 = \sqrt{2}$ and with equal amplitudes, and one in which they move in opposite directions with frequency $\omega_2 = 2$ and with equal amplitudes.

CHAPTER 9

NONLINEAR SYSTEMS AND PHENOMENA

SECTION 9.1

STABILITY AND THE PHASE PLANE

1. The only solution of the homogeneous system $2x - y = 0$, $x - 3y = 0$ is the origin $(0, 0)$. The only figure among Figs. 9.1.11 through 9.1.18 showing a single critical point at the origin is Fig. 9.1.13. Thus the only critical point of the given autonomous system is the saddle point $(0, 0)$ shown in Figure 9.1.13 in the text.

3. The only solution of the system $x - 2y + 3 = 0$, $x - y + 2 = 0$ is the point $(-1, 1)$. The only figure among Figs. 9.1.11 through 9.1.18 showing a single critical point at $(-1, 1)$ is Fig. 9.1.18. Thus the only critical point of the given autonomous system is the stable center $(-1, 1)$ shown in Figure 9.1.18 in the text.

5. The first equation $1 - y^2 = 0$ gives $y = 1$ or $y = -1$ at a critical point. Then the second equation $x + 2y = 0$ gives $x = -2$ or $x = 2$, respectively. The only figure among Figs. 9.1.11 through 9.1.18 showing two critical points at $(-2, 1)$ and $(2, -1)$ is Fig. 9.1.11. Thus the critical points of the given autonomous system are the spiral point $(-2, 1)$ and the saddle point $(2, 1)$ shown in Figure 9.1.11 in the text.

7. The first equation $4x - x^3 = 0$ gives $x = -2$, $x = 0$, or $x = 2$ at a critical point. Then the second equation $x - 2y = 0$ gives $y = -1$, $y = 0$, or $y = 1$, respectively. The only figure among Figs. 9.1.11 through 9.1.18 showing three critical points at $(-2, -1)$, $(0, 0)$, and $(2, 1)$ is Fig. 9.1.14. Thus the critical points of the given autonomous system are the spiral point $(0, 0)$ and the saddle points $(-2, 1)$ and $(2, 1)$ shown in Figure 9.1.14 in the text.

In each of Problems 9–12 we need only set $x' = x'' = 0$ and solve the resulting equation for x.

9. The equation $4x - x^3 = x(4 - x^2) = 0$ has the three solutions $x = 0, \pm 2$. This gives the three equilibrium solutions $x(t) \equiv 0$, $x(t) \equiv 2$, $x(t) \equiv -2$ of the given 2nd-order differential equation.

11. The equation $4\sin x = 0$ is satisfied by $x = n\pi$ for any integer n. Thus the given 2nd-order equation has infinitely many equilibrium solutions: $x(t) \equiv n\pi$ for any integer n.

In Problems 13–16, the given x- and y-equations are independent exponential differential equations that we can solve immediately by inspection.

13. Solution: $x(t) = x_0 e^{-2t}$, $y(t) = y_0 e^{-2t}$

Then $y = (y_0/x_0)x = kx$, so the trajectories are straight lines through the origin. Clearly $x(t), y(t) \to 0$ as $t \to +\infty$, so the origin is a stable proper node like the one shown in Figure 9.1.4 in the text.

15. Solution: $x(t) = x_0 e^{-2t}$, $y(t) = y_0 e^{-t}$

Then $x = (x_0/y_0^2)(y_0 e^{-t})^2 = ky^2$, so the trajectories are parabolas of the form $x = ky^2$, and clearly $x(t), y(t) \to 0$ as $t \to +\infty$. Thus the origin is a stable improper node similar to the one shown in Figure 9.1.5 in the text.

17. Differentiation of the first equation and substitution using the second one gives

$$x'' = y' = x, \text{ so } x'' + x = 0.$$

We therefore get the general solution

$$x(t) = A \cos t + B \sin t$$
$$y(t) = B \cos t - A \sin t \quad (y = x').$$

Then

$$x^2 + y^2 = (A\cos t + B\sin t)^2 + (B\cos t - A\sin t)^2$$
$$= (A^2 + B^2)\cos^2 t + (A^2 + B^2)\sin^2 t = A^2 + B^2.$$

Therefore the trajectories are clockwise-oriented circles centered at the origin, and the origin is a stable center.

19. Elimination of y as in Problem 17 gives $x'' + 4x = 0$, so we get the general solution

$$x(t) = A \cos 2t + B \sin 2t,$$
$$y(t) = B \cos 2t - A \sin 2t \quad (y = \tfrac{1}{2}x').$$

Then $x^2 + y^2 = A^2 + B^2$, so the origin is a stable center, and the trajectories are clockwise-oriented circles centered at (0, 0).

21. We want to solve the system

$$-ky + x(1 - x^2 - y^2) = 0$$
$$kx + y(1 - x^2 - y^2) = 0.$$

If we multiply the first equation by $-y$ and the second one by x, then add the two results, we get $k(x^2+y^2) = 0$. It therefore follows that $x = y = 0$.

23. The equation $dy/dx = -x/y$ separates to $x\,dx + y\,dy = 0$, so $x^2 + y^2 = C$. Thus the trajectories consist of the origin $(0, 0)$ and the circles $x^2 + y^2 = C > 0$.

25. The equation $dy/dx = -x/4y$ separates to $x\,dx + 4y\,dy = 0$, so $x^2 + 4y^2 = C$. Thus the trajectories consist of the origin $(0, 0)$ and the ellipses $x^2 + 4y^2 = C > 0$.

27. If $\phi(t) = x(t + \gamma)$ and $\psi(t) = y(t + \gamma)$ then

$$\phi'(t) = x'(t + \gamma) = y(t + \gamma) = \psi(t),$$

but

$$\psi(t) = y'(t + \gamma) = x(t + \gamma)\cdot(t + \gamma) = t\,\phi(t) + \gamma\,\phi(t) \neq t\,\phi(t).$$

SECTION 9.2

LINEAR AND ALMOST LINEAR SYSTEMS

In Problems 1–10 we first find the roots λ_1 and λ_2 of the characteristic equation of the coefficient matrix of the given linear system. We can then read the type and stability of the critical point (0,0) from Theorem 1 and the table of Figure 9.2.9 in the text.

1. The roots $\lambda_1 = -1$ and $\lambda_2 = -3$ of the characteristic equation $\lambda^2 + 4\lambda + 3 = 0$ are both negative, so (0,0) is an asymptotically stable node.

3. The roots $\lambda_1 = -1$ and $\lambda_2 = 3$ of the characteristic equation $\lambda^2 - 2\lambda - 3 = 0$ have different signs, so (0,0) is an unstable saddle point.

5. The roots $\lambda_1 = \lambda_2 = -1$ of the characteristic equation $\lambda^2 + 2\lambda + 1 = 0$ are negative and equal, so (0,0) is an asymptotically stable node.

7. The roots $\lambda_1, \lambda_2 = 1 \pm 2i$ of the characteristic equation $\lambda^2 - 2\lambda + 5 = 0$ are complex conjugates with positive real part, so (0,0) is an unstable spiral point.

9. The roots $\lambda_1, \lambda_2 = \pm 2i$ of the characteristic equation $\lambda^2 + 4 = 0$ are pure imaginary, so (0,0) is a stable (but not asymptotically stable) center.

11. The substitution $u = x - 2$, $v = y - 1$ transforms the given system to the system

$$u' = u - 2v, \qquad v' = 3u - 4v$$

with characteristic roots $\lambda_1 = -1$, $\lambda_2 = -2$ that are both negative. Hence (2, 1) is an asymptotically stable node.

13. The substitution $u = x - 2$, $v = y - 2$ transforms the given system to the system

$$u' = 2u - v, \qquad v' = 3u - 2v$$

with characteristic roots $\lambda_1 = 1$, $\lambda_2 = -1$ that are real with different signs. Hence (2, 2) is an unstable saddle point.

15. The substitution $u = x - 1$, $v = y - 1$ transforms the given system to the system

$$u' = u - v, \qquad v' = 5u - 3v$$

with characteristic roots $\lambda_1, \lambda_2 = -1 \pm i$ that are complex conjugates with negative real part. Hence (1, 1) is an asymptotically stable spiral point.

17. The substitution $u = x - 5/2$, $v = y + 1/2$ transforms the given system to the system

$$u' = u - 5v, \qquad v' = u - v$$

with pure imaginary characteristic roots $\lambda_1, \lambda_2 = \pm 2i$. Hence $(5/2, -1/2)$ is a stable (but not asymptotically stable) center.

In each of Problems 19–28 the associated linear system is obtained simply by deleting the nonlinear terms in the given almost linear system. We first find the characteristic roots λ_1 and λ_2 of the coefficient matrix of the associated linear system, and then apply Theorem 2 to determine as much as we can about the type and stability of the critical point (0,0) of the given almost linear system.

19. The roots $\lambda_1 = -2$ and $\lambda_2 = -3$ of the characteristic equation $\lambda^2 + 5\lambda + 6 = 0$ of the associated linear system are real and both negative. Hence (0,0) is an asymptotically stable node of the given almost linear system.

21. The roots $\lambda_1 = -3$ and $\lambda_2 = 2$ of the characteristic equation $\lambda^2 + \lambda - 6 = 0$ of the associated linear system are real and have different signs. Hence (0,0) is an unstable saddle point of the given almost linear system.

23. The roots $\lambda_1, \lambda_2 = -2 \pm 2i$ of the characteristic equation $\lambda^2 + 4\lambda + 8 = 0$ of the associated linear system are complex conjugates with negative real part. Hence (0,0) is an asymptotically stable spiral point of the given almost linear system.

25. The roots $\lambda_1 = \lambda_2 = -1$ of the characteristic equation $\lambda^2 + 2\lambda + 1 = 0$ of the associated linear system are real, equal, and negative. Hence (0,0) is an asymptotically stable node or spiral point of the given almost linear system.

27. The roots $\lambda_1, \lambda_2 = \pm i$ of the characteristic equation $\lambda^2 + 1 = 0$ of the associated linear system are pure imaginary. Hence (0,0) is either a center or a spiral point of the given almost linear system, but its stability is not determined by Theorem 2.

29. The critical points of the given system are (0,0) and (1, 1). At (0,0) the characteristic roots are $\lambda_1 = -1, \lambda_2 = 1$ so (0,0) is an unstable saddle point. The substitution $u = x - 1, v = y - 1$ transforms the given system to the almost linear system

$$u' = u - v, \qquad v' = 2u + v + u^2$$

whose linearization has characteristic roots $\lambda_1, \lambda_2 = \pm i$. Hence (1, 1) is either a center or a spiral point, but its stability is indeterminate.

31. The critical points of the given system are (1, 1) and (−1,−1). The substitution $u = x - 1, v = y - 1$ transforms it to the almost linear system

$$u' = 2v + v^2, \qquad v' = 3u - v + 3u^2 + u^3$$

whose linearization has characteristic roots $\lambda_1 = -3, \lambda_2 = 2$. Hence (1, 1) is an unstable saddle point of the given system. The substitution $u = x + 1, v = y + 1$ transforms it to the almost linear system

$$u' = -2v + v^2, \qquad v' = 3u - v - 3u^2 + u^3$$

whose linearization has characteristic roots $\lambda_1, \lambda_2 = (-1 \pm i\sqrt{23})/2$. Hence (−1,−1) is an asymptotically stable spiral point.

33. The characteristic equation of the given linear system is

$$(\lambda - \varepsilon)^2 + 1 = 0$$

with characteristic roots $\lambda_1, \lambda_2 = \varepsilon \pm i$.

(a) So if $\varepsilon < 0$ then λ_1, λ_2 are complex conjugates with negative real part, and hence (0, 0) is an asymptotically stable spiral point.

(b) If $\varepsilon = 0$ then $\lambda_1, \lambda_2 = \pm i$ (pure imaginary), so (0,0) is a stable center.

(c) If $\varepsilon > 0$, the situation is the same as in (a) except that the real part is positive, so (0, 0) is an unstable spiral point.

35. **(a)** If $h = 0$ we have the familiar system $x' = y$, $y' = -x$ with circular trajectories about the origin, which is therefore a center.

(b) The change to polar coordinates as in Example 6 of Section 9.1 is routine, yielding $r' = hr^3$ and $\theta' = -1$.

(c) If $h = -1$, then $r' = -r^3$ integrates to give $2r^2 = 1/(t + C)$ where C is a positive constant, so clearly $r \to 0$ as $t \to +\infty$, and thus the origin is a stable spiral point.

(d) If $h = +1$, then $r' = r^3$ integrates to give $2r^2 = -1/(t + C)$ where $C = -B$ is a positive constant. It follows that $2r^2 = 1/(B - t)$, so now r increases as t starts at 0 and increases.

37. The substitution $y = vx$ in the homogeneous first-order equation

$$\frac{dy}{dx} = \frac{y(2x^3 - y^3)}{x(x^3 - 2y^3)}$$

yields

$$x\frac{dv}{dx} = -\frac{v^4 + v}{2v^3 - 1}.$$

Separating the variables and integrating by partial fractions, we get

$$\int \left(-\frac{1}{v} + \frac{1}{v+1} + \frac{2v-1}{v^2 - v + 1}\right) dv = -\int \frac{dx}{x}$$

$$\ln((v+1)(v^2 - v + 1)) = \ln v - \ln x + \ln C$$

$$(v+1)(v^2 - v + 1) = \frac{Cv}{x}$$

$$v^3 + 1 = \frac{Cv}{x}.$$

Finally, the replacement $v = y/x$ yields $x^3 + y^3 = Cxy$.

SECTION 9.3

ECOLOGICAL APPLICATIONS: PREDATORS AND COMPETITORS

1. If $x = u + b/q$, $y = v + a/p$ then $x' = u'$ and $y' = v'$ so the first predator-prey equation in

$$x' = ax - pxy, \quad y' = -by + qxy \tag{1}$$

yields

$$\begin{aligned} u' &= a(u + b/q) - p(u + b/q)(v + a/p) \\ &= au + ab/q - puv - au - bpv/q - ab/q = -bpv/q - puv. \end{aligned}$$

upon cancellation of a couple of pairs of terms. In a quite similar way the second predator-prey equation transforms to $v' = aqu/p + quv$. The coefficient matrix of the linearization of this transformed system is

$$\mathbf{A} = \begin{bmatrix} 0 & -bp/a \\ aq/p & 0 \end{bmatrix}.$$

Its characteristic values $\lambda = \pm i\sqrt{bq}$ are pure imaginary, so the linear system has a stable center at the origin.

3. The effect of using the insecticide is to replace b by $b + f$ and a by $a - f$ in the predator-prey equations, while leaving p and q unchanged. Hence the new harmful population is $(b + f)/q > b/q = x_E$, and the new benign population is $(a - f)/p < a/p = y_E$.

Problems 4–7 deal with the competition system

$$x' = 60x - 4x^2 - 3xy, \quad y' = 42y - 2y^2 - 3xy. \tag{2}$$

In each problem we denote by $\mathbf{A}$ the coefficient matrix of the linearized system.

5. The substitution $x = u$, $y = v + 21$ gives the linearization $u' = -3u$, $v' = -63u - 42v$ at $(0, 21)$. Then the characteristic equation

$$|\mathbf{A} - \lambda \mathbf{I}| = \begin{vmatrix} -3 - \lambda & 0 \\ -63 & -42 - \lambda \end{vmatrix} = (\lambda + 3)(\lambda + 42) = 0$$

has negative roots $\lambda_1 = -3$ and $\lambda_2 = -42$ indicating a nodal sink.

7. The substitution $x = u+6$, $y = v+12$ gives the linearization $u' = -24u - 18v$, $v' = -36u - 24v$ at $(6, 12)$. Then the characteristic equation

$$|A - \lambda I| = \begin{vmatrix} -24 - \lambda & -18 \\ -36 & -24 - \lambda \end{vmatrix} = (-24 - \lambda)^2 - 2(18)^2 = 0$$

has real roots $\lambda_1 = -24 + 18\sqrt{2} > 0$, $\lambda_2 = -24 - 18\sqrt{2} < 0$ of opposite sign, indicating a saddle point.

Problems 8–10 deal with the competition system

$$x' = 60x - 3x^2 - 4xy, \quad y' = 42y - 3y^2 - 2xy \tag{3}$$

9. The substitution $x = u+20$, $y = v$ gives the linearization $u' = -60u - 80v$, $v' = 2v$ at $(20, 0)$. Then the characteristic equation

$$|A - \lambda I| = \begin{vmatrix} -60 - \lambda & -80 \\ 0 & 2 - \lambda \end{vmatrix} = (\lambda + 60)(\lambda - 2) = 0$$

has real roots $\lambda_1 = -60$ and $\lambda_2 = 2$ of opposite sign, indicating a saddle point.

Problems 11–13 deal with the predator-prey system

$$x' = 5x - x^2 - xy, \quad y' = -2y + xy. \tag{4}$$

11. We simply delete the quadratic terms in (4) to linearize at $(0,0)$. Then it is quite obvious that the coefficient matrix $A = \begin{bmatrix} 5 & 0 \\ 0 & -2 \end{bmatrix}$ has real eigenvalues $\lambda_1 = 5$ and $\lambda_2 = -2$ of opposite sign, indicating a saddle point.

13. The substitution $x = u+2$, $y = v+3$ gives the linearization $u' = -2u - 2v$, $v' = 3u$. at $(2, 3)$. Then the characteristic equation

$$|A - \lambda I| = \begin{vmatrix} -2 - \lambda & -2 \\ 3 & -\lambda \end{vmatrix} = \lambda^2 + 2\lambda + 6 = 0$$

has complex conjugate roots $\lambda = -1 \pm i\sqrt{5}$ with negative real part, indicating a spiral sink.

Problems 14–17 deal with the predator-prey system

$$x' = x^2 - 2x - xy, \quad y' = y^2 - 4y + xy. \tag{5}$$

15. The substitution $x = u$, $y = v + 4$ gives the linearization $u' = -6u$, $v' = 4u + 4v$ at $(0, 4)$. Then the characteristic equation

$$|A - \lambda I| = \begin{vmatrix} -6 - \lambda & 0 \\ 4 & 4 - \lambda \end{vmatrix} = (\lambda + 6)(\lambda - 4) = 0$$

has negative roots $\lambda_1 = -6$ and $\lambda_2 = 4$ of opposite sign, indicating a **saddle point**.

17. The substitution $x = u + 3$, $y = v + 1$ gives the linearization $u' = 3u - 3v$, $v' = u + v$ at $(3, 1)$. Then the characteristic equation

$$|A - \lambda I| = \begin{vmatrix} 3 - \lambda & -3 \\ 1 & 1 - \lambda \end{vmatrix} = \lambda^2 - 4\lambda + 6 = 0$$

has complex conjugate roots $\lambda = 2 \pm i\sqrt{2}$ with positive real part, indicating a **spiral source**.

Problems 18 and 19 deal with the predator-prey system

$$x' = 2x - xy, \quad y' = -5y + xy \tag{7}$$

19. The substitution $x = u + 5$, $y = v + 2$ gives the linearization $u' = -5v$, $v' = 2u$ at $(5, 2)$. Then the characteristic equation

$$|A - \lambda I| = \begin{vmatrix} -\lambda & -5 \\ 2 & -\lambda \end{vmatrix} = \lambda^2 + 10 = 0$$

has pure imaginary roots $\lambda = \pm i\sqrt{10}$, so the origin is a stable center for the **linearized** system. This is the indeterminate case, but the figure in the text suggests that $(5, 2)$ is also a stable center for the original system in (7).

Problems 20–22 deal with the predator-prey system

$$x' = -3x + x^2 - xy, \quad y' = -5y + xy \tag{8}$$

21. The substitution $x = u + 3$, $y = v$ gives the linearization $u' = 3u - 3v$, $v' = -2v$ at $(3, 0)$. Then the characteristic equation

$$|\mathbf{A}-\lambda\mathbf{I}| = \begin{vmatrix} 3-\lambda & -3 \\ 0 & -2-\lambda \end{vmatrix} = (\lambda-3)(\lambda+2) = 0$$

has real roots $\lambda_1 = 3$ and $\lambda_2 = -2$ of opposite sign, indicating a saddle point.

Problems 23–25 deal with the predator-prey system

$$x' = 7x - x^2 - xy, \quad y' = -5y + xy \tag{9}$$

23. We simply delete the quadratic terms in (9) to linearize at (0,0). Then it is quite obvious that the coefficient matrix $\mathbf{A} = \begin{bmatrix} 7 & 0 \\ 0 & -5 \end{bmatrix}$ has real eigenvalues $\lambda_1 = 7$ and $\lambda_2 = -5$ of opposite sign, indicating a saddle point.

25. The substitution $x = u+5$, $y = u+2$ gives the linearization $u' = -5u - 5v$, $v' = 2u$ at $(5, 2)$. Then the characteristic equation

$$|\mathbf{A}-\lambda\mathbf{I}| = \begin{vmatrix} -5-\lambda & -5 \\ 2 & -\lambda \end{vmatrix} = \lambda^2 + 5\lambda + 10 = 0$$

has complex conjugate roots $\lambda = \frac{1}{2}\left(-5 \pm i\sqrt{15}\right)$ with negative real part, indicating a spiral sink.

SECTION 9.4

NONLINEAR MECHANICAL SYSTEMS

In each of Problems 1–4 we need only substitute the familiar power series for the exponential, sine, and cosine functions, and then discard all higher-order terms. For each problem we give the corresponding linear system, the eigenvalues λ_1 and λ_2, and the type of this critical point.

1. $x' = 1 - \left(1 + x + \frac{1}{2}x^2 + \cdots\right) + 2y \approx -x + 2y$

$y' = -x - 4\left(y - \frac{1}{6}y^3 + \cdots\right) \approx -x - 4y$

The coefficient matrix $\mathbf{A} = \begin{bmatrix} -1 & 2 \\ -1 & -4 \end{bmatrix}$ has negative eigenvalues $\lambda_1 = -2$ and $\lambda_2 = -3$ indicating a stable nodal sink.

3. $x' = \left(1 + x + \tfrac{1}{2}x^2 + \cdots\right) + 2y - 1 \approx x + 2y$

$y' = 8x + \left(1 + y + \tfrac{1}{2}y^2 + \cdots\right) - 1 \approx 8x + y$

The coefficient matrix $\mathbf{A} = \begin{bmatrix} 1 & 2 \\ 8 & 1 \end{bmatrix}$ has real eigenvalues $\lambda_1 = -3$ and $\lambda_2 = 5$ of opposite sign, indicating an unstable saddle point.

5. The critical points are of the form $(0, n\pi)$ where n is an integer, so we substitute $x = u$, $y = v + n\pi$. Then

$$u' = x' = -u + \sin(v + n\pi) = -u + (\cos n\pi)v = -u + (-1)^n v.$$

Hence the linearized system at $(0, n\pi)$ is

$$u' = -u \pm v, \quad v' = 2u$$

where we take the plus sign if n is even, the minus sign if n is odd. If n is even the eigenvalues are $\lambda_1 = 1$ and $\lambda_2 = -2$, so $(0, n\pi)$ is an unstable saddle point. If n is odd the eigenvalues are $\lambda_1, \lambda_2 = (-1 \pm i\sqrt{7})/2$, so $(0, n\pi)$ is a stable spiral point.

7. The critical points are of the form $(n\pi, n\pi)$ where n is an integer, so we substitute $x = u + n\pi$, $y = v + n\pi$. Then

$$u' = x' = 1 - e^{u-v} = 1 - \left(1 + (u-v) + \tfrac{1}{2}(u-v)^2 + \cdots\right) \approx -u + v,$$

$$v' = y' = 2\sin(u + n\pi) = 2\sin u \cos n\pi \approx 2(-1)^n u.$$

Hence the linearized system at $(n\pi, n\pi)$ is

$$u' = -u + v, \quad v' = \pm 2u$$

and has coefficient matrix $\mathbf{A} = \begin{bmatrix} -1 & 1 \\ \pm 2 & 0 \end{bmatrix}$, where we take the plus sign if n is even, the minus sign if n is odd. With n even, The characteristic equation $\lambda^2 + \lambda - 2 = 0$ has real roots $\lambda_1 = 1$ and $\lambda_2 = -2$ of opposite sign, so $(n\pi, n\pi)$ is an unstable saddle point. With n odd, the characteristic equation $\lambda^2 + \lambda + 2 = 0$ has complex conjugate eigenvalues are $\lambda_1, \lambda_2 = (-1 \pm i\sqrt{7})/2$ with negative real part, so $(n\pi, n\pi)$ is a stable spiral point.

For Problems 9–11 the linearization of the damped pendulum system at the critical point $(n\pi, 0)$ is obtained by substituting $x = u + n\pi$, $y = v$. This gives the linear system

$$u' = v,$$
$$v' = -\omega^2 \sin(u + n\pi) - cv = -\omega^2 \sin u \cos n\pi - cv \approx -\omega^2(-1)^n u - cv$$

with coefficient matrix $\mathbf{A} = \begin{bmatrix} 0 & 1 \\ -(-1)^n \omega^2 & -c \end{bmatrix}$, where we take the plus sign if n is odd, the minus sign if n is even.

9. If n is odd then the characteristic equation $\lambda^2 + c\lambda - \omega^2 = 0$ has real roots

$$\lambda_1, \lambda_2 = \frac{-c \pm \sqrt{c^2 + 4\omega^2}}{2}$$

with opposite signs, so $(n\pi, 0)$ is an unstable saddle point.

11. If $c^2 < 4\omega^2$ then the two eigenvalues

$$\lambda_1, \lambda_2 = \frac{-c \pm \sqrt{c^2 - 4\omega^2}}{2} = -\frac{c}{2} \pm \frac{i}{2}\sqrt{4\omega^2 - c^2}$$

are complex conjugates with negative real part, so $(n\pi, 0)$ is a stable spiral point.

Problems 12–16 call for us to find and classify the critical points of the first order-system $x' = y$, $y' = -f(x, y)$ that corresponds to the given equation $x'' + f(x, x') = 0$.

13. The critical points are $(0, 0)$ and $(\pm 2, 0)$. At $(0, 0)$, the substitution $x = u$, $y = v$ gives the linear system $u' = v$, $v' = -20u - 2v$. The coefficient matrix $\mathbf{A} = \begin{bmatrix} 0 & 1 \\ -20 & -2 \end{bmatrix}$ has complex conjugate eigenvalues $\lambda = -1 \pm i\sqrt{19}$ with negative real part, which indicates a spiral sink.

At $(-2, 0)$, the substitution $x = u - 2$, $y = v$ gives the linear system $u' = v$, $v' = 40u - 2v$. The coefficient matrix $\mathbf{A} = \begin{bmatrix} 0 & 1 \\ 40 & -2 \end{bmatrix}$ has real eigenvalues $\lambda = -1 \pm \sqrt{41}$ of opposite sign, which indicates a saddle point.

At $(2, 0)$, the substitution $x = u + 2$, $y = v$ gives the same linear system $u' = v$, $v' = 40u - 2v$ as at $(-2, 0)$, so a saddle point is indicated here also.

15. The critical points are (0, 0) and (4, 0). At (0, 0), the substitution $x = u$, $y = v$ gives the linear system $u' = v$, $v' = -4u$. The coefficient matrix $\mathbf{A} = \begin{bmatrix} 0 & 1 \\ -4 & 0 \end{bmatrix}$ has pure imaginary eigenvalues $\lambda = \pm 2i$ with opposite signs, which indicates a stable center.

At (4,0), the substitution $x = u + 4$, $y = v$ gives the linear system $u' = v$, $v' = 4u$. The coefficient matrix $\mathbf{A} = \begin{bmatrix} 0 & 1 \\ 4 & 0 \end{bmatrix}$ has real eigenvalues $\lambda = \pm 2$ of opposite sign, which indicates a saddle point.

The statements of Problems 17–22 in the text include their answers and rather fully outline their solutions, which therefore are omitted here.

CHAPTER 10

LAPLACE TRANSFORM METHODS

SECTION 10.1

LAPLACE TRANSFORMS AND INVERSE TRANSFORMS

The objectives of this section are especially clearcut. They include familiarity with the definition of the Laplace transform $\mathcal{L}\{f(t)\} = F(s)$ that is given in Equation (1) in the textbook, the direct application of this definition to calculate Laplace transforms of simple functions (as in Examples 1–3), and the use of known transforms (those listed in Figure 10.1.2) to find Laplace transforms and inverse transforms (as in Examples 4–6). Perhaps students need to be told explicitly to memorize the transforms that are listed in the short table that appears in Figure 10.1.2.

1. $\mathcal{L}\{t\} = \int_0^\infty e^{-st} t \, dt \qquad (u = -st, \quad du = -s \, dt)$

$= \int_0^{-\infty} \left[\frac{1}{s^2}\right] u e^u \, du = \frac{1}{s^2}\left[(u-1)e^u\right]_0^{-\infty} = \frac{1}{s^2}$

3. $\mathcal{L}\{e^{3t+1}\} = \int_0^\infty e^{-st} e^{3t+1} \, dt = e \int_0^\infty e^{-(s-3)t} \, dt = \dfrac{e}{s-3}$

5. $\mathcal{L}\{\sinh t\} = \tfrac{1}{2}\mathcal{L}\{e^t - e^{-t}\} = \tfrac{1}{2}\int_0^\infty e^{-st}\left(e^t - e^{-t}\right) dt = \tfrac{1}{2}\int_0^\infty \left(e^{-(s-1)t} - e^{-(s+1)t}\right) dt$

$= \dfrac{1}{2}\left[\dfrac{1}{s-1} - \dfrac{1}{s+1}\right] = \dfrac{1}{s^2 - 1}$

7. $\mathcal{L}\{f(t)\} = \int_0^1 e^{-st} \, dt = \left[-\dfrac{1}{s}e^{-st}\right]_0^1 = \dfrac{1-e^{-s}}{s}$

9. $\mathcal{L}\{f(t)\} = \int_0^1 e^{-st} t \, dt = \dfrac{1 - e^{-s} - se^{-s}}{s^2}$

11. $\mathcal{L}\{\sqrt{t} + 3t\} = \dfrac{\Gamma(3/2)}{s^{3/2}} + 3 \cdot \dfrac{1}{s^2} = \dfrac{\sqrt{\pi}}{2s^{3/2}} + \dfrac{3}{s^2}$

13. $\mathcal{L}\{t - 2e^{3t}\} = \dfrac{1}{s^2} - \dfrac{2}{s-3}$

15. $\mathcal{L}\{1 + \cosh 5t\} = \dfrac{1}{s} + \dfrac{s}{s^2 - 25}$

17. $\mathcal{L}\{\cos^2 2t\} = \dfrac{1}{2}\mathcal{L}\{1+\cos 4t\} = \dfrac{1}{2}\left(\dfrac{1}{s} + \dfrac{s}{s^2+16}\right)$

19. $\mathcal{L}\{(1+t)^3\} = \mathcal{L}\{1+3t+3t^2+t^3\} = \dfrac{1}{s} + 3\cdot\dfrac{1!}{s^2} + 3\cdot\dfrac{2!}{s^3} + \dfrac{3!}{s^4} = \dfrac{1}{s} + \dfrac{3}{s^2} + \dfrac{6}{s^3} + \dfrac{6}{s^4}$

21. Integration by parts with $u = t$ and $dv = e^{-st}\cos 2t\, dt$ yields

$$\mathcal{L}\{t \cos 2t\} = \int_0^\infty t e^{-st}\cos 2t\, dt = -\dfrac{1}{s^2+4}\int_0^\infty e^{-st}(-s\cos 2t + 2\sin 2t)\, dt$$

$$= -\dfrac{1}{s^2+4}\left[-s\mathcal{L}\{\cos 2t\} + 2\mathcal{L}\{\sin 2t\}\right]$$

$$= -\dfrac{1}{s^2+4}\left[\dfrac{-s^2}{s^2+4} + \dfrac{4}{s^2+4}\right] = \dfrac{s^2-4}{(s^2+4)^2}.$$

23. $\mathcal{L}^{-1}\left\{\dfrac{3}{s^4}\right\} = \mathcal{L}^{-1}\left\{\dfrac{1}{2}\cdot\dfrac{6}{s^4}\right\} = \dfrac{1}{2}t^3$

25. $\mathcal{L}^{-1}\left\{\dfrac{1}{s} - \dfrac{2}{s^{5/2}}\right\} = \mathcal{L}^{-1}\left\{\dfrac{1}{s} - \dfrac{2}{\Gamma(5/2)}\cdot\dfrac{\Gamma(5/2)}{s^{5/2}}\right\} = 1 - \dfrac{2}{\frac{3}{2}\cdot\frac{1}{2}\sqrt{\pi}}\cdot t^{3/2} = 1 - \dfrac{8t^{3/2}}{3\sqrt{\pi}}$

27. $\mathcal{L}^{-1}\left\{\dfrac{3}{s-4}\right\} = 3\cdot\mathcal{L}^{-1}\left\{\dfrac{1}{s-4}\right\} = 3e^{4t}$

29. $\mathcal{L}^{-1}\left\{\dfrac{5-3s}{s^2+9}\right\} = \dfrac{5}{3}\cdot\mathcal{L}^{-1}\left\{\dfrac{3}{s^2+9}\right\} - 3\cdot\mathcal{L}^{-1}\left\{\dfrac{s}{s^2+9}\right\} = \dfrac{5}{3}\sin 3t - 3\cos 3t$

31. $\mathcal{L}^{-1}\left\{\dfrac{10s-3}{25-s^2}\right\} = -10\cdot\mathcal{L}^{-1}\left\{\dfrac{s}{s^2-25}\right\} + \dfrac{3}{5}\cdot\mathcal{L}^{-1}\left\{\dfrac{5}{s^2-25}\right\} = -10\cosh 5t + \dfrac{3}{5}\sinh 5t$

33. $\mathcal{L}\{\sin kt\} = \mathcal{L}\left\{\dfrac{e^{ikt} - e^{-ikt}}{2i}\right\} = \dfrac{1}{2i}\left(\dfrac{1}{s-ik} - \dfrac{1}{s+ik}\right)$

$= \dfrac{1}{2i}\cdot\dfrac{2ik}{(s-ik)(s-ik)} = \dfrac{k}{s^2+k^2}$ (because $i^2 = -1$)

35. Using the given tabulated integral with $a = -s$ and $b = k$, we find that

Section 10.1

215

$$\mathcal{L}\{\cos kt\} = \int_0^\infty e^{-st}\cos kt\, dt = \left[\frac{e^{-st}}{s^2+k^2}(-s\cos kt + k\sin kt)\right]_{t=0}^\infty$$

$$= \lim_{t\to\infty}\left(\frac{e^{-st}}{s^2+k^2}(-s\cos kt + k\sin kt)\right) - \frac{e^0}{s^2+k^2}(-s\cdot 1 + k\cdot 0) = \frac{s}{s^2+k^2}.$$

37. $f(t) = 1 - u_a(t) = 1 - u(t-a)$ so $\mathcal{L}\{f(t)\} = \mathcal{L}\{1\} - \mathcal{L}\{u_a(t)\} = \dfrac{1}{s} - \dfrac{e^{-as}}{s}$

39. Use of the geometric series gives

$$\mathcal{L}\{f(t)\} = \sum_{n=0}^\infty \mathcal{L}\{u(t-n)\} = \sum_{n=0}^\infty \frac{e^{-ns}}{s} = \frac{1}{s}\left(1 + e^{-s} + e^{-2s} + e^{-3s} + \cdots\right)$$

$$= \frac{1}{s}\left(1 + (e^{-s}) + (e^{-s})^2 + (e^{-s})^3 + \cdots\right) = \frac{1}{s}\cdot\frac{1}{1-e^{-s}} = \frac{1}{s(1-e^{-s})}.$$

41. By checking values at sample points, you can verify that $g(t) = 2f(t) - 1$ in terms of the square wave function $f(t)$ of Problem 40. Hence

$$\mathcal{L}\{g(t)\} = \mathcal{L}\{2f(t) - 1\} = \frac{2}{s(1+e^{-s})} - \frac{1}{s} = \frac{1}{s}\left(\frac{2}{1+e^{-s}} - 1\right) = \frac{1}{s}\cdot\frac{1-e^{-s}}{1+e^{-s}}$$

$$= \frac{1}{s}\cdot\frac{1-e^{-s}}{1+e^{-s}}\cdot\frac{e^{s/2}}{e^{s/2}} = \frac{1}{s}\cdot\frac{e^{s/2}-e^{-s/2}}{e^{s/2}+e^{-s/2}} = \frac{1}{s}\cdot\frac{\frac{1}{2}(e^{s/2}-e^{-s/2})}{\frac{1}{2}(e^{s/2}+e^{-s/2})}$$

$$= \frac{1}{s}\cdot\frac{\sinh(s/2)}{\cosh(s/2)} = \frac{1}{s}\tanh\frac{s}{2}.$$

SECTION 10.2

TRANSFORMATION OF INITIAL VALUE PROBLEMS

The focus of this section is on the use of transforms of derivatives (Theorem 1) to solve initial value problems (as in Examples 1 and 2). Transforms of integrals (Theorem 2) appear less frequently in practice, and the extension of Theorem 1 at the end of Section 10.2 may be considered entirely optional (except perhaps for electrical engineering students).

In Problems 1–10 we give first the transformed differential equation, then the transform $X(s)$ of the solution, and finally the inverse transform $x(t)$ of $X(s)$.

1. $[s^2 X(s) - 5s] + 4\{X(s)\} = 0$

$$X(s) = \frac{5s}{s^2+4} = 5 \cdot \frac{s}{s^2+4}$$

$x(t) = \mathcal{L}^{-1}\{X(s)\} = 5\cos 2t$

3. $[s^2 X(s) - 2] - [sX(s)] - 2[X(s)] = 0$

$$X(s) = \frac{2}{s^2-s-2} = \frac{2}{(s-2)(s+1)} = \frac{2}{3}\left(\frac{1}{s-2} - \frac{1}{s+1}\right)$$

$x(t) = (2/3)(e^{2t} - e^{-t})$

5. $[s^2 X(s)] + [X(s)] = 2/(s^2+4)$

$$X(s) = \frac{2}{(s^2+1)(s^2+4)} = \frac{2}{3} \cdot \frac{1}{s^2+1} - \frac{1}{3} \cdot \frac{2}{s^2+4}$$

$x(t) = (2\sin t - \sin 2t)/3$

7. $[s^2 X(s) - s] + [X(s)] = s/s^2+9)$

$(s^2 + 1)X(s) = s + s/(s^2+9) = (s^3 + 10s)/(s^2+9)$

$$X(s) = \frac{s^2+10s}{(s^2+1)(s^2+9)} = \frac{9}{9} \cdot \frac{s}{s^2+1} - \frac{1}{8} \cdot \frac{s}{s^2+9}$$

$x(t) = (9\cos t - \cos 3t)/8$

9. $s^2 X(s) + 4sX(s) + 3X(s) = 1/s$

$$X(s) = \frac{1}{s(s^2+4s+3)} = \frac{1}{s(s+1)(s+3)} = \frac{1}{3} \cdot \frac{1}{s} - \frac{1}{2} \cdot \frac{1}{s+1} + \frac{1}{6} \cdot \frac{1}{s+3}$$

$x(t) = (2 - 3e^{-t} + e^{-3t})/6$

11. The transformed equations are

$$sX(s) - 1 = 2X(s) + Y(s)$$
$$sY(s) + 2 = 6X(s) + 3Y(s).$$

We solve for the Laplace transforms

$$X(s) = \frac{s-5}{s(s-5)} = \frac{1}{s}$$

$$Y(s) = X(s) = \frac{-2s+10}{s(s-5)} = -\frac{2}{s}.$$

Section 10.2   217

Hence the solution is given by

$$x(t) = 1, \qquad y(t) = -2.$$

13. The transformed equations are

$$sX(s) + 2[sY(s) - 1] + X(s) = 0$$
$$sX(s) - [sY(s) - 1] + Y(s) = 0,$$

which we solve for the transforms

$$X(s) = -\frac{2}{3s^2 - 1} = -\frac{2}{3} \cdot \frac{1}{s^2 - 1/3} = -\frac{2}{\sqrt{3}} \cdot \frac{1/\sqrt{3}}{s^2 - (1/\sqrt{3})^2}$$

$$X(s) = \frac{3s+1}{3s^2-1} = \frac{s+1/3}{s^2-1/3} = \frac{s}{s^2-(1/\sqrt{3})^2} + \frac{1}{\sqrt{3}} \cdot \frac{1/\sqrt{3}}{s^2-(1/\sqrt{3})^2}.$$

Hence the solution is

$$x(t) = -(2/\sqrt{3})\sinh(t/\sqrt{3})$$
$$y(t) = \cosh(t/\sqrt{3}) + (1/\sqrt{3})\sinh(t/\sqrt{3}).$$

15. The transformed equations are

$$[s^2X - s] + [sX - 1] + [sY - 1] + 2X - Y = 0$$
$$[s^2Y - s] + [sX - 1] + [sY - 1] + 4X - 2Y = 0,$$

which we solve for

$$X(s) = \frac{s^2+3s+2}{s^3+3s^2+3s} = \frac{1}{3}\left(\frac{2}{s} + \frac{s+3}{s^2+3s+3}\right) = \frac{1}{3}\left(\frac{2}{s} + \frac{s+3}{(s+3/2)^2+(3/4)}\right)$$

$$= \frac{1}{3}\left(\frac{2}{s} + \frac{s+3/2}{(s+3/2)^2+(\sqrt{3}/2)^2} + \sqrt{3} \cdot \frac{\sqrt{3}/2}{(s+3/2)^2+(\sqrt{3}/2)^2}\right)$$

$$Y(s) = \frac{-s^3-2s^2+2s+4}{s^3+3s^2+3s} = \frac{1}{21}\left(\frac{28}{s} - \frac{9}{s-1} + \frac{2s+15}{s^2+3s+3}\right)$$

$$= \frac{1}{21}\left(\frac{28}{s} - \frac{9}{s-1} + \frac{2s+15}{(s+3/2)^2+3/4}\right)$$

$$= \frac{1}{21}\left(\frac{28}{s} - \frac{9}{s-1} + 2 \cdot \frac{s+3/2}{(s+3/2)^2+(\sqrt{3}/2)^2} + 8\sqrt{3} \cdot \frac{\sqrt{3}/2}{(s+3/2)^2+(\sqrt{3}/2)^2}\right).$$

Here we've used some fairly heavy-duty partial fractions (Section 10.3). The transforms

$$\mathcal{L}\{e^{at}\cos kt\} = \frac{s-a}{(s-a)^2+k^2}, \quad \mathcal{L}\{e^{at}\sin kt\} = \frac{k}{(s-a)^2+k^2}$$

from the inside-front-cover table (with $a = -3/2$, $k = \sqrt{3}/2$) finally yield

$$x(t) = \frac{1}{3}\left\{2 + e^{-3t/2}\left[\cos\left(\sqrt{3}t/2\right) + \sqrt{3}\sin\left(\sqrt{3}t/2\right)\right]\right\}$$

$$y(t) = \frac{1}{21}\left\{28 - 9e^t + e^{-3t/2}\left[2\cos\left(\sqrt{3}t/2\right) + 8\sqrt{3}\sin\left(\sqrt{3}t/2\right)\right]\right\}.$$

17. $f(t) = \int_0^t e^{3\tau}\,d\tau = \left[\frac{1}{3}e^{3\tau}\right]_{\tau=0}^t = \frac{1}{3}(e^{3t}-1)$

19. $f(t) = \int_0^t \frac{1}{2}\sin 2\tau\,d\tau = \left[-\frac{1}{4}\cos 2\tau\right]_{\tau=0}^t = \frac{1}{4}(1-\cos 2t)$

21. $f(t) = \int_0^t\left[\int_0^\tau \sin t\,dt\right]d\tau = \int_0^t (1-\cos\tau)\,d\tau = \left[\tau - \sin\tau\right]_{\tau=0}^t = t - \sin t$

23. $f(t) = \int_0^t\left[\int_0^\tau \sinh t\,dt\right]d\tau = \int_0^t (\cosh\tau - 1)\,d\tau = \left[\sinh\tau - \tau\right]_{\tau=0}^t = \sinh t - t$

25. With $f(t) = \cos kt$ and $F(s) = s/(s^2+k^2)$, Theorem 1 in this section yields

$$\mathcal{L}\{-k\sin kt\} = \mathcal{L}\{f'(t)\} = sF(s) - 1 = s\cdot\frac{s}{s^2+k^2} - 1 = -\frac{k^2}{s^2+k^2},$$

so division by $-k$ yields $\mathcal{L}\{\sin kt\} = k/(s^2+k^2)$.

27. (a) With $f(t) = t^n e^{at}$ and $f'(t) = nt^{n-1}e^{at} + at^n e^{at}$, Theorem 1 yields

$$\mathcal{L}\{nt^{n-1}e^{at} + at^n e^{at}\} = s\mathcal{L}\{t^n e^{at}\}$$

so

$$n\mathcal{L}\{t^{n-1}e^{at}\} = (s-a)\mathcal{L}\{t^n e^{at}\}$$

and hence

$$\mathcal{L}\{t^n e^{at}\} = \frac{n}{s-a}\mathcal{L}\{t^{n-1}e^{at}\}.$$

(b) $n=1$: $\mathcal{L}\{te^{at}\} = \dfrac{1}{s-a}\mathcal{L}\{e^{at}\} = \dfrac{1}{s-a}\cdot\dfrac{1}{s-a} = \dfrac{1}{(s-a)^2}$

$n=2$: $\mathcal{L}\{t^2 e^{at}\} = \dfrac{2}{s-a}\mathcal{L}\{te^{at}\} = \dfrac{2}{s-a}\cdot\dfrac{1}{(s-a)^2} = \dfrac{2!}{(s-a)^3}$

$n=3$: $\mathcal{L}\{t^3 e^{at}\} = \dfrac{3}{s-a}\mathcal{L}\{t^2 e^{at}\} = \dfrac{3}{s-a}\cdot\dfrac{2!}{(s-a)^3} = \dfrac{3!}{(s-a)^4}$

And so forth.

29. Let $f(t) = t\sinh kt$, so $f(0) = 0$. Then

$$f'(t) = \sinh kt + kt\cosh kt$$
$$f''(t) = 2k\cosh kt + k^2 t\sinh kt,$$

and thus $f'(0) = 0$, so Formula (5) in this section yields

$$\mathcal{L}\{2k\cosh kt + k^2 t\sinh kt\} = s^2\mathcal{L}\{\sinh kt\},$$

$$2k\cdot\dfrac{s}{s^2-k^2} + k^2 F(s) = s^2 F(s).$$

We readily solve this last equation for

$$\mathcal{L}\{t\cosh kt\} = F(s) = \dfrac{2ks}{\left(s^2-k^2\right)^2}.$$

31. Using the known transform of $\sin kt$ and the Problem 28 transform of $t\cos kt$, we obtain

$$\mathcal{L}\left\{\dfrac{1}{2k^3}(\sin kt - kt\cos kt)\right\} = \dfrac{1}{2k^3}\cdot\dfrac{k}{s^2+k^2} - \dfrac{k}{2k^3}\cdot\dfrac{s^2-k^2}{\left(s^2+k^2\right)^2}$$

$$= \dfrac{1}{2k^2}\left[\dfrac{1}{s^2+k^2} - \dfrac{s^2-k^2}{\left(s^2+k^2\right)^2}\right] = \dfrac{1}{2k^2}\cdot\dfrac{2k^2}{\left(s^2+k^2\right)^2} = \dfrac{1}{\left(s^2+k^2\right)^2}$$

33. $f(t) = u_a(t) - u_b(t) = u(t-a) - u(t-b)$, so the result of Problem 32 gives

$$\mathcal{L}\{f(t)\} = \mathcal{L}\{u(t-a)\} - \mathcal{L}\{u(t-b)\} = \dfrac{e^{-as}}{s} - \dfrac{e^{-bs}}{s} = \dfrac{e^{-as} - e^{-bs}}{s}.$$

35. Let's write $g(t)$ for the on-off function of this problem to distinguish it from the square wave function of Problem 34. Then comparison of Figures 10.2.6 and 10.2.7 makes it clear that $g(t) = \tfrac{1}{2}(1 + f(t))$, so (using the result of Problem 34) we obtain

$$G(s) = \frac{1}{2s} + \frac{1}{2}F(s) = \frac{1}{2s} + \frac{1}{2s}\tanh\frac{s}{2} = \frac{1}{2s}\left(1 + \frac{e^{s/2} - e^{-s/2}}{e^{s/2} + e^{-s/2}} \cdot \frac{e^{-s/2}}{e^{-s/2}}\right)$$

$$= \frac{1}{2s}\left(1 + \frac{1 - e^{-s}}{1 + e^{-s}}\right) = \frac{1}{2s} \cdot \frac{2}{1 + e^{-s}} = \frac{1}{s(1 + e^{-s})}.$$

37. We observe that $f(0) = 0$ and that the sawtooth function has jump -1 at each of the points $t_n = n = 1, 2, 3, \cdots$. Also, $f'(t) \equiv 1$ wherever the derivative is defined. Hence Eq. (21) in this section gives

$$\frac{1}{s} = sF(s) + \sum_{n=1}^{\infty} e^{-ns} = sF(s) - 1 + \sum_{n=0}^{\infty} e^{-ns} = sF(s) - 1 + \frac{1}{1 - e^{-ns}},$$

using the geometric series $\sum_{n=0}^{\infty} x^n = 1/(1-x)$ with $x = e^{-s}$. Solution for $F(s)$ gives

$$F(s) = \frac{1}{s^2} + \frac{1}{s} - \frac{1}{s(1 - e^{-s})} = \frac{1}{s^2} - \frac{e^{-s}}{s(1 - e^{-s})}.$$

SECTION 10.3

TRANSLATION AND PARTIAL FRACTIONS

This section is devoted to the computational nuts and bolts of the staple technique for the inversion of Laplace transforms — partial fraction decompositions. If time does not permit going further in this chapter, Sections 10.1–10.3 provide a self-contained introduction to Laplace transforms that suffices for the most common elementary applications.

1. $\mathcal{L}\{t^4\} = \dfrac{24}{s^5}$, so $\mathcal{L}\{t^4 e^{\pi t}\} = \dfrac{24}{(s-\pi)^5}$

3. $\mathcal{L}\{\sin 3\pi t\} = \dfrac{3\pi}{s^2 + 9\pi^2}$, so $\mathcal{L}\{e^{-2t}\sin 3\pi t\} = \dfrac{3\pi}{(s+2)^2 + 9\pi^2}$.

5. $F(s) = \dfrac{3}{2s-4} = \dfrac{3}{2} \cdot \dfrac{1}{s-2}$, so $f(t) = \dfrac{3}{2}e^{2t}$

7. $F(s) = \dfrac{1}{(s+2)^2}$, so $f(t) = te^{-2t}$

9. $F(s) = 3 \cdot \dfrac{s-3}{(s-3)^2+16} + \dfrac{7}{2} \cdot \dfrac{4}{(s-3)^2+16}$, so $f(t) = e^{3t}[3\cos 4t + (7/2)\sin 4t]$

$s^2 + 6x + 9 + 16$

11. $F(s) = \dfrac{1}{4} \cdot \dfrac{1}{s-2} - \dfrac{1}{4} \cdot \dfrac{1}{s+2}$, so $f(t) = \dfrac{1}{4}(e^{2t} - e^{-2t}) = \dfrac{1}{2}\sinh 2t$

13. $F(s) = 3 \cdot \dfrac{1}{s+2} - 5 \cdot \dfrac{1}{s+5}$, so $f(t) = 3e^{-2t} - 5e^{-5t}$

15. $F(s) = \dfrac{1}{25}\left(-1 \cdot \dfrac{1}{s} - 5 \cdot \dfrac{1}{s^2} + \dfrac{1}{s-5}\right)$, so $f(t) = \dfrac{1}{25}\left(-1 - 5t + e^{5t}\right)$

17. $F(s) = \dfrac{1}{8}\left(\dfrac{1}{s^2-4} - \dfrac{1}{s^2+4}\right) = \dfrac{1}{16}\left(\dfrac{2}{s^2-4} - \dfrac{2}{s^2+4}\right)$

$f(t) = \dfrac{1}{16}(\sinh 2t - \sin 2t)$

19. $F(s) = \dfrac{s^2 - 2s}{(s^2+1)(s^2+4)} = \dfrac{1}{3}\left(\dfrac{-2s-1}{s^2+1} + \dfrac{2s+4}{s^2+4}\right)$

$f(t) = \dfrac{1}{3}(-2\cos t - \sin t + 2\cos 2t + 2\sin 2t)$

21. First we need to find A, B, C, D so that

$$\dfrac{s^2+3}{(s^2+2s+2)^2} = \dfrac{As+B}{s^2+2s+2} + \dfrac{Cs+D}{(s^2+2s+2)^2}.$$

When we multiply both sides by the quadratic factor s^2+2s+2 and collect coefficients, we get the linear equations

$$-2B - D + 3 = 0$$
$$-2A - 2B - C = 0$$
$$-2A - B + 1 = 0$$
$$-A = 0$$

which we solve for $A=0, B=1, C=-2, D=1$. Thus

$$F(s) = \frac{1}{(s+1)^2+1} + \frac{-2s+1}{\left[(s+1)^2+1\right]^2} = \frac{1}{(s+1)^2+1} - 2 \cdot \frac{s+1}{\left[(s+1)^2+1\right]^2} + 3 \cdot \frac{1}{\left[(s+1)^2+1\right]^2}.$$

We now use the inverse Laplace transforms given in Eq. (16) and (17) of Section 10.3 — supplying the factor e^{-t} corresponding to the translation $s \to s+1$ — and get

$$f(t) = e^{-t}\left[\sin t - 2 \cdot \frac{1}{2} t \sin t + 3 \cdot \frac{1}{2}(\sin t - t\cos t)\right] = \frac{1}{2} e^{-t}(5\sin t - 2t \sin t - 3t \cos t).$$

23. $\quad \dfrac{s^3}{s^4+4a^4} = \dfrac{1}{2}\left(\dfrac{s-a}{s^2-2as+2a^2} + \dfrac{s+a}{s^2+2as+2a^2}\right),$

and $s^2 \pm 2as + 2a^2 = (s \pm a)^2 + a^2$, so it follows that

$$\mathcal{L}^{-1}\left\{\frac{s^3}{s^4+4a^4}\right\} = \frac{1}{2}\left(e^{at} + e^{-at}\right)\cos at = \cosh at \cos at.$$

25. $\quad \dfrac{s}{s^4+4a^4} = \dfrac{1}{4a}\left(\dfrac{s}{s^2-2as+2a^2} - \dfrac{s}{s^2+2as+2a^2}\right)$

$$= \frac{1}{4a}\left(\frac{s-a}{s^2-2as+2a^2} + \frac{a}{s^2-2as+2a^2} - \frac{s+a}{s^2+2as+2a^2} + \frac{a}{s^2+2as+2a^2}\right),$$

and $s^2 \pm 2as + 2a^2 = (s \pm a)^2 + a^2$, so it follows that

$$\mathcal{L}^{-1}\left\{\frac{s}{s^4+4a^4}\right\} = \frac{1}{4a}\left[e^{at}(\cos at + \sin at) - e^{-at}(\cos at - \sin at)\right]$$

$$= \frac{1}{2a}\left[\frac{1}{2}\left(e^{at}+e^{-at}\right)\sin at + \frac{1}{2}\left(e^{at}-e^{-at}\right)\cos at\right]$$

$$= \frac{1}{2a}(\cosh at \sin at + \sinh at \cos at).$$

In Problems 27–40 we give first the transformed equation, then the Laplace transform $X(s)$ of the solution, and finally the desired solution $x(t)$.

27. $\quad [s^2 X(s) - 2s - 3] + 6[sX(s) - 2] + 25X(s) = 0$

$$X(s) = \frac{2s+15}{s^2+6s+25} = 2 \cdot \frac{s+3}{(s+3)^2+16} + \frac{9}{4} \cdot \frac{4}{(s+3)^2+16}$$

$x(t) = e^{-3t}[2\cos 4t + (9/4)\sin 4t]$

Section 10.3

29. $s^2 X(s) - 4X(s) = \dfrac{3}{s^2}$

$X(s) = \dfrac{3}{s^2(s^2-4)} = \dfrac{3}{4}\left(\dfrac{1}{s^2-4} - \dfrac{1}{s^2}\right)$

$x(t) = \dfrac{3}{8}\sinh 2t - \dfrac{3}{4}t = \dfrac{3}{8}(\sinh 2t - 2t)$

31. $[s^3 X(s) - s - 1] + [s^2 X(s) - 1] - 6[sX(s)] = 0$

$X(s) = \dfrac{s+2}{s^3 + s^2 - 6s} = \dfrac{1}{15}\left(-\dfrac{5}{s} - \dfrac{1}{s+3} + \dfrac{6}{s-2}\right)$

$x(t) = \dfrac{1}{15}\left(-5 - e^{-3t} + 6e^{2t}\right)$

33. $[s^4 X(s) - 1] + X(s) = 0$

$X(s) = \dfrac{1}{s^4 + 1}$

It therefore follows from Problem 26 with $a = \sqrt[4]{1/4} = 1/\sqrt{2}$ that

$x(t) = \dfrac{1}{\sqrt{2}}\left(\cosh\dfrac{t}{\sqrt{2}}\sin\dfrac{t}{\sqrt{2}} - \sinh\dfrac{t}{\sqrt{2}}\cos\dfrac{t}{\sqrt{2}}\right).$

35. $\left[s^4 X(s) - 1\right] + 8s^2 X(s) + 16 X(s) = 0$

$X(s) = \dfrac{1}{s^4 + 8s^2 + 16} = \dfrac{1}{(s^2+4)^2}$

$x(t) = \dfrac{1}{16}(\sin 2t - 2t\cos 2t)$ (by Eq. (17) in Section 10.3)

37. $\left[s^2 X(s) - 2\right] + 4sX(s) + 13X(s) = \dfrac{1}{(s+1)^2}$

$X(s) = \dfrac{2 + 1/(s+1)^2}{s^2 + 4s + 13} = \dfrac{2s^2 + 4s + 13}{(s+1)^2(s^2 + 4s + 13)}$

$= \dfrac{1}{50}\left[-\dfrac{1}{s+1} + \dfrac{5}{(s+1)^2} + \dfrac{s+98}{(s+2)^2 + 9}\right]$

$$= \frac{1}{50}\left[-\frac{1}{s+1}+\frac{5}{(s+1)^2}+\frac{s+2}{(s+2)^2+9}+32\cdot\frac{3}{(s+2)^2+9}\right]$$

$$x(t) = \frac{1}{50}\left[(-1+5t)e^{-t}+e^{-2t}(\cos 3t+32\sin 3t)\right]$$

39. $x''+9x = 6\cos 3t, \quad x(0) = x'(0) = 0$

$$s^2 X(s)+9X(s) = \frac{6s}{s^2+9}$$

$$X(s) = \frac{6s}{\left(s^2+9\right)^2}$$

$$x(t) = 6\cdot\frac{1}{2\cdot 3}t\sin 3t = t\sin 3t \qquad \text{(by Eq. (16) in Section 10.3)}$$

SECTION 10.4

DERIVATIVES, INTEGRALS, AND PRODUCTS OF TRANSFORMS

This section completes the presentation of the standard "operational properties" of Laplace transforms, the most important one here being the convolution property $\mathcal{L}\{f*g\} = \mathcal{L}\{f\}\cdot\mathcal{L}\{g\}$, where the **convolution** $f*g$ is defined by

$$f*g(t) = \int_0^t f(x)g(t-x)\,dx.$$

Here we use x rather than τ as the variable of integration; compare with Eq. (3) in Section 10.4 of the textbook

1. With $f(t)=t$ and $g(t)=1$ we calculate

$$t*1 = \int_0^t x\cdot 1\,dx = \left[\frac{1}{2}x^2\right]_{x=0}^{x=t} = \frac{1}{2}t^2.$$

3. To compute $(\sin t)*(\sin t) = \int_0^t \sin x\sin(t-x)\,dx,$ we first apply the identity $\sin A \sin B = [\cos(A-B) - \cos(A+B)]/2.$ This gives

$$(\sin t) * (\sin t) = \int_0^t \sin x \sin(t-x)\,dx$$

$$= \frac{1}{2}\int_0^t [\cos(2x-t) - \cos t]\,dx$$

$$= \frac{1}{2}\left[\frac{1}{2}\sin(2x-t) - x\cos t\right]_{x=0}^{x=t}$$

$$(\sin t)*(\sin t) = \frac{1}{2}(\sin t - t\cos t).$$

5. $\quad e^{at} * e^{at} = \int_0^t e^{ax}e^{a(t-x)}\,dx = \int_0^t e^{at}\,dx = e^{at}[x]_{x=0}^{x=t} = te^{at}$

7. $\quad f(t) = 1 * e^{3t} = e^{3t} * 1 = \int_0^t e^{3x}\cdot 1\,dx = \frac{1}{3}(e^{3t}-1)$

9. $\quad f(t) = \frac{1}{9}\sin 3t * \sin 3t = \frac{1}{9}\int_0^t \sin 3x \sin 3(t-x)\,dx$

$$= \frac{1}{9}\int_0^t \sin 3x [\sin 3t \cos 3x - \cos 3t \sin 3x]\,dx$$

$$= \frac{1}{9}\sin 3t \int_0^t \sin 3x \cos 3x\,dx - \frac{1}{9}\cos 3t \int_0^t \sin^2 3x\,dx$$

$$= \frac{1}{9}\sin 3t \left[\frac{1}{6}\sin^2 3x\right]_{x=0}^{x=t} - \frac{1}{9}\cos 3t \left[\frac{1}{2}\left(x - \frac{1}{6}\sin 6x\right)\right]_{x=0}^{x=t}$$

$$f(t) = \frac{1}{54}(\sin 3t - 3t\cos 3t)$$

11. $\quad f(t) = \cos 2t * \cos 2t = \int_0^t \cos 2x \cos 2(t-x)\,dx$

$$= \int_0^t \cos 2x(\cos 2t \cos 2x + \sin 2t \sin 2x)\,dx$$

$$= (\cos 2t)\int_0^t \cos^2 2x\,dx + (\sin 2t)\int_0^t \cos 2x \sin 2x\,dx$$

$$= (\cos 2t)\left[\frac{1}{2}\left(x + \frac{1}{4}\sin 4x\right)\right]_{x=0}^{x=t} + (\sin 2t)\left[\frac{1}{4}\sin^2 2x\right]_{x=0}^{x=t}$$

$$f(t) = \frac{1}{4}(\sin 2t + 2t \cos 2t)$$

13. $\quad f(t) = e^{3t} * \cos t = \int_0^t (\cos x)e^{3(t-x)}\,dx$

$$= e^{3t}\int_0^t e^{-3x}\cos x\,dx$$

$$= e^{3t} \left[\frac{e^{-3x}}{10} (-3\cos x + \sin x) \right]_{x=0}^{x=t} \qquad \text{(by integral formula \#50)}$$

$$f(t) = \frac{1}{10}\left(3e^{3t} - 3\cos t + \sin t\right)$$

15. $\mathcal{L}\{t \sin t\} = -\dfrac{d}{ds}(\mathcal{L}\{\sin t\}) = -\dfrac{d}{ds}\left(\dfrac{3}{s^2+9}\right) = \dfrac{6s}{(s^2+9)^2}$

17. $\mathcal{L}\{e^{2t}\cos 3t\} = (s-2)/(s^2 - 4s + 13)$
    $\mathcal{L}\{te^{2t}\cos 3t\} = -(d/ds)[(s-2)/(s^2 - 4s + 13)] = (s^2 - 4s - 5)/(s^2 - 4s + 13)^2$

19. $\mathcal{L}\left\{\dfrac{\sin t}{t}\right\} = \displaystyle\int_s^\infty \dfrac{ds}{s^2+1} = \left[\tan^{-1} s\right]_s^\infty = \dfrac{\pi}{2} - \tan^{-1} s = \tan^{-1}\left(\dfrac{1}{s}\right)$

21. $\mathcal{L}\{e^{3t} - 1\} = \dfrac{1}{s-3} - \dfrac{1}{s}$, so

$$\mathcal{L}\left\{\dfrac{e^{3t}-1}{t}\right\} = \int_s^\infty \left(\dfrac{1}{s-3} - \dfrac{1}{s}\right) ds = \left[\ln\left(\dfrac{s-3}{s}\right)\right]_s^\infty = \ln\left(\dfrac{s}{s-3}\right)$$

23. $f(t) = -\dfrac{1}{t}\mathcal{L}^{-1}\{F'(s)\} = -\dfrac{1}{t}\mathcal{L}^{-1}\left\{\dfrac{1}{s-2} - \dfrac{1}{s+2}\right\} = -\dfrac{1}{t}\left(e^{2t} - e^{-2t}\right) = -\dfrac{2\sinh 2t}{t}$

25. $f(t) = -\dfrac{1}{t}\mathcal{L}^{-1}\{F'(s)\} = -\dfrac{1}{t}\mathcal{L}^{-1}\left\{\dfrac{2s}{s^2+1} - \dfrac{1}{s+2} - \dfrac{1}{s-3}\right\} = \dfrac{1}{t}\left(e^{-2t} + e^{3t} - 2\cos t\right)$

27. $f(t) = -\dfrac{1}{t}\mathcal{L}^{-1}\{F'(s)\} = -\dfrac{1}{t}\mathcal{L}^{-1}\left\{\dfrac{-2/s^3}{1+1/s^2}\right\}$

$$= \dfrac{2}{t}\mathcal{L}^{-1}\left\{\dfrac{1}{s^3+s}\right\} = \dfrac{2}{t}\mathcal{L}^{-1}\left\{\dfrac{1}{s} - \dfrac{s}{s^2+1}\right\} = \dfrac{2}{t}(1 - \cos t)$$

29. $-[s^2 X(s) - x'(0)]' - [s\, X(s)]' - 2[s\, X(s)] + X(s) = 0$
    $s(s+1)X'(s) + 4s\, X(s) = 0$ \qquad (separable)
    $X(s) = \dfrac{A}{(s+1)^4}$ with $A \neq 0$
    $x(t) = Ct^3 e^{-t}$ with $C \neq 0$

31. $-[s^2 X(s) - x'(0)]' + 4[s X(s)]' - [s X(s)] - 4[X(s)]' + 2X(s) = 0$

$(s^2 - 4s + 4)X'(s) + (3s - 6)X(s) = 0$ (separable)

$(s - 2)X'(s) + 3X(s) = 0$

$X(s) = \dfrac{A}{(s-2)^3}$ with $A \neq 0$

$x(t) = Ct^2 e^{2t}$ with $C \neq 0$

33. $-[s^2 X(s) - x(0)]' - 2[s X(s)] - [X(s)]' = 0$

$(s^2 + 1)X'(s) + 4s X(s) = 0$ (separable)

$X(s) = \dfrac{A}{(s^2 + 1)^2}$ with $A \neq 0$

$x(t) = C(\sin t - t \cos t)$ with $C \neq 0$

35. $\mathcal{L}^{-1}\left\{\dfrac{1}{(s-1)\sqrt{s}}\right\} = e^t * \dfrac{1}{\sqrt{\pi t}} = \int_0^t \dfrac{1}{\sqrt{\pi x}} \cdot e^{t-x} dx$

$= \dfrac{e^t}{\sqrt{\pi}} \int_0^{\sqrt{t}} \dfrac{1}{u} \cdot e^{-u^2} \cdot 2u\, du = \dfrac{2e^t}{\sqrt{\pi}} \int_0^{\sqrt{t}} e^{-u^2} du = e^t \operatorname{erf}\left(\sqrt{t}\right)$

37. $s^2 X(s) + 2s X(s) + X(s) = F(s)$

$X(s) = F(s) \cdot \dfrac{1}{(s+1)^2}$

$x(t) = te^{-t} * f(t) = \int_0^t \tau e^{-\tau} f(t-\tau)\, d\tau$

SECTION 10.5

PERIODIC AND PIECEWISE CONTINUOUS FORCING FUNCTIONS

1. $F(s) = e^{-3s} \mathcal{L}\{t\}$ so Eq. (3b) in Theorem 1 gives

$$f(t) = u(t-3) \cdot (t-3) = \begin{cases} 0 & \text{if } t < 3, \\ t-3 & \text{if } t \geq 3. \end{cases}$$

The graph of f is shown below.

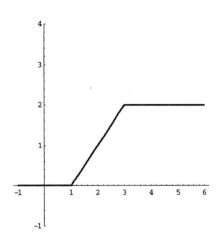

3. $F(s) = e^{-s}\mathcal{L}\{e^{-2t}\}$ so $f(t) = u(t-1)\cdot e^{-2(t-1)} = \begin{cases} 0 & \text{if } t<1, \\ e^{-2(t-1)} & \text{if } t\geq 1. \end{cases}$

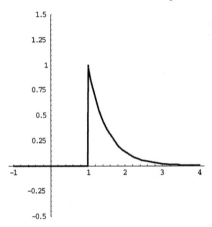

5. $F(s) = e^{-\pi s}\mathcal{L}\{\sin t\}$ so
$$f(t) = u(t-\pi)\cdot\sin(t-\pi) = -u(t-\pi)\sin t = \begin{cases} 0 & \text{if } t<\pi, \\ -\sin t & \text{if } t\geq \pi. \end{cases}$$

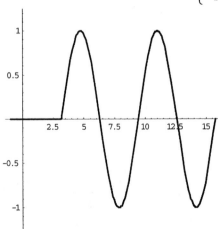

7. $F(s) = \mathcal{L}\{\sin t\} - e^{-2\pi s}\mathcal{L}\{\sin t\}$ so

$$f(t) = \sin t - u(t-2\pi)\sin(t-2\pi) = [1-u(t-2\pi)]\sin t = \begin{cases} \sin t & \text{if } t < 2\pi, \\ 0 & \text{if } t \geq 2\pi. \end{cases}$$

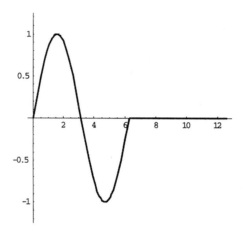

9. $F(s) = \mathcal{L}\{\cos \pi t\} + e^{-3s}\mathcal{L}\{\cos \pi t\}$ so

$$f(t) = \cos \pi t + u(t-3)\cos \pi(t-3) = [1-u(t-3)]\cos \pi t = \begin{cases} \cos \pi t & \text{if } t < 3, \\ 0 & \text{if } t \geq 3. \end{cases}$$

11. $f(t) = 2 - u(t-3) \cdot 2$ so $F(s) = \dfrac{2}{s} - e^{-3s}\dfrac{2}{s} = \dfrac{2}{s}\left(1 - e^{-3s}\right).$

13. $f(t) = [1 - u(t-2\pi)]\sin t = \sin t - u(t-2\pi)\sin(t-2\pi)$ so

$$F(s) = \dfrac{1}{s^2+1} - e^{-2\pi s} \cdot \dfrac{1}{s^2+1} = \dfrac{1 - e^{-2\pi s}}{s^2+1}.$$

15. $f(t) = [1 - u(t-3\pi)]\sin t = \sin t + u(t-3\pi)\sin(t-3\pi)$ so

$$F(s) = \dfrac{1}{s^2+1} + \dfrac{e^{-3\pi s}}{s^2+1} = \dfrac{1 + e^{-3\pi s}}{s^2+1}.$$

17. $f(t) = [u(t-2) - u(t-3)]\sin \pi t = u(t-2)\sin \pi(t-2) + u(t-3)\sin \pi(t-3)$ so

$$F(s) = \left(e^{-2s} + e^{-3s}\right) \cdot \dfrac{\pi}{s^2 + \pi^2} = \dfrac{\pi\left(e^{-2s} + e^{-3s}\right)}{s^2 + \pi^2}.$$

19. If $g(t) = t + 1$ then $f(t) = u(t-1) \cdot t = u(t-1) \cdot g(t-1)$ so

$$F(s) = e^{-s}G(s) = e^{-s}L\{t+1\} = e^{-s} \cdot \left(\dfrac{1}{s^2} + \dfrac{1}{s}\right) = \dfrac{e^{-s}(s+1)}{s^2}.$$

21. If $g(t) = t+1$ and $h(t) = t+2$ then

$$f(t) = t[1-u(t-1)] + (2-t)[u(t-1) - u(t-2)]$$
$$= t - 2tu(t-1) + 2u(t-1) - 2u(t-2) + tu(t-2)$$
$$= t - 2u(t-1)g(t-1) + 2u(t-1) - 2u(t-2) + u(t-2)h(t-2)$$

so

$$F(s) = \frac{1}{s^2} - 2e^{-s}\left(\frac{1}{s^2} + \frac{1}{s}\right) + \frac{2e^{-s}}{s} - \frac{2e^{-2s}}{s} + e^{-2s}\left(\frac{1}{s^2} + \frac{2}{s}\right) = \frac{(1-e^{-s})^2}{s^2}.$$

23. With $f(t) = 1$ and $p = 1$, Formula (6) in the text gives

$$\mathcal{L}\{1\} = \frac{1}{1-e^{-s}} \int_0^1 e^{-st} \cdot 1 \, dt = \frac{1}{1-e^{-s}} \left[-\frac{e^{-st}}{s}\right]_{t=0}^{t=1} = \frac{1}{s}.$$

25. With $p = 2a$ and $f(t) = 1$ if $0 \leq t \leq a$, $f(t) = 0$ if $a < t \leq 2a$, Formula (6) gives

$$\mathcal{L}\{f(t)\} = \frac{1}{1-e^{-2as}} \int_0^a e^{-st} \cdot 1 \, dt = \frac{1}{1-e^{-2as}} \left[-\frac{e^{-st}}{s}\right]_{t=0}^{t=a}$$

$$= \frac{1-e^{-as}}{s(1-e^{-as})(1+e^{-as})} = \frac{1}{s(1+e^{-as})}.$$

27. $G(s) = \mathcal{L}\{t/a - f(t)\} = (1/as^2) - F(s)$. Now substitution of the result of Problem 26 in place of $F(s)$ immediately gives the desired transform.

29. With $p = 2\pi/k$ and $f(t) = \sin kt$ for $0 \leq t \leq \pi/k$ while $f(t) = 0$ for $\pi/k \leq t \leq 2\pi/k$, Formula (6) and the integral formula

$$\int e^{at} \sin bt \, dt = e^{at}\left[\frac{a\sin bt - b\cos bt}{a^2 + b^2}\right] + C$$

give

$$\mathcal{L}\{f(t)\} = \frac{1}{1-e^{-2\pi s/k}} \int_0^{\pi/k} e^{-st} \cdot \sin kt \, dt$$

$$= \frac{1}{1-e^{-2\pi s/k}} \left[e^{-st}\left(\frac{-s\sin kt - k\cos kt}{s^2 + k^2}\right)\right]_{t=0}^{t=\pi/k}$$

$$= \frac{1}{1-e^{-2\pi s/k}} \left[\frac{e^{-\pi s/k}(k) - (-k)}{s^2 + k^2}\right]$$

$$= \frac{k\left(1+e^{-\pi s/k}\right)}{\left(1-e^{-\pi s/k}\right)\left(1+e^{-\pi s/k}\right)\left(s^2+k^2\right)} = \frac{k}{\left(s^2+k^2\right)\left(1-e^{-\pi s/k}\right)}.$$

In Problems 31–37, we first write and transform the appropriate differential equation. Then we solve for the transform of the solution, and finally inverse transform to find the desired solution.

31. $x'' + 4x = 1 - u(t - \pi)$

$$s^2 X(s) + 4X(s) = \frac{1-e^{-\pi s}}{s}$$

$$X(s) = \frac{1-e^{-\pi s}}{s(s^2+4)} = \frac{1}{4}(1-e^{-\pi s})\left(\frac{1}{s} - \frac{s}{s^2+4}\right)$$

$x(t) = (1/4)[1 - u(t-\pi)][1 - \cos 2(t-\pi)] = (1/2)[1 - u(t-\pi)]\sin^2 t$

33. $x'' + 9x = [1 - u(t - 2\pi)]\sin t$

$$X(s) = \frac{1-e^{-2\pi s}}{(s^2+1)(s^2+4)} = \frac{1}{8}(1-e^{-2\pi s})\left(\frac{1}{s^2+1} - \frac{1}{s^2+9}\right)$$

$$x(t) = \frac{1}{8}[1-u(t-2\pi)]\left(\sin t - \frac{1}{3}\sin 3t\right)$$

35. $x'' + 4x' + 4x = [1 - u(t - 2)]t = t - u(t - 2)g(t - 2)$ where $g(t) = t + 2$

$$(s+2)^2 X(s) = \frac{1}{s^2} - e^{-2s}\left(\frac{2}{s} + \frac{1}{s^2}\right)$$

$$X(s) = \frac{1}{s^2(s+2)^2} - e^{-2s}\frac{2s+1}{s^2(s+2)^2}$$

$$= \frac{1}{4}\left(-\frac{1}{s} + \frac{1}{s^2} + \frac{1}{s+2} + \frac{1}{(s+2)^2}\right) - \frac{1}{4}e^{-2s}\left(\frac{1}{s} + \frac{1}{s^2} - \frac{1}{s+2} - \frac{3}{(s+2)^2}\right)$$

$x(t) = (1/4)\{-1 + t + (1+t)e^{-2t} + u(t-2)[1 - t + (3t-5)e^{-2(t-2)}]\}$

37. $x'' + 2x' + 10x = f(t)$, $\quad x(0) = x'(0) = 0$

As in the solution of Example 6 we find first that

$$\left(s^2 + 2s + 10\right)X(s) = \frac{10}{s} + \frac{20}{s}\sum_{n=1}^{\infty}(-1)^n e^{-n\pi s},$$

so

$$X(s) = \frac{10}{s(s^2+2s+10)} + 2\sum_{n=1}^{\infty}\frac{10(-1)^n e^{-n\pi s}}{s(s^2+2s+10)}.$$

If

$$g(t) = \mathcal{L}^{-1}\left\{\frac{10}{s\left[(s+1)^2+9\right]}\right\} = 1 - \frac{1}{3}e^{-t}(3\cos 3t + \sin t),$$

then it follows that

$$x(t) = g(t) + 2\sum_{n=1}^{\infty}(-1)^n u_{n\pi}(t)g(t-n\pi).$$

The following graph of $x(t)$ exhibits the resulting resonance.

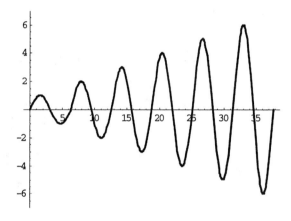

CHAPTER 11

POWER SERIES METHODS

SECTION 11.1

INTRODUCTION AND REVIEW OF POWER SERIES

The power series method consists of substituting a series $y = \Sigma c_n x^n$ into a given differential equation in order to determine what the coefficients $\{c_n\}$ must be in order that the power series will satisfy the equation. It might be pointed out that, if we find a recurrence relation in the form $c_{n+1} = \phi(n)c_n$, then we can determine the radius of convergence ρ of the series solution directly from the recurrence relation,

$$\rho = \lim_{n \to \infty} \left| \frac{c_n}{c_{n+1}} \right| = \lim_{n \to \infty} \left| \frac{1}{\phi(n)} \right|.$$

In Problems 1–10 we give first that recurrence relation that can be used to find the radius of convergence and to calculate the succeeding coefficients $c_1, c_2, c_3, \cdots$ in terms of the arbitrary constant c_0. Then we give the series itself

1. $c_{n+1} = \dfrac{c_n}{n+1}$; it follows that $c_n = \dfrac{c_0}{n!}$ and $\rho = \lim_{n \to \infty}(n+1) = \infty$.

$$y(x) = c_0\left(1 + x + \frac{x^2}{2} + \frac{x^3}{6} + \frac{x^4}{24} + \cdots\right) = c_0\left(1 + \frac{x}{1!} + \frac{x^2}{2!} + \frac{x^3}{3!} + \frac{x^4}{4!} + \cdots\right) = c_0 e^x$$

3. $c_{n+1} = -\dfrac{3c_n}{2(n+1)}$; it follows that $c_n = \dfrac{(-1)^n 3^n c_0}{2^n n!}$ and $\rho = \lim_{n \to \infty}\dfrac{2(n+1)}{3} = \infty$.

$$y(x) = c_0\left(1 - \frac{3x}{2} + \frac{9x^2}{8} - \frac{9x^3}{16} + \frac{27x^4}{128} - \cdots\right)$$

$$= c_0\left(1 - \frac{3x}{1!2} + \frac{3^2 x^2}{2!2^2} - \frac{3^3 x^3}{3!2^3} + \frac{3^4 x^4}{4!2^4} - \cdots\right) = c_0 e^{-3x/2}$$

5. When we substitute $y = \Sigma c_n x^n$ into the equation $y' = x^2 y$, we find that

$$c_1 + 2c_2 x + \sum_{n=0}^{\infty}\left[(n+3)c_{n+3} - c_n\right]x^{n+1} = 0.$$

Hence $c_1 = c_2 = 0$ — which we see by equating constant terms and x-terms on the two sides of this equation — and $c_3 = \dfrac{c_n}{n+3}$. It follows that

$$c_{3k+1} = c_{3k+2} = 0 \quad \text{and} \quad c_{3k} = \dfrac{c_0}{3 \cdot 6 \cdots (3k)} = \dfrac{c_0}{k! 3^k}.$$

Hence

$$y(x) = c_0\left(1 + \dfrac{x^3}{3} + \dfrac{x^6}{18} + \dfrac{x^9}{162} + \cdots\right) = c_0\left(1 + \dfrac{x^3}{1!3} + \dfrac{x^6}{2!3^2} + \dfrac{x^9}{3!3^3} + \cdots\right) = c_0 e^{(x^3/3)}.$$

and $\rho = \infty$.

7. $c_{n+1} = 2c_n$; it follows that $c_n = 2^n c_0$ and $\rho = \lim_{n \to \infty} \dfrac{1}{2} = \dfrac{1}{2}$.

$$y(x) = c_0\left(1 + 2x + 4x^2 + 8x^3 + 16x^4 + \cdots\right)$$
$$= c_0\left[1 + (2x) + (2x)^2 + (2x)^3 + (2x)^4 + \cdots\right] = \dfrac{c_0}{1-2x}$$

9. $c_{n+1} = \dfrac{(n+2)c_n}{n+1}$; it follows that $c_n = (n+1)c_0$ and $\rho = \lim_{n \to \infty} \dfrac{n+1}{n+2} = 1$.

$$y(x) = c_0\left(1 + 2x + 3x^2 + 4x^3 + 5x^4 + \cdots\right)$$

Separation of variables gives $y(x) = \dfrac{c_0}{(1-x)^2}$.

In Problems 11–14 the differential equations are second-order, and we find that the two initial coefficients c_0 and c_1 are both arbitrary. In each case we find the even-degree coefficients in terms of c_0 and the odd-degree coefficients in terms of c_1. The solution series in these problems are all recognizable power series that have infinite radii of convergence.

11. $c_{n+1} = \dfrac{c_n}{(n+1)(n+2)}$; it follows that $c_{2k} = \dfrac{c_0}{(2k)!}$ and $c_{2k+1} = \dfrac{c_1}{(2k+1)!}$.

$$y(x) = c_0\left(1 + \dfrac{x^2}{2!} + \dfrac{x^4}{4!} + \dfrac{x^6}{6!} + \cdots\right) + c_1\left(x + \dfrac{x^3}{3!} + \dfrac{x^5}{5!} + \dfrac{x^7}{7!} + \cdots\right) = c_0 \cosh x + c_1 \sinh x$$

13. $c_{n+1} = -\dfrac{9c_n}{(n+1)(n+2)}$; it follows that $c_{2k} = \dfrac{(-1)^k 3^{2k} c_0}{(2k)!}$ and $c_{2k+1} = \dfrac{(-1)^k 3^{2k} c_1}{(2k+1)!}$.

$$y(x) = c_0\left(1 - \dfrac{9x^2}{2} + \dfrac{27x^4}{8} - \dfrac{81x^6}{80} + \cdots\right) + c_1\left(x - \dfrac{3x^3}{2} + \dfrac{27x^5}{40} - \dfrac{81x^7}{560} + \cdots\right)$$

$$= c_0 \left(1 - \frac{(3x)^2}{2!} + \frac{(3x)^4}{4!} - \frac{(3x)^6}{6!} + \cdots \right) + \frac{c_1}{3}\left((3x) - \frac{(3x)^3}{3!} + \frac{(3x)^5}{5!} - \frac{(3x)^7}{7!} + \cdots \right)$$

$$= c_0 \cos 3x + \frac{c_1}{3} \sin x$$

15. Assuming a power series solution of the form $y = \Sigma c_n x^n$, we substitute it into the differential equation $xy' + y = 0$ and find that $(n+1)c_n = 0$ for all $n \geq 0$. This implies that $c_n = 0$ for all $n \geq 0$, which means that the only power series solution of our differential equation is the trivial solution $y(x) \equiv 0$. Therefore the equation has no *non-trivial* power series solution.

17. Assuming a power series solution of the form $y = \Sigma c_n x^n$, we substitute it into the differential equation $x^2 y' + y = 0$. We find that $c_0 = c_1 = 0$ and that $c_{n+1} = -nc_n$ for $n \geq 1$, so it follows that $c_n = 0$ for all $n \geq 0$. Just as in Problems 15 and 16, this means that the equation has no *non-trivial* power series solution.

In Problems 19–22 we first give the recurrence relation that results upon substitution of an assumed power series solution $y = \Sigma c_n x^n$ into the given second-order differential equation. Then we give the resulting general solution, and finally apply the initial conditions $y(0) = c_0$ and $y'(0) = c_1$ to determine the desired particular solution.

19. $c_{n+2} = -\dfrac{2^2 c_n}{(n+1)(n+2)}$ for $n \geq 0$, so $c_{2k} = \dfrac{(-1)^k 2^{2k} c_0}{(2k)!}$ and $c_{2k+1} = \dfrac{(-1)^k 2^{2k} c_1}{(2k+1)!}$.

$$y(x) = c_0 \left(1 - \frac{2^2 x^2}{2!} + \frac{2^4 x^4}{4!} - \frac{2^6 x^6}{6!} + \cdots \right) + c_1 \left(x - \frac{2^2 x^3}{3!} + \frac{2^4 x^5}{5!} - \frac{2^6 x^7}{7!} + \cdots \right)$$

$c_0 = y(0) = 0$ and $c_1 = y'(0) = 3$, so

$$y(x) = 3\left(x - \frac{2^2 x^3}{3!} + \frac{2^4 x^5}{5!} - \frac{2^6 x^7}{7!} + \cdots \right)$$

$$= \frac{3}{2}\left[(2x) - \frac{(2x)^3}{3!} + \frac{(2x)^5}{5!} - \frac{(2x)^7}{7!} + \cdots \right] = \frac{3}{2} \sin 2x.$$

21. $c_{n+1} = \dfrac{2nc_n - c_{n-1}}{n(n+1)}$ for $n \geq 1$; with $c_0 = y(0) = 0$ and $c_1 = y'(0) = 1$, we obtain

$c_2 = 1$, $c_3 = \dfrac{1}{2}$, $c_4 = \dfrac{1}{6} = \dfrac{1}{3!}$, $c_5 = \dfrac{1}{24} = \dfrac{1}{4!}$, $c_6 = \dfrac{1}{120} = \dfrac{1}{5!}$. Evidently $c_n = \dfrac{1}{(n-1)!}$, so

$$y(x) = x + x^2 + \frac{x^3}{2!} + \frac{x^4}{3!} + \frac{x^5}{4!} + \cdots = x\left(1 + x + \frac{x^2}{2!} + \frac{x^3}{3!} + \frac{x^4}{4!} + \cdots \right) = xe^x.$$

23. $c_0 = c_1 = 0$ and the recursion relation

$$(n^2 - n + 1)c_n + (n-1)c_{n-1} = 0$$

for $n \geq 2$ imply that $c_n = 0$ for $n \geq 0$. Thus any assumed power series solution $y = \Sigma c_n x^n$ must reduce to the trivial solution $y(x) \equiv 0$.

25. This problem is pretty fully outlined in the textbook. The only hard part is squaring the power series:

$$\begin{aligned}\left(1 + c_3 x^3 + c_5 x^5 + c_7 x^7 + c_9 x^9 + c_{11} x^{11} + \cdots\right)^2 \\ = x^2 + 2c_3 x^4 + \left(c_3^2 + 2c_5\right)x^6 + \left(2c_3 c_5 + 2c_7\right)x^8 + \\ \left(c_5^2 + 2c_3 c_7 + 2c_9\right)x^{10} + \left(2c_5 c_7 + 2c_3 c_9 + 2c_{11}\right)x^{12} + \cdots\end{aligned}$$

SECTION 11.2

POWER SERIES SOLUTIONS

Instead of deriving in detail the recurrence relations and solution series for Problems 1 through 15, we indicate where some of these problems and answers originally came from. Each of the differential equations in Problems 1–10 is of the form

$$(Ax^2 + B)y'' + Cxy' + Dy = 0$$

with selected values of the constants A, B, C, D. When we substitute $y = \Sigma c_n x^n$, shift indices where appropriate, and collect coefficients, we get

$$\sum_{n=0}^{\infty}\left[An(n-1)c_n + B(n+1)(n+2)c_{n+2} + Cnc_n + Dc_n\right]x^n = 0.$$

Thus the recurrence relation is

$$c_{n+2} = -\frac{An^2 + (C-A)n + D}{B(n+1)(n+2)}c_n \quad \text{for } n \geq 0.$$

It yields a solution of the form

$$y = c_0 \, y_{\text{even}} + c_1 \, y_{\text{odd}}$$

where y_{even} and y_{odd} denote series with terms of even and odd degrees, respectively. The even-degree series $c_0 + c_2 x^2 + c_4 x^4 + \cdots$ converges (by the ratio test) provided that

$$\lim_{n\to\infty}\left|\frac{c_{n+2}x^{n+2}}{c_n x^n}\right| = \left|\frac{Ax^2}{B}\right| < 1.$$

Hence its radius of convergence is at least $\rho = \sqrt{|B/A|}$, as is that of the odd-degree series $c_1 x + c_3 x^3 + c_5 x^4 + \cdots$. (See Problem 6 for an example in which the radius of convergence is, surprisingly, greater than $\sqrt{|B/A|}$.)

In Problems 1–15 we give first the recurrence relation and the radius of convergence, then the resulting power series solution.

1. $c_{n+2} = c_n;$ $\rho = 1;$ $c_0 = c_2 = c_4 = \cdots;$ $c_1 = c_3 = c_4 = \cdots$

$$y(x) = c_0 \sum_{n=0}^{\infty} x^{2n} + c_1 \sum_{n=0}^{\infty} x^{2n+1} = \frac{c_0 + c_1 x}{1 - x^2}$$

3. $c_{n+2} = -\dfrac{c_n}{(n+2)};$ $\rho = \infty;$

$$c_{2n} = \frac{(-1)^n c_0}{(2n)(2n-2)\cdots 4\cdot 2} = \frac{(-1)^n c_0}{n!\, 2^n}; \qquad c_{2n+1} = \frac{(-1)^n c_1}{(2n+1)(2n-1)\cdots 5\cdot 3} = \frac{(-1)^n c_1}{(2n+1)!!}$$

$$y(x) = c_0 \sum_{n=0}^{\infty} (-1)^n \frac{x^{2n}}{n!\, 2^n} + c_1 \sum_{n=0}^{\infty} (-1)^n \frac{x^{2n+1}}{(2n+1)!!}$$

5. $c_{n+2} = \dfrac{nc_n}{3(n+2)};$ $\rho = 3;$ $c_2 = c_4 = c_6 = \cdots = 0$

$$c_{2n+1} = \frac{2n-1}{3(2n+1)}\cdot\frac{2n-3}{3(2n-1)}\cdots\frac{3}{3(5)}\cdot\frac{1}{3(3)}c_1 = \frac{c_1}{(2n+1)3^n}$$

$$y(x) = c_0 + c_1 \sum_{n=0}^{\infty} \frac{x^{2n+1}}{(2n+1)3^n}$$

7. $c_{n+2} = -\dfrac{(n-4)^2}{3(n+1)(n+2)}c_n;$ $\rho \geq \sqrt{3}$

The factor $(n-4)$ yields $c_6 = c_8 = c_{10} = \cdots = 0$, so y_{even} is a 4th-degree polynomial. We find first that $c_3 = -c_1/2$ and $c_5 = c_1/120$, and then for $n \geq 3$ that

$$c_{2n+1} = \left(-\frac{(2n-5)^2}{3(2n)(2n+1)}\right)\left(-\frac{(2n-7)^2}{3(2n-2)(2n-1)}\right)\cdots\left(-\frac{1^2}{3(6)(7)}\right)c_5 =$$

$$= (-1)^{n-2}\frac{[(2n-5)!!]^2}{3^{n-2}(2n+1)(2n-1)\cdots 7\cdot 6}\cdot\frac{c_1}{120} = 9\cdot(-1)^n\frac{[(2n-5)!!]^2}{3^n(2n+1)!}c_1$$

$$y(x) = c_0\left(1 - \frac{8}{3}x^2 + \frac{8}{27}x^4\right) + c_1\left[x - \frac{1}{2}x^3 + \frac{1}{120}x^5 + 9\sum_{n=3}^{\infty}\frac{[(2n-5)!!]^2(-1)^n}{(2n+1)!\,3^n}x^{2n+1}\right]$$

9. $\quad c_{n+2} = \frac{(n+3)(n+4)}{(n+1)(n+2)}c_n; \qquad \rho = 1$

$$c_{2n} = \frac{(2n+1)(2n+2)}{(2n-1)(2n)}\cdot\frac{(2n-1)(2n)}{(2n-3)(2n-2)}\cdots\frac{3\cdot 4}{1\cdot 2}c_0 = \frac{1}{2}(n+1)(2n+1)c_0$$

$$c_{2n+1} = \frac{(2n+2)(2n+3)}{(2n)(2n+1)}\cdot\frac{(2n)(2n+1)}{(2n-2)(2n-1)}\cdots\frac{4\cdot 5}{2\cdot 3}c_1 = \frac{1}{3}(n+1)(2n+3)c_1$$

$$y(x) = c_0\sum_{n=0}^{\infty}(n+1)(2n+1)x^{2n} + \frac{1}{3}c_1\sum_{n=0}^{\infty}(n+1)(2n+3)x^{2n+1}$$

11. $\quad c_{n+2} = \frac{2(n-5)}{5(n+1)(n+2)}c_n; \qquad \rho = \infty$

The factor $(n-5)$ yields $c_7 = c_9 = c_{11} = \cdots = 0$, so y_{odd} is a 5th-degree polynomial. We find first that $c_2 = -c_1$, $c_4 = c_0/10$ and $c_6 = c_0/750$, and then for $n \geq 4$ that

$$c_{2n} = \frac{2(2n-7)}{5(2n)(2n-1)}\cdot\frac{2(2n-5)}{5(2n-2)(2n-3)}\cdots\frac{2(1)}{5(8)(7)}c_6$$

$$= \frac{2^{n-3}(2n-7)!!}{5^{n-3}(2n)(2n-1)\cdots(8)(7)}\cdot\frac{c_0}{750} =$$

$$= \frac{5^3\cdot 6!}{2^3\cdot 750}\cdot\frac{2^n(2n-7)!!}{5^n(2n)(2n)\cdots(8)(7)\cdot 6!}\cdot c_1 = 15\cdot\frac{2^n(2n-7)!!}{5^n(2n)!}c_0$$

$$y(x) = c_1\left(x - \frac{4x^3}{15} + \frac{4x^5}{375}\right) + c_0\left[1 - x^2 + \frac{x^4}{10} + \frac{x^6}{750} + 15\sum_{n=4}^{\infty}\frac{(2n-7)!!\,2^n}{(2n)!\,5^n}x^{2n}\right]$$

13. $\quad c_{n+3} = -\frac{c_n}{n+3}; \qquad \rho = \infty$

When we substitute $y = \Sigma c_n x^n$ into the given differential equation, we find first that $c_2 = 0$, so the recurrence relation yields $c_5 = c_8 = c_{11} = \cdots = 0$ also.

$$y(x) = c_0 \sum_{n=0}^{\infty} \frac{(-1)^n x^{3n}}{n! \, 3^n} + c_1 \sum_{n=0}^{\infty} \frac{(-1)^n x^{3n+1}}{1 \cdot 4 \cdots (3n+1)}$$

15. $c_{n+4} = -\dfrac{c_n}{(n+3)(n+4)}; \qquad \rho = \infty$

When we substitute $y = \Sigma c_n x^n$ into the given differential equation, we find first that $c_2 = c_3 = 0$, so the recurrence relation yields $c_6 = c_{10} = \cdots = 0$ and $c_7 = c_{11} = \cdots = 0$ also. Then

$$c_{4n} = \frac{-1}{(4n)(4n-1)} \cdot \frac{-1}{(4n-4)(4n-5)} \cdots \frac{-1}{4 \cdot 3} c_0 = \frac{(-1)^n c_0}{4^n n! \, (4n-1)(4n-5) \cdots 5 \cdot 3},$$

$$c_{3n+1} = \frac{-1}{(4n+1)(4n)} \cdot \frac{-1}{(4n-3)(4n-4)} \cdots \frac{-1}{5 \cdot 4} c_1 = \frac{(-1)^n c_1}{4^n n! \, (4n+1)(4n-3) \cdots 9 \cdot 5}.$$

$$y(x) = c_0 \left[1 + \sum_{n=1}^{\infty} \frac{(-1)^n x^{4n}}{4^n n! \, 3 \cdot 7 \cdots (4n-1)} \right] + c_1 \left[x + \sum_{n=1}^{\infty} \frac{(-1)^n x^{4n+1}}{4^n n! \, 5 \cdot 9 \cdots (4n+1)} \right]$$

17. The recurrence relation

$$c_{n+2} = -\frac{(n-2)c_n}{(n+1)(n+2)}$$

yields $c_2 = c_0 = y(0) = 1$ and $c_4 = c_6 = \cdots = 0$. Because $c_1 = y'(0) = 0$, it follows also that $c_1 = c_3 = c_5 = \cdots = 0$. Thus the desired particular solution is $y(x) = 1 + x^2$.

19. The substitution $t = x - 1$ yields $(1 - t^2)y'' - 6ty' - 4y = 0$, where primes now denote differentiation with respect to t. When we substitute $y = \Sigma c_n t^n$ we get the recurrence relation

$$c_{n+2} = \frac{n+4}{n+2} c_n.$$

for $n \geq 0$, so the solution series has radius of convergence $\rho = 1$, and therefore converges if $-1 < t < 1$. The initial conditions give $c_0 = 0$ and $c_1 = 1$, so $c_{\text{even}} = 0$ and

$$c_{2n+1} = \frac{2n+3}{2n+1} \cdot \frac{2n+1}{2n-1} \cdots \frac{7}{5} \cdot \frac{5}{3} c_1 = \frac{2n+3}{3}.$$

Thus

$$y = \frac{1}{3} \sum_{n=0}^{\infty} (2n+3) t^{2n+1} = \frac{1}{3} \sum_{n=0}^{\infty} (2n+3)(x-1)^{2n+1},$$

and the x-series converges if $0 < x < 2$.

21. The substitution $t = x+2$ yields $(4t^2 + 1)y'' = 8y$, where primes now denote differentiation with respect to t. When we substitute $y = \Sigma c_n t^n$ we get the recurrence relation

$$c_{n+2} = -\frac{4(n-2)}{(n+2)}c_n$$

for $n \geq 0$. The initial conditions give $c_0 = 1$ and $c_1 = 0$. It follows that $c_{odd} = 0$, $c_2 = 4$ and $c_4 = c_6 = \cdots = 0$, so the solution reduces to

$$y = 2 + 4t^2 = 1 + 4(x+2)^2.$$

In Problems 23–26 we first derive the recurrence relation, and then calculate the solution series $y_1(x)$ with $c_0 = 1$ and $c_1 = 0$, the solution series $y_2(x)$ with $c_0 = 0$ and $c_1 = 1$.

23. Substitution of $y = \Sigma c_n x^n$ yields

$$c_0 + 2c_2 + \sum_{n=1}^{\infty}\left[c_{n-1} + c_n + (n+1)(n+2)c_{n+2}\right]x^n = 0,$$

so

$$c_2 = -\frac{1}{2}c_0, \qquad c_{n+2} = -\frac{c_{n-1} + c_n}{(n+1)(n+2)} \quad \text{for } n \geq 1.$$

$$y_1(x) = 1 - \frac{x^2}{2} - \frac{x^3}{6} + \frac{x^4}{24} + \cdots; \qquad y_2(x) = x - \frac{x^3}{6} - \frac{x^4}{12} + \frac{x^5}{120} + \cdots$$

25. Substitution of $y = \Sigma c_n x^n$ yields

$$2c_2 + 6c_3 x + \sum_{n=2}^{\infty}\left[c_{n-2} + (n-1)c_{n-1} + (n+1)(n+2)c_{n+2}\right]x^n = 0,$$

so

$$c_2 = c_3 = 0, \qquad c_{n+2} = -\frac{c_{n-2} + (n-1)c_{n-1}}{(n+1)(n+2)} \quad \text{for } n \geq 2.$$

$$y_1(x) = 1 - \frac{x^4}{12} + \frac{x^7}{126} + \frac{x^8}{672} + \cdots; \qquad y_2(x) = x - \frac{x^4}{12} - \frac{x^5}{20} + \frac{x^7}{126} + \cdots$$

27. Substitution of $y = \Sigma c_n x^n$ yields

$$c_0 + 2c_2 + (2c_1 + 6c_3)x + \sum_{n=2}^{\infty}\left[2c_{n-2} + (n+1)c_n + (n+1)(n+2)c_{n+2}\right]x^n = 0,$$

so

$$c_2 = -\frac{c_0}{2}, \qquad c_3 = -\frac{c_1}{3}, \qquad c_{n+2} = -\frac{2c_{n-2} + (n+1)c_n}{(n+1)(n+2)} \quad \text{for } n \geq 2.$$

With $c_0 = y(0) = 1$ and $c_1 = y'(0) = -1$, we obtain

$$y(x) = 1 - x - \frac{x^2}{2} + \frac{x^3}{3} - \frac{x^4}{24} + \frac{x^5}{30} + \frac{29x^6}{720} - \frac{13x^7}{630} - \frac{143x^8}{40320} + \frac{31x^9}{22680} + \cdots.$$

Finally, $x = 0.5$ gives

$$y(0.5) = 1 - 0.5 - 0.125 + 0.041667 - 0.002604 + 0.001042$$
$$+ 0.000629 - 0.000161 - 0.000014 + 0.000003 + \cdots$$

$$y(0.5) \approx 0.415562 \approx 0.4156.$$

29. When we substitute $y = \Sigma c_n x^n$ and $\cos x = \Sigma (-1)^n x^{2n}/(2n)!$ and then collect coefficients of the terms involving $1, x, x^2, \cdots, x^6$, we obtain the equations

$$c_0 + 2c_2 = 0, \quad c_1 + 6c_3 = 0, \quad 12c_4 = 0, \quad -2c_3 + 20c_5 = 0,$$

$$\frac{1}{12}c_2 - 5c_4 + 30c_6 = 0, \quad \frac{1}{4}c_3 - 9c_5 + 42c_6 = 0,$$

$$-\frac{1}{360}c_2 + \frac{1}{2}c_4 - 14c_6 + 56c_8 = 0.$$

Given c_0 and c_1, we can solve easily for $c_2, c_3, \cdots, c_8$ in turn. With the choices $c_0 = 1, c_1 = 0$ and $c_0 = 0, c_1 = 1$ we obtain the two series solutions

$$y_1(x) = 1 - \frac{x^2}{2} + \frac{x^6}{720} + \frac{13x^8}{40320} + \cdots \quad \text{and} \quad y_2(x) = x - \frac{x^3}{6} - \frac{x^5}{60} - \frac{13x^7}{5040} + \cdots.$$

SECTION 11.3

FROBENIUS SERIES SOLUTIONS

1. Upon division of the given differential equation by x we see that $P(x) = 1 - x^2$ and $Q(x) = (\sin x)/x$. Because both are analytic at $x = 0$ — in particular, $(\sin x)/x \to 1$ as $x \to 0$ because

$$\frac{\sin x}{x} = \frac{1}{x}\sum_{n=0}^{\infty} \frac{(-1)^n x^{2n+1}}{(2n+1)!} = \sum_{n=1}^{\infty} \frac{(-1)^n x^{2n}}{(2n+1)!} = 1 - \frac{x^2}{3!} + \frac{x^4}{5!} - \frac{x^6}{7!} + \cdots$$

— it follows that $x = 0$ is an ordinary point.

3. When we rewrite the given equation in the standard form of Equation (3) in this section, we see that $p(x) = (\cos x)/x$ and $q(x) = x$. Because $(\cos x)/x \to \infty$ as $x \to 0$ it follows that $p(x)$ is not analytic, so $x = 0$ is an irregular singular point.

5. In the standard form of Equation (3) we have $p(x) = 2/(1 + x)$ and $q(x) = 3x^2/(1 + x)$. Both are analytic, so $x = 0$ is a regular singular point. The indicial equation is

$$r(r-1) + 2r = r^2 + r = r(r+1) = 0,$$

so the exponents are $r_1 = 0$ and $r_2 = -1$.

7. In the standard form of Equation (3) we have $p(x) = (6 \sin x)/x$ and $q(x) = 6$, so $x = 0$ is a regular singular point with $p_0 = q_0 = 6$. The indicial equation is $r^2 + 5r + 6 = 0$, so the exponents are $r_1 = -2$ and $r_2 = -3$.

9. The only singular point of the differential equation $y'' + \dfrac{x}{1-x} y' + \dfrac{x^2}{1-x} y = 0$ is $x = 1$.
Upon substituting $t = x - 1$, $x = t + 1$ we get the transformed equation
$$y'' - \dfrac{t+1}{t} y' - \dfrac{(t+1)^2}{t} y = 0,$$ where primes now denote differentiation with respect to t.
In the standard form of Equation (3) we have $p(t) = -(1+t)$ and $q(t) = -t(1+t)^2$. Both these functions are analytic, so it follows that $x = 1$ is a regular singular point of the original equation.

11. The only singular points of the differential equation $y'' - \dfrac{2x}{1-x^2} y' + \dfrac{12}{1-x^2} y = 0$ are $x = +1$ and $x = -1$.

   $x = +1$: Upon substituting $t = x - 1$, $x = t + 1$ we get the transformed equation
   $$y'' + \dfrac{2(t+1)}{t(t+2)} y' - \dfrac{12}{t(t+2)} y = 0,$$ where primes now denote differentiation with respect to t. In the standard form of Equation (3) we have $p(t) = \dfrac{2(t+1)}{t+2}$ and $q(t) = -\dfrac{12t}{t+2}$.
   Both these functions are analytic at $t = 0$, so it follows that $x = +1$ is a regular singular point of the original equation.

   $x = -1$: Upon substituting $t = x + 1$, $x = t - 1$ we get the transformed equation
   $$y'' + \dfrac{2(t-1)}{t(t-2)} y' - \dfrac{12}{t(t-2)} y = 0,$$ where primes now denote differentiation with respect to t. In the standard form of Equation (3) we have $p(t) = \dfrac{2(t-1)}{t-2}$ and $q(t) = -\dfrac{12t}{t-2}$.
   Both these functions are analytic at $t = 0$, so it follows that $x = -1$ is a regular singular point of the original equation.

13. The only singular points of the differential equation $y'' + \dfrac{1}{x-2} y' + \dfrac{1}{x+2} y = 0$ are $x = +2$ and $x = -2$.

$x = +2$: Upon substituting $t = x - 2$, $x = t + 2$ we get the transformed equation $y'' + \dfrac{1}{t+4}y' + \dfrac{1}{t}y = 0$, where primes now denote differentiation with respect to t. In the standard form of Equation (3) we have $p(t) = \dfrac{t}{t+4}$ and $q(t) = t$. Both these functions are analytic at $t = 0$, so it follows that $x = +2$ is a regular singular point of the original equation.

$x = -2$: Upon substituting $t = x + 2$, $x = t - 2$ we get the transformed equation $y'' + \dfrac{1}{t}y' + \dfrac{1}{t-4}y = 0$, where primes now denote differentiation with respect to t. In the standard form of Equation (3) we have $p(t) \equiv 1$ and $q(t) = \dfrac{t^2}{t-4}$. Both these functions are analytic at $t = 0$, so it follows that $x = -2$ is a regular singular point of the original equation.

15. The only singular point of the differential equation $y'' - \dfrac{x^2 - 4}{(x-2)^2}y' + \dfrac{x+2}{(x-2)^2}y = 0$ is $x = 2$. Upon substituting $t = x - 2$, $x = t + 2$ we get the transformed equation $y'' - \dfrac{t+4}{t}y' + \dfrac{t+4}{t^2}y = 0$, where primes now denote differentiation with respect to t. In the standard form of Equation (3) we have $p(t) = -(t+4)$ and $q(t) = t+4$. Both these functions are analytic, so it follows that $x = 2$ is a regular singular point of the original equation.

Each of the differential equations in Problems 17–20 is of the form

$$Axy'' + By' + Cy = 0$$

with indicial equation $Ar^2 + (B - A)r = 0$. Substitution of $y = \Sigma c_n x^{n+r}$ into the differential equation yields the recurrence relation

$$c_n = -\dfrac{C c_{n-1}}{A(n+r)^2 + (B-A)(n+r)}$$

for $n \geq 1$. In these problems the exponents $r_1 = 0$ and $r_2 = (A-B)/A$ do *not* differ by an integer, so this recurrence relation yields two linearly independent Frobenius series solutions when we apply it separately with $r = r_1$ and with $r = r_2$.

17. With exponent $r_1 = 0$: $c_n = -\dfrac{c_{n-1}}{4n^2 - 2n}$

$$y_1(x) = x^0\left(1 - \frac{x}{2} + \frac{x^2}{24} - \frac{x^3}{720} + \cdots\right) = \sum_{n=0}^{\infty} \frac{(-1)^n \left(\sqrt{x}\right)^{2n}}{(2n)!} = \cos\sqrt{x}$$

With exponent $r_2 = \dfrac{1}{2}$: $c_n = -\dfrac{c_{n-1}}{4n^2 + 2n}$

$$y_2(x) = x^{1/2}\left(1 - \frac{x}{6} + \frac{x^2}{120} - \frac{x^3}{5040} + \cdots\right) = \sum_{n=0}^{\infty} \frac{(-1)^n \left(\sqrt{x}\right)^{2n+1}}{(2n+1)!} = \sin\sqrt{x}$$

19. With exponent $r_1 = 0$: $c_n = \dfrac{c_{n-1}}{2n^2 - 3n}$

$$y_1(x) = x^0\left(1 - x - \frac{x^2}{2} - \frac{x^3}{18} - \frac{x^4}{360} - \cdots\right) = 1 - x - \sum_{n=2}^{\infty} \frac{x^n}{n!(2n-3)!!}$$

With exponent $r_2 = \dfrac{3}{2}$: $c_n = \dfrac{c_{n-1}}{2n^2 + 3n}$

$$y_2(x) = x^{3/2}\left(1 + \frac{x}{5} + \frac{x^2}{70} + \frac{x^3}{1890} + \frac{x^4}{83160} + \cdots\right) = x^{3/2}\left[1 + 3\sum_{n=1}^{\infty} \frac{x^n}{n!(2n+3)!!}\right]$$

The differential equations in Problems 21–24 are all of the form

$$Ax^2 y'' + Bxy' + (C + Dx^2)y = 0 \tag{1}$$

with indical equation

$$\phi(r) = Ar^2 + (B - A)r + C = 0. \tag{2}$$

Substitution of $y = \Sigma c_n x^{n+r}$ into the differential equation yields

$$\phi(r)c_0 x^r + \phi(r+1)c_1 x^{r+1} + \sum_{n=2}^{\infty}\left[\phi(r+n)c_n + Dc_{n-2}\right]x^{n+r} = 0. \tag{3}$$

In each of Problems 21–24 the exponents r_1 and r_2 do *not* differ by an integer. Hence when we substitute either $r = r_1$ or $r = r_2$ into Equation (*) above, we find that c_0 is arbitrary because $\phi(r)$ is then zero, that $c_1 = 0$ — because its coefficient $\phi(r+1)$ is then nonzero — and that

$$c_n = -\frac{Dc_{n-2}}{\phi(r+n)} = -\frac{Dc_{n-2}}{A(n+r)^2 + (B-A)(n+r) + C} \tag{4}$$

for $n \geq 2$. Thus this recurrence formula yields two linearly independent Frobenius series solutions when we apply it separately with $r = r_1$ and with $r = r_2$.

21. With exponent $r_1 = 1$: $c_1 = 0$, $c_n = \dfrac{2c_{n-2}}{n(2n+3)}$

$$y_1(x) = x^1\left(1 + \frac{x^2}{7} + \frac{x^4}{154} + \frac{x^6}{6930} + \cdots\right) = x\left[1 + \sum_{n=1}^{\infty} \frac{x^{2n}}{n!\cdot 7\cdot 11 \cdots (4n+3)}\right]$$

With exponent $r_2 = -\dfrac{1}{2}$: $c_1 = 0$, $c_n = \dfrac{2c_{n-2}}{n(2n-3)}$

$$y_2(x) = x^{-1/2}\left(1 + x^2 + \frac{x^4}{10} + \frac{x^6}{270} + \cdots\right) = \frac{1}{\sqrt{x}}\left[1 + \sum_{n=1}^{\infty} \frac{x^{2n}}{n!\cdot 1\cdot 5 \cdots (4n-3)}\right]$$

23. With exponent $r_1 = \dfrac{1}{2}$: $c_1 = 0$, $c_n = \dfrac{c_{n-2}}{n(6n+7)}$

$$y_1(x) = x^{1/2}\left(1 + \frac{x^2}{38} + \frac{x^4}{4712} + \frac{x^6}{1215696} + \cdots\right) = \sqrt{x}\left[1 + \sum_{n=1}^{\infty} \frac{x^{2n}}{2^n n!\cdot 19\cdot 31 \cdots (12n+7)}\right]$$

With exponent $r_2 = -\dfrac{2}{3}$: $c_1 = 0$, $c_n = \dfrac{c_{n-2}}{n(6n-7)}$

$$y_2(x) = x^{-2/3}\left(1 + \frac{x^2}{10} + \frac{x^4}{680} + \frac{x^6}{118320} + \cdots\right) = x^{-2/3}\left[1 + \sum_{n=1}^{\infty} \frac{x^{2n}}{2^n n!\cdot 5\cdot 17 \cdots (12n-7)}\right]$$

25. With exponent $r_1 = \dfrac{1}{2}$: $c_n = -\dfrac{c_{n-1}}{2n}$

$$y_1(x) = x^{1/2}\left(1 - \frac{x}{2} + \frac{x^2}{8} - \frac{x^3}{48} + \frac{x^4}{384} - \cdots\right) = \sqrt{x}\sum_{n=0}^{\infty} \frac{(-1)^n x^n}{n!\, 2^n} = \sqrt{x}\, e^{-x/2}$$

With exponent $r_2 = 0$: $c_n = -\dfrac{c_{n-1}}{2n-1}$

$$y_2(x) = x^0\left(1 - x + \frac{x^2}{3} - \frac{x^3}{15} + \frac{x^4}{105} - \cdots\right) = 1 + \sum_{n=1}^{\infty} \frac{(-1)^n x^n}{(2n-1)!!}$$

The differential equations in Problems 27–29 (after multiplication by x) and the one in Problem 31 are of the same form (1) above as those in Problems 21–24. However, now the exponents r_1 and $r_2 = r_1 - 1$ *do* differ by an integer. Hence when we substitute the smaller exponent $r = r_2$ into Equation (3), we find that c_0 and c_1 are *both* arbitrary, and that c_n is given (for $n \geq 2$) by the recurrence relation in (4). Thus the *smaller* exponent r_2 yields the general solution $y(x) = c_0 y_1(x) + c_1 y_2(x)$ in terms of the two linearly independent Frobenius series solutions $y_1(x)$ and $y_2(x)$.

27. Exponents $r_1 = 0$ and $r_2 = -1$; with $r = -1$: $c_n = -\dfrac{9c_{n-2}}{n(n-1)}$

$$y(x) = \frac{c_0}{x}\left(1 - \frac{9x^2}{2} + \frac{27x^4}{8} - \frac{81x^6}{80} + \cdots\right) + \frac{c_1}{x}\left(x - \frac{3x^3}{2} + \frac{27x^5}{40} - \frac{81x^7}{560} + \cdots\right)$$

$$= \frac{c_0}{x}\left(1 - \frac{9x^2}{2} + \frac{81x^4}{24} - \frac{729x^6}{720} + \cdots\right) + \frac{c_1}{3x}\left(3x - \frac{27x^3}{6} + \frac{243x^5}{120} - \frac{2187x^7}{5040} + \cdots\right)$$

$$y(x) = c_0 \frac{\cos 3x}{x} + \frac{1}{3}c_1 \frac{\sin 3x}{x}$$

29. Exponents $r_1 = 0$ and $r_2 = -1$; with $r = -1$: $c_n = -\dfrac{c_{n-2}}{4n(n-1)}$

$$y(x) = \frac{c_0}{x}\left(1 - \frac{x^2}{8} + \frac{x^4}{384} - \frac{x^6}{46080} + \cdots\right) + \frac{c_1}{x}\left(x - \frac{x^3}{24} + \frac{x^5}{1920} - \frac{x^7}{322560} + \cdots\right)$$

$$= \frac{c_0}{x}\left(1 - \frac{x^2}{2^2 \cdot 2} + \frac{x^4}{2^4 \cdot 24} - \frac{x^6}{2^6 \cdot 720} + \cdots\right) + \frac{2c_1}{x}\left(\frac{x}{2} - \frac{x^3}{2^3 \cdot 6} + \frac{x^5}{2^5 \cdot 120} - \frac{x^7}{2^7 \cdot 5040} + \cdots\right)$$

$$y(x) = \frac{c_0}{x}\cos\frac{x}{2} + \frac{2c_1}{x}\sin\frac{x}{2}$$

31. The given differential equation $4x^2 y'' - 4xy' + (3 - 4x^2)y = 0$ has indicial equation $4r^2 - 8r + 3 = (2r - 3)(2r - 1) = 0$, so its exponents are $r_1 = 3/2$ and $r_2 = 1/2$. With $r = 3/2$, the recurrence relation $c_n = c_{n-2}/n(n-1)$ yields the general solution

$$y(x) = c_0 x^{1/2}\left(1 + \frac{x^2}{2} + \frac{x^4}{24} + \frac{x^6}{720} + \cdots\right) + c_1 x^{1/2}\left(x + \frac{x^3}{6} + \frac{x^5}{120} + \frac{x^7}{5040} + \cdots\right)$$

$$y(x) = c_0 \sqrt{x}\cosh x + c_1 \sqrt{x}\sinh x.$$

33. Exponents $r_1 = 1/2$ and $r_2 = -1$. With each exponent we find that c_0 is arbitrary and we can solve recursively for c_n in terms of c_{n-1}.

$$y_1(x) = \sqrt{x}\left(1 + \frac{11x}{20} - \frac{11x^2}{224} + \frac{671x^3}{24192} - \frac{9577x^4}{387072} + \cdots\right)$$

$$y_2(x) = \frac{1}{x}\left(1 + 10x + 5x^2 + \frac{10x^3}{9} - \frac{7x^4}{18} + \cdots\right)$$

35. Substitution of $y = x^r \sum c_n x^n$ into the differential equation yields a result of the form

$$-rc_0 x^{r-1} + (\cdots)x^r + (\cdots)x^{r+1} + \cdots = 0,$$

so we see immediately that $c_0 \neq 0$ implies that $r = 0$. Then substitution of the power series $y = \sum c_n x^n$ yields

$$(c_0 - c_1) + (4c_1 - 2c_2)x + (9c_2 - 3c_3)x^2 + (16c_3 - 4c_4)x^4 + \cdots = 0$$

Evidently $c_n = nc_{n-1}$, so if $c_0 = 1$ it follows that $c_n = n!$ for $n \geq 1$. But the series $\sum n! x^n$ has zero radius of convergence, and hence converges only if $x = 0$. We therefore conclude that the given differential equation has *no* nontrivial Frobenius series solution.

37. Substitution of $y = x^r \sum c_n x^n$ into the differential equation $x^3 y'' - y' + y = 0$ yields a result of the form

$$(r-1)^2 c_0 x^r + (\cdots)x^{r+1} + (\cdots)x^{r+2} + \cdots = 0,$$

so it follows that $r = 1$. But then substitution of $y = x \sum c_n x^n$ into the differential equation yields

$$c_1 x^2 + 4c_2 x^3 + 9c_3 x^4 + 16c_4 x^5 + 25c_5 x^6 + \cdots = 0,$$

so it follows that $c_1 = c_2 = c_3 = c_4 = \cdots = 0$. Hence $y(x) = c_0 x$.

39. Exponents $r_1 = 1$ and $r_2 = -1$; with $r = +1$: $c_1 = 0$, $c_n = -\dfrac{c_{n-2}}{n(n+2)}$

$$y(x) = c_0 x \left(1 - \frac{x^2}{8} + \frac{x^4}{192} - \frac{x^6}{9216} + \frac{x^8}{737280} - \cdots \right)$$

$$= c_0 x \left(1 - \frac{x^2}{2^2 1! 2!} + \frac{x^4}{2^4 2! 3!} - \frac{x^6}{2^6 3! 4!} + \frac{x^8}{2^8 4! 5!} - \cdots \right)$$

If $c_0 = 1/2$, then

$$y(x) = J_1(x) = \frac{x}{2} \sum_{n=0}^{\infty} \frac{(-1)^n}{n!(n+1)} \left(\frac{x}{2}\right)^{2n}$$

Now, consider the smaller exponent $r_2 = -1$. A Frobenius series with $r = -1$ is of the form $y = x^{-1} \sum_{n=0}^{\infty} c_n x^n$ with $c_0 \neq 0$. However, substitution of this series into Bessel's equation of order 1 gives

$$-c_1 + c_0 x + (c_1 + 3c_3)x^2 + (c_2 + 8c_4)x^3 + (c_3 + 15c_5)x^5 + \cdots = 0,$$

so it follows that $c_0 = 0$, after all. Thus Bessel's equation of order 1 does not have a Frobenius series solution with leading term $c_0 x^{-1}$. However, there is a little more here that meets the eye. We see further that c_2 is arbitrary and that $c_1 = 0$ and $c_n = c_{n-2}/n(n-2)$ for $n > 2$. It follows that our assumed Frobenius series

$$y = x^{-1}\sum_{n=0}^{\infty} c_n x^n \quad \text{actually reduces to}$$

$$y(x) = c_2 x\left(1 - \frac{x^2}{8} + \frac{x^4}{192} - \frac{x^6}{9216} + \frac{x^8}{737280} - \cdots\right).$$

But this is the same as our series solution obtained above using the larger exponent $r = +1$ (calling the arbitrary constant c_2 rather than c_0).

SECTION 11.4

BESSEL'S EQUATION

Of course Bessel's equation is the most important special ordinary differential equation in mathematics, and every student should be exposed at least to Bessel functions of the first kind.

1. $\quad J_0'(x) = D_x\left(1 + \sum_{m=1}^{\infty} \frac{(-1)^m x^{2m}}{2^{2m}(m!)^2}\right) = \sum_{m=1}^{\infty} \frac{(-1)^m 2m\, x^{2m-1}}{2^{2m}(m!)^2}$

$\quad = \sum_{m=1}^{\infty} \frac{(-1)^m x^{2m-1}}{2^{2m-1}(m-1)!(m!)} = \sum_{m=0}^{\infty} \frac{(-1)^{m+1} x^{2m+1}}{2^{2m+1}(m)!(m+1)!}$

$\quad = -\sum_{m=0}^{\infty} \frac{(-1)^m x^{2m+1}}{2^{2m+1}(m)!(m+1)!} = -J_1(x)$

3. **(a)** $\Gamma\left(m + \frac{2}{3}\right) = \Gamma\left(\frac{3m+2}{3}\right) = \frac{3m-1}{3} \cdot \frac{3m-4}{3} \cdot \Gamma\left(\frac{3m-4}{3}\right)$

$\quad = \frac{3m-1}{3} \cdot \frac{3m-4}{3} \cdots \cdots \frac{5}{3} \cdot \frac{2}{3} \cdot \Gamma\left(\frac{2}{3}\right) = \frac{2 \cdot 5 \cdot 8 \cdots (3m-1)}{3^m} \Gamma\left(\frac{2}{3}\right)$

(b) $J_{-1/3}(x) = \sum_{m=0}^{\infty} \frac{(-1)^m}{m!\,\Gamma(m+2/3)} \left(\frac{x}{2}\right)^{2m-1/3} = \frac{(x/2)^{-1/3}}{\Gamma(2/3)} \sum_{m=0}^{\infty} \frac{(-1)^m 3^m x^{2m}}{m!\, 2 \cdot 3 \cdot 8 \cdots (3m-1)}$

5. Starting with $p = 3$ in Equation (26) we get

$$J_4(x) = \frac{6}{x} J_3(x) - J_2(x) = \frac{6}{x}\left[\frac{4}{x} J_2(x) - J_1(x)\right] - J_2(x)$$

$$= \left(\frac{24}{x^2} - 1\right)\left[\frac{2}{x} J_1(x) - J_0(x)\right] - \frac{6}{x} J_1(x) = \frac{x^2-24}{x^2} J_0(x) + \frac{8(6-x^2)}{x^3} J_1(x)$$

9. $\Gamma(p + m + 1) = (p + m)(p + m - 1)\cdots(p + 2)(p + 1)\Gamma(p + 1)$, so

$$J_p(x) = \sum_{m=0}^{\infty} \frac{(-1)^m}{m!\Gamma(p+m+1)} \left(\frac{x}{2}\right)^{2m+p}$$

$$= \frac{(x/2)^p}{\Gamma(p+1)} \sum_{m=0}^{\infty} \frac{(-1)^m}{m!(p+1)(p+2)\cdots(p+m)} \left(\frac{x}{2}\right)^{2m}.$$

In Problems 11–18 we use a conspicuous dot • to indicate our choice of u and dv in the integration by parts formula $\int u \cdot dv = uv - \int v\, du$. We use repeatedly the facts (from Example 1) that $\int xJ_0(x)\, dx = xJ_1(x) + C$ and $\int J_1(x)\, dx = -J_0(x) + C$.

11. $\int x^2 J_0(x)\, dx = \int x \cdot xJ_0(x)\, dx$

$\qquad = x^2 J_1(x) - \int x \cdot J_1(x)\, dx$

$\qquad = x^2 J_1(x) - \left(-xJ_0(x) + \int J_0(x)\, dx\right)$

$\qquad = x^2 J_1(x) + xJ_0(x) - \int J_0(x)\, dx + C$

13. $\int x^4 J_0(x)\, dx = \int x^3 \cdot xJ_0(x)\, dx$

$\qquad = x^4 J_1(x) - 3\int x^3 \cdot J_1(x)\, dx$

$\qquad = x^4 J_1(x) - 3\left(-x^3 J_0(x) + 3\int x \cdot xJ_0(x)\, dx\right)$

$\qquad = x^4 J_1(x) + 3x^3 J_0(x) - 9\left(x^2 J_1(x) - \int x \cdot J_1(x)\, dx\right)$

$\qquad = x^4 J_1(x) + 3x^3 J_0(x) - 9x^2 J_1(x) + 9\left(-xJ_0(x) + \int J_0(x)\, dx\right)$

$\qquad = (x^4 - 9x^2)J_1(x) + (3x^3 - 9x)J_0(x) + 9\int J_0(x)\, dx + C$

15. $\int x^2 J_1(x)\, dx = \int x^2 \cdot J_1(x)\, dx$

$\qquad = -x^2 J_0(x) + 2\int xJ_0(x)\, dx = -x^2 J_0(x) + 2xJ_1(x) + C$

17. $\int x^4 J_1(x)\, dx = \int x^4 \cdot J_1(x)$

$\qquad = -x^4 J_0(x) + 4\int x^2 \cdot xJ_0(x)\, dx$

$\qquad = -x^4 J_0(x) + 4\left(x^3 J_1(x) - 2\int x^2 \cdot J_1(x)\, dx\right)$

$\qquad = -x^4 J_0(x) + 4x^3 J_1(x) - 8\left(-x^2 J_0(x) + 2\int xJ_0(x)\, dx\right)$

$\qquad = (-x^4 + 8x^2)J_0(x) + (4x^3 - 16x)J_1(x) + C$

Problems 19–30 are routine applications of the theorem in this section. In each case it is necessary only to identify the coefficients A, B, C and the exponent q in the differential equation

$$x^2 y'' + Axy' + (B + Cx^q)y = 0. \qquad (1)$$

Then we can calculate the values

$$\alpha = \frac{1-A}{2}, \quad \beta = \frac{q}{2}, \quad k = \frac{2\sqrt{C}}{q}, \quad p = \frac{\sqrt{(1-A)^2 - 4B}}{q} \qquad (2)$$

and finally write the general solution

$$y(x) = x^\alpha \left[c_1 J_p(kx^\beta) + c_2 J_{-p}(kx^\beta) \right] \qquad (3)$$

specified in Theorem 1 on solutions in terms of Bessel functions. This is a "template procedure" that we illustrate only in a couple of problems.

19. We have $A = -1, B = 1, C = 1, q = 2$ so

$$\alpha = \frac{1-(-1)}{2} = 1, \quad \beta = \frac{2}{2} = 1, \quad k = \frac{2\sqrt{1}}{2} = 1, \quad p = \frac{\sqrt{(1-(-1))^2 - 4(1)}}{2} = 0,$$

so our general solution is $y(x) = x[c_1 J_0(x) + c_2 Y_0(x)]$, using $Y_0(x)$ because $p = 0$ is an integer.

21. $y(x) = x[c_1 J_{1/2}(3x^2) + c_2 J_{-1/2}(3x^2)]$

23. To match the given equation with Eq. (1) above, we first divide through by the leading coefficient 16 to obtain the equation

$$x^2 y'' + \frac{5}{3} xy' + \left(-\frac{5}{36} + \frac{1}{4} x^3 \right) y = 0$$

with $A = 5/3, B = -5/36, C = 1/4,$ and $q = 3$. Then

$$\alpha = \frac{1 - 5/3}{3} = -\frac{1}{3}, \quad \beta = \frac{3}{2}, \quad k = \frac{2\sqrt{1/4}}{3} = \frac{1}{3}, \quad p = \frac{\sqrt{(1-5/3)^2 - 4(-5/36)}}{3} = \frac{1}{3},$$

so our general solution is $y(x) = x^{-1/3}[c_1 J_{1/3}(x^{3/2}/3) + c_2 J_{-1/3}(x^{3/2}/3)]$.

25. $y(x) = x^{-1}[c_1 J_0(x) + c_2 Y_0(x)]$

27. $y(x) = x^{1/2}[c_1 J_{1/2}(2x^{3/2}) + c_2 J_{-1/2}(2x^{3/2})]$

29. $y(x) = x^{1/2}[c_1 J_{1/6}(x^3/3) + c_2 J_{-1/6}(x^3/3)]$

31. We want to solve the equation $xy'' + 2y' + xy = 0$. If we rewrite it as

$$x^2 y'' + 2xy' + x^2 y = 0$$

then we have the form in Equation (1) with $A = 2$, $B = 0$, $C = 1$, and $q = 2$. Then Equation (2) gives $\alpha = -1/2$, $\beta = 1$, $k = 1$, and $p = 1/2$, so by Equation (3) the general solution is

$$y(x) = x^{-1/2}[c_1 J_{1/2}(x) + c_1 J_{-1/2}(x)]$$

$$= x^{-1/2}\left[c_1 \sqrt{\frac{2}{\pi x}} \cos x + c_2 \sqrt{\frac{2}{\pi x}} \sin x\right]$$

$$= \frac{1}{x}(a_1 \cos x + a_2 \sin x)$$

(with $a_i = c_i \sqrt{2/\pi}$), using Equations (19) in Section 11.4.

33. The substitution

$$y = -\frac{u'}{u}, \quad y' = \frac{(u')^2}{u^2} - \frac{u''}{u}$$

immediately transforms $y' = x^2 + y^2$ to $u'' + x^2 u = 0$. The equivalent equation

$$x^2 u'' + x^4 u = 0$$

is of the form in (1) with $A = B = 0$, $C = 1$, and $q = 4$. Equations (2) give $\alpha = 1/2$, $\beta = 2$, $k = 1/2$, and $p = 1/4$, so the general solution is

$$u(x) = x^{1/2}[c_1 J_{1/4}(x^2/2) + c_2 J_{-1/4}(x^2/2)].$$

To compute $u'(x)$, let $z = x^2/2$ so $x = 2^{1/2} z^{1/2}$. Then Equation (22) in Section 11.4 with $p = 1/4$ yields

$$\frac{d}{dx}\left(x^{1/2} J_{1/4}(x^2/2)\right) = \frac{d}{dz}\left(2^{1/4} z^{1/4} J_{1/4}(z)\right) \cdot \frac{dz}{dx}$$

$$= 2^{1/4} z^{1/4} J_{-3/4}(z) \cdot \frac{dz}{dx}$$

$$= 2^{1/4} \cdot \frac{x^{1/2}}{2^{1/4}} J_{-3/4}(x^2/2) \cdot x = x^{3/2} J_{-3/4}(x^2/2).$$

Similarly, Equation (23) in Section 11.4 with $p = -1/4$ yields

$$\frac{d}{dx}\left(x^{1/2}J_{-1/4}(x^2/2)\right) = \frac{d}{dz}\left(2^{1/4}z^{1/4}J_{-1/4}(z)\right)\cdot\frac{dz}{dx} = -x^{3/2}J_{3/4}(x^2/2).$$

Therefore
$$u'(x) = x^{3/2}[c_1 J_{-3/4}(x^2/2) - c_2 J_{3/4}(x^2/2)].$$

It follows finally that the general solution of the Riccati equation $y' = x^2 + y^2$ is

$$y(x) = -\frac{u'}{u} = x\cdot\frac{J_{3/4}(\tfrac{1}{2}x^2) - cJ_{-3/4}(\tfrac{1}{2}x^2)}{cJ_{1/4}(\tfrac{1}{2}x^2) + J_{-1/4}(\tfrac{1}{2}x^2)}$$

where the arbitrary constant is $c = c_1/c_2$.

34. Substitution of the series expressions for the Bessel functions in the formula for $y(x)$ in Problem 33 yields

$$y(x) = x\cdot\frac{A\left(\tfrac{1}{2}x^2\right)^{3/4}(1+\cdots) - cB\left(\tfrac{1}{2}x^2\right)^{-3/4}(1+\cdots)}{cC\left(\tfrac{1}{2}x^2\right)^{1/4}(1+\cdots) + D\left(\tfrac{1}{2}x^2\right)^{-1/4}(1+\cdots)}$$

where each pair of parentheses encloses a power series in x with constant term 1, and

$$A = 2^{-3/4}/\Gamma(7/4) \qquad B = 2^{3/4}/\Gamma(1/4)$$
$$C = 2^{-1/4}/\Gamma(5/4) \qquad D = 2^{1/4}/\Gamma(3/4).$$

Multiplication of numerator and denominator by $x^{1/2}$ and a bit of simplification gives

$$y(x) = \frac{2^{-3/4}Ax^3(1+\cdots) - 2^{3/4}cB(1+\cdots)}{2^{-1/4}cCx(1+\cdots) + 2^{1/4}D(1+\cdots)}.$$

It now follows that

$$y(0) = \frac{-2^{3/4}cB}{2^{1/4}D} = \frac{-2^{1/2}\left(2^{3/4}/\Gamma(1/4)\right)}{2^{1/4}/\Gamma(3/4)} = -2c\cdot\frac{\Gamma(3/4)}{\Gamma(1/4)}. \qquad (*)$$

(a) If $y(0) = 0$ then (*) gives $c = 0$ in the general solution formula of Problem 33.

(b) If $y(0) = 1$ then (*) gives $c = -\Gamma(1/4)/2\Gamma(3/4)$. More generally, (*) yields the formula

$$y(x) = x\cdot\frac{2\Gamma\left(\tfrac{3}{4}\right)J_{3/4}(\tfrac{1}{2}x^2) + y_0\Gamma\left(\tfrac{1}{4}\right)J_{-3/4}(\tfrac{1}{2}x^2)}{2\Gamma\left(\tfrac{3}{4}\right)J_{-1/4}(\tfrac{1}{2}x^2) - y_0\Gamma\left(\tfrac{1}{4}\right)J_{1/4}(\tfrac{1}{2}x^2)}$$

for the solution of the initial value problem $y' = x^2 + y^2$, $y(0) = y_0$.

APPENDIX A

EXISTENCE AND UNIQUENESS OF SOLUTIONS

In Problems 1–12 we apply the iterative formula

$$y_{n+1} = b + \int_a^x f(t, y_n(t))\, dt$$

to compute successive approximations $\{y_n(x)\}$ to the solution of the initial value problem

$$y' = f(x, y), \qquad y(a) = b.$$

starting with $y_0(x) = b$.

1. $y_0(x) = 3$

     $y_1(x) = 3 + 3x$

     $y_2(x) = 3 + 3x + 3x^2/2$

     $y_3(x) = 3 + 3x + 3x^2/2 + x^3/2$

     $y_4(x) = 3 + 3x + 3x^2/2 + x^3/2 + x^4/8$

     $y(x) \;= 3 - 3x + 3x^2/2 + x^3/2 + x^4/8 + \cdots = 3e^x$

3. $y_0(x) = 1$

     $y_1(x) = 1 - x^2$

     $y_2(x) = 1 - x^2 + x^4/2$

     $y_3(x) = 1 - x^2 + x^4/2 - x^6/6$

     $y_4(x) = 1 - x^2 + x^4/2 - x^6/6 + x^8/24$

     $y(x) \;= 1 - x^2 + x^4/2 - x^6/6 + x^8/24 - \cdots = \exp(-x^2)$

5. $y_0(x) = 0$

     $y_1(x) = 2x$

     $y_2(x) = 2x + 2x^2$

     $y_3(x) = 2x + 2x^2 + 4x^3/3$

     $y_4(x) = 2x + 2x^2 + 4x^3/3 + 2x^4/3$

     $y(x) \;= 2x + 2x^2 + 4x^3/3 + 2x^4/3 + \cdots = e^{2x} - 1$

7. $y_0(x) = 0$
$y_1(x) = x^2$
$y_2(x) = x^2 + x^4/2$
$y_3(x) = x^2 + x^4/2 + x^6/6$
$y_4(x) = x^2 + x^4/2 + x^6/6 + x^8/24$
$y(x) = x^2 + x^4/2 + x^6/6 + x^8/24 + \cdots = \exp(x^2) - 1$

9. $y_0(x) = 1$
$y_1(x) = (1 + x) + x^2/2$
$y_2(x) = (1 + x + x^2) + x^3/6$
$y_3(x) = (1 + x + x^2 + x^3/3) + x^4/24$
$y(x) = 1 + x + x^2 + x^3/3 + x^4/12 + \cdots = 2e^x - 1 - x$

11. $y_0(x) = 1$
$y_1(x) = 1 + x$
$y_2(x) = (1 + x + x^2) + x^3/3$
$y_3(x) = (1 + x + x^2 + x^3) + 2x^4/3 + x^5/3 + x^6/9 + x^7/63$
$y(x) = 1 + x + x^2 + x^3 + x^4 + \cdots = 1/(1 - x)$

13. $\begin{bmatrix} x_0(t) \\ y_0(t) \end{bmatrix} = \begin{bmatrix} 1 \\ -1 \end{bmatrix}$

$\begin{bmatrix} x_1(t) \\ y_1(t) \end{bmatrix} = \begin{bmatrix} 1 + 3t \\ -1 + 5t \end{bmatrix}$

$\begin{bmatrix} x_2(t) \\ y_2(t) \end{bmatrix} = \begin{bmatrix} 1 + 3t + \frac{1}{2}t^2 \\ -1 + 5t - \frac{1}{2}t^2 \end{bmatrix}$

$\begin{bmatrix} x_3(t) \\ y_3(t) \end{bmatrix} = \begin{bmatrix} 1 + 3t + \frac{1}{2}t^2 + \frac{1}{3}t^3 \\ -1 + 5t - \frac{1}{2}t^2 + \frac{5}{6}t^3 \end{bmatrix}$